Numerical analysis

Numerical analysis

A M Cohen, J F Cutts, R Fielder,
D E Jones, J Ribbans, E Stuart

A HALSTED PRESS BOOK

John Wiley and Sons
New York

Published in the U.S.A.
by **Halsted Press**
A Division of **John Wiley and Sons, Inc., New York**

Library of Congress Cataloging in Publication Data
Main entry under title:
Numerical analysis.
"A Halsted Press book."
Bibliography: p.
1. Numerical analysis. I. Cohen, Alan M.
QA297.N82 519.4 73-282
ISBN 0-470-16423-9

PRINTED AND BOUND IN GREAT BRITAIN

Contents

Preface

This book started life as an extension of a short course given in the University of Wales Institute of Science and Technology by the editor and his colleagues during the winter of 1967. It contains the material which now forms the basis of undergraduate honours courses in the College and only assumes an elementary knowledge of real and complex variable theory. It aims to form a foundation for an appreciation of numerical work and to pave the way to understanding more advanced treatises and research papers.

In recent years it has become fashionable to resort to computers for all types of mathematical calculations – and with good reason. For most mathematical calculations can be re-formulated in a manner which gives rise to an algebraic problem such as the evaluation of a polynomial, or the solution of a system of linear equations, etc. The calculations in these algebraic problems can be performed iteratively and the great speed of computer arithmetic can thus be used to advantage. Or can it? As we shall see in chapter 1, correct mathematical re-formulation and correct programming are often insufficient to produce answers to our problems which can be classified as being 'correct'. Thus more work must be attached to problem solving than just re-formulation and coding for a computer, as we hope will be made clear by examples throughout the book.

The plan of presentation of most chapters has been that of step by step. We start with an elementary method and then proceed to develop this or give alternative, more sophisticated methods. At each step we have tried to concern ourselves with points of numerical detail, and with an analysis of the effects of errors arising in the calculation, although this may not be apparent in some chapters. In chapter 3, for example, we are effectively calculating polynomial functions and inner products, for which error analyses are given earlier, and so they are omitted here. However, considerable attention is paid in this chapter to truncation errors. Again, we have tried to impress on the reader the fact that our initial data is in general inexact. This means that not only do we have to trace the error effect through the calculation with the given data but also through all possible admissible variations in the data. An analysis of this type is termed an interval analysis and the basic ideas are sketched in chapter 1 and are supplemented in the exercises. A full list of the topics discussed in the book is given in the table of contents.

The editor wishes to acknowledge the many helpful comments received, particularly those from Professor L. Fox, Professor E. E. Jones, Professor M. B. Priestley, Mr. D. M. Ellis, Professor J. Leech, Mr. F. J. Wilson, and McGraw-Hill and their advisers. He is also grateful to Dr. D. Davies for supplying him with a copy of I.C.I.'s notes on optimization and to Mr. M. Cottam and Mr. T. Pindard for writing programmes to check some of the tables and exercises. Thanks are also due to Mrs. S. M. Cousins, Mrs. S. M. Bennett, and Mrs. J. Ribbans for typing the greater part of this book. A number of exercises have their origins in the examination papers of the University of Wales Institute, the University of Manchester, and the Mathematical Association (Diploma Examination). The editor is grateful to these bodies for permission to reproduce their examination questions. Further, the editor wishes to thank the many publishers and journal editors who have granted permission for extracts of books and research papers to appear in this work.

Finally, we dedicate this book to all who made it possible. We have benefited much by sitting at the feet of the wise, and we hope that, through this book, it may be possible to transmit a spark from their fire to all our readers.

Alan M. Cohen
January 1972

Contributors

The contributors to this book, in addition to the editor, are J. F. Cutts, R. Fielder, D. E. Jones, J. Ribbans, and E. Stuart. A break-down of the individual contributions is

A. M. Cohen	(chapters 1, 2, 4, 6, 7, 9, and 13)
J. F. Cutts	(chapter 8)
R. Fielder	(chapter 10)
D. E. Jones	(chapters 11 and 12)
J. Ribbans	(chapters 3 and 4)
E. Stuart	(chapter 14)

1 Introduction

1.1 Mathematical and physical problems

The subject of numerical analysis is concerned with finding best possible numerical answers to two types of problems. In the first type, termed by many authors a mathematical problem, the data is given exactly and it would be reasonable to ask for answers with a high degree of precision, provided the problem is well-conditioned. In the second type which is often called a physical problem, the data has been obtained as a result of measurement and is inaccurate to some degree. We now have to decide the worthwhile precision of the answers, i.e., the number of digits we can quote in our answer. One cannot quote more digits than the number remaining unchanged however the data is varied within its limits of tolerance.

Let us consider a simple example to illustrate these remarks. Suppose that l_t, the length of a metal bar at temperature t, is related to l_0, the length of the bar at zero temperature, by the hypothetical formula

$$l_t = l_0\,(1 + \alpha t + \beta t^2)$$

where α and β are constants. If we know that the exact values of α, β, l_0 are 0·001 253, 0·000 068, and 1 respectively then

$$l_{10} = 1 + (0{\cdot}001\,253)10 + (0{\cdot}000\,068)100 = 1{\cdot}019\,330$$

This is an example of a mathematical problem and our result is exact. Now let us assume that our data has been obtained experimentally and that $l_0 = 1$ (exactly), $\alpha = 0{\cdot}001\,253 \pm 10^{-6}$, $\beta = 0{\cdot}000\,068 \pm 10^{-6}$, i.e.,

$$0{\cdot}001\,252 \leqslant \alpha \leqslant 0{\cdot}001\,254,$$
$$0{\cdot}000\,067 \leqslant \beta \leqslant 0{\cdot}000\,069.$$

All that we can say about l_{10} is

$$1 + (0{\cdot}001\,252)10 + (0{\cdot}000\,067)100 \leqslant l_{10} \leqslant 1 + (0{\cdot}001\,254)10 + (0{\cdot}000\,069)100$$

or

$$1{\cdot}019\,220 \leqslant l_{10} \leqslant 1{\cdot}019\,440.$$

Clearly, in this physical problem, we cannot retain more than three decimal places in our answer for l_{10}.

The above physical problem is one of a class of problems where the final result is less accurate than the original data. One must not however jump to the conclusion that this is true in general. Consider, for instance,

$$y = x^p, \qquad 0 < p < 1, \qquad x > 1.$$

If δx is a small error in x and δy is the associated error in y we know that, neglecting second order terms,

$$\delta y = p x^{p-1}\, \delta x,$$

so that

$$|\delta y| < |\delta x|.$$

Thus the absolute value of the error in y is less than that in x, and hence our result is more accurate than our original data.

1.2 Representation of numbers by a computer

Having mentioned the two types of problems which a numerical analyst encounters it is only natural to say something about the tools he employs to obtain his end product. We shall concern ourselves here with just one such tool, the electronic digital computer, and will assume that our computer represents numbers in binary form.

Our computer is almost certain to use two main modes of operation, namely, a fixed point mode and a floating point mode. In the fixed point mode every computed binary number x has a fixed number of digits and satisfies

$$-1 \leqslant x \leqslant 1$$

while in the floating point mode every number x is of the form

$$x = a \,.\, 2^b,$$

where the index b is an integer (positive or negative) lying in some finite range and the mantissa a is a binary number with a fixed number of digits and $\frac{1}{2} \leqslant |a| < 1$ unless $a = 0$. From these definitions we see that every number in our computer is the ratio of two integers. We know, however, from the theory of numbers, that irrational numbers, such as $\sqrt{2}$ and π, cannot be represented exactly as the ratio of two integers and even simple fractions like 1/3 and 1/10 cannot be represented exactly in binary form. It follows that nearly all the problems we tackle by means of a computer can be classed as physical problems since, in almost all cases, we will have a representational error.

It has been tacitly assumed that, given the data, no errors would be incurred in our calculation. With our computer this is not the case as we now demonstrate for our floating point mode. As it would probably fog the issue if we used binary arithmetic for this demonstration, we shall use for this, and later illustrations, four decimal digit floating point numbers

$$a \,.\, 10^b,$$

where, unless $|a| = 0$,

$$0{\cdot}1000 \leqslant |a| \leqslant 0{\cdot}9999, \qquad -10 \leqslant b \leqslant 10.$$

Examples of 'Computer Arithmetic'.

(*a*) $10^3(0{\cdot}5971) + 10^{-5}(0{\cdot}4268)$
$\rightarrow 10^3(0{\cdot}5971) + 10^3(0{\cdot}0000)$
$\rightarrow 10^3(0{\cdot}5971)$

(*b*) $10^3(0{\cdot}5971) + 10^0(0{\cdot}4268)$
$\rightarrow 10^3(0{\cdot}5971) + 10^3(0{\cdot}0004)$
$\rightarrow 10^3(0{\cdot}5975)$

(*c*) $10^3(0{\cdot}5971) + 10^3(0{\cdot}4268)$
$\rightarrow 10^3(1{\cdot}0239)$
$\rightarrow 10^4(0{\cdot}1024)$

(*d*) $10^{-1}(0{\cdot}5971) - 10^{-1}(0{\cdot}5956)$
$\rightarrow 10^{-1}(0{\cdot}0015)$
$\rightarrow 10^{-3}(0{\cdot}1500)$

(*e*) $10^{-1}(0{\cdot}1008) - 10^{-2}(0{\cdot}9994)$
$\rightarrow 10^{-1}(0{\cdot}1008 - 0{\cdot}0999)$
$\rightarrow 10^{-4}(0{\cdot}9000)$

(*f*) $10^3(0{\cdot}5971) \times 10^{-5}(0{\cdot}4268)$
$\rightarrow 10^{-2}(0{\cdot}5971 \times 0{\cdot}4268) \rightarrow 10^{-2}(0{\cdot}2548\,4228)$
$\rightarrow 10^{-2}(0{\cdot}2548)$

(*g*) $10^3(0{\cdot}5971) \div 10^{-5}(0{\cdot}4268)$
$\rightarrow 10^8(0{\cdot}5971) \div (0{\cdot}4268)$
$\rightarrow 10^9(0{\cdot}0597\,1000) \div (0{\cdot}4268)$
$\rightarrow 10^9(0{\cdot}1399\,0) \rightarrow 10^9(0{\cdot}1399)$

(*h*) $10^{-5}(0{\cdot}4268) \div 10^3(0{\cdot}5971)$
$\rightarrow 10^{-8}(0{\cdot}4268\,0000) \div (0{\cdot}5971)$
$\rightarrow 10^{-8}(0{\cdot}7147\,9) \rightarrow 10^{-8}(0{\cdot}7148)$.

To explain the above results we consider first examples (*a*) – (*e*). In all these cases the indices are compared and the mantissa of the smaller number is shifted down until its exponent agrees with that of the larger number. This shifted mantissa is rounded to four digits and the operation of addition (or subtraction) performed, our final answer being adjusted as necessary to conform with our convention for four decimal digit floating point numbers. In (*f*) we add indices and multiply mantissas rounding the mantissa (after possibly shifting) to four digits. In (*g*) and (*h*) we subtract indices, extend the dividend to eight digits, and then carry out the division.

Clearly, the arithmetic performed by the computer is, in general, not exact. This, coupled with the fact that our data may be inexact, leads us, in any problem, to speculate about the magnitude of the accumulated error in our final result. This result may be considered satisfactory if, for some specified small positive number ϵ, one of the following two conditions holds

(*a*) $$\|\text{output} - f(\text{input})\| \leqslant \epsilon,$$
(*b*) $$\|f^{-1}(\text{output}) - \text{input}\| \leqslant \epsilon;$$

where we are measuring in (*a*) the difference, in some sense, between the 'true output' for the given input and the answer outputted by the computer; and in (*b*) the difference, in some sense, between the 'true input' corresponding to the actual computer output and the given input. An analysis of a given problem which establishes a result of the form (*a*) is called a *forward error analysis*, while an analysis producing a result of the form (*b*) is called a *backward error analysis*.

Suppose, for instance, we obtained as the result of calculation the value $10^1(0{\cdot}1414)$ for $\sqrt{2}$ and ϵ is 10^{-3}. Then this result may be considered satisfactory in both the 'forward' and 'backward' senses since

$$|10^1(0{\cdot}1414) - \sqrt{2}| < 10^{-3},$$

and

$$|[10^1(0{\cdot}1414)]^2 - 2| < 10^{-3}.$$

We now look into some of the deficiencies of computer arithmetic and describe how these deficiencies may be overcome.

1.3 Subtraction of nearly equal numbers

We take as an example the evaluation of the roots of the quadratic polynomial

$$ax^2 + bx + c = 0. \qquad (1.1)$$

The roots are given by the formula

$$x = [-b \pm \sqrt{(b^2 - 4ac)}]/2a. \qquad (1.2)$$

Let $a = 1$, $b = -18$, $c = 1$ in (1.1) so that we are finding the roots x_1, x_2 of the equation

$$x^2 - 18x + 1 = 0.$$

Equation (1.2) gives

$$x_1 = 9 + \sqrt{80}, \qquad x_2 = 9 - \sqrt{80},$$

so that on our four decimal digit machine taking $\sqrt{80}$ as $10^1(0{\cdot}8944)$, we obtain

$$x_1 = 10^2(0{\cdot}1794), \qquad x_2 = 10^{-1}(0{\cdot}5600).$$

Here x_2 is obtained as the small difference of two relatively large numbers, 9 and $\sqrt{80}$, and two significant digits are lost. On the other hand if x_2 is obtained not from (1.2) but from the relation

$$x_1 x_2 = c/a \quad (=1), \tag{1.3}$$

we find $x_2 = 10^{-1}(0{\cdot}5574)$, with an error of only one unit. Thus, while the formulae $x_2 = 9 - \sqrt{80}$ and $x_2 = 1/(9 + \sqrt{80})$ are equivalent, the latter is far superior from the point of view of numerical calculation.

Another example in which significant digits are lost is the computation of

$$\frac{1}{3} - \frac{1}{4},$$

or, more generally,

$$\frac{1}{x} - \frac{1}{x+1}, \qquad 3 \leqslant x \leqslant 1000, \text{ say.}$$

However, if we carry out the computation of $1/12$, or in the more general case $1/[x(x+1)]$, we get as many significant digits as the representation permits.

The moral from both these examples is that subtraction of nearly equal numbers should be avoided if possible.

1.4 Addition in descending order of magnitude

Suppose we are required to evaluate e^4 using the series expansion

$$e^x = 1 + x + \frac{x^2}{2!} + \frac{x^3}{3!} + \ldots + \frac{x^{n-1}}{(n-1)!} + \ldots, \tag{1.4}$$

and to estimate the error in our answer. Note the correct answer to four significant digits is 54·60.

There are three types of errors occurring in this problem. Two types, the errors due to representation and computation, have already been mentioned. The third arises because it is impossible to enumerate all the terms in the series. We shall assume that the series is curtailed at $x^{n-1}/(n-1)!$, where $n > x > 0$, so that the cumulative error in neglecting the terms $x^n/n!$, $x^{n+1}/(n+1)!$, etc., called the truncation error, is

$$\frac{x^n}{n!}\left(1 + \frac{x}{n+1} + \frac{x^2}{(n+1)(n+2)} + \ldots\right) < \frac{x^n}{n!}\left(1 + \frac{x}{n} + \frac{x^2}{n^2} + \ldots\right) < \frac{x^n}{n!}\left(1 - \frac{x}{n}\right)^{-1} \tag{1.5}$$

When $x = 4$, the truncated series consisting of the first fourteen terms, the first three being exact and the remainder correctly rounded, is

$$\begin{aligned}&10^1(0{\cdot}1000 + 0{\cdot}4000 + 0{\cdot}8000) + 10^2(0{\cdot}1067 + 0{\cdot}1067)\\ &+ 10^1(0{\cdot}8534 + 0{\cdot}5689 + 0{\cdot}3251 + 0{\cdot}1625) + 10^0(0{\cdot}7224 + 0{\cdot}2890 + 0{\cdot}1051)\\ &+ 10^{-1}(0{\cdot}3503 + 0{\cdot}1078).\end{aligned}$$

The truncation error is less than

$$\frac{4^{14}}{14!}\left(1-\frac{4}{14}\right)^{-1} < 0{\cdot}0044,$$

while the representation error in the rounded terms is at most 0·5 in the fourth significant digit. It follows that the representation error is at most 0·012 16. Addition of the two largest terms $10^2(0{\cdot}1067)$, $10^2(0{\cdot}1067)$ gives $10^2(0{\cdot}2134)$ without any computational error. Adding the next largest number to the partial sum gives $10^2(0{\cdot}2987)$ with a computational error of 0·004. Carrying on in this manner yields a total of 54·61 and a maximum computational error of almost 0·02 giving a total error of at most 0·03656, say 0·04. Thus

$$|e^4 - 54{\cdot}61| < 0{\cdot}04.$$

If the terms are added up in reverse order, i.e., starting with the smallest term first, we find

$$|e^4 - 54{\cdot}60| < 0{\cdot}03. \tag{1.6}$$

Here our truncation and representation errors are the same as previously but our computational error is now only 0·007 31.

The moral from this example is that numbers should be added in ascending order of magnitude wherever possible. We remark, too, that (1.4) can be used to obtain a value for e^{-4}. It turns out that our answer is in the region of 0·02 and our estimated error is of the same magnitude. In other words, it is possible that not a single digit in our final answer is correct! However, if we estimate e^{-4} by the formula

$$e^{-4} = 1/e^4,$$

using our computed result of 54·60 for e^4, then it follows from (1.6) that our result is correct to three significant digits.

The function e^x is often required in computational problems and, after the foregoing example, one would naturally be tempted to ask 'How do we know that we will obtain a satisfactory answer for e^x when using our computer?' The answer here is that in all computers the standard functions are evaluated by methods which minimize the effects of errors. For example, a method of evaluating e^x by computer is as follows: We assume x is represented in the form

$$x = a.2^b.$$

Then

$$e^x = 2^{x \log_2 e}.$$

We carry out the floating point multiplication of $x \log_2 e$ and evaluate

$$\alpha = [x \log_2 e] = \text{the integer part of } x \log_2 e,$$
$$\beta = \{x \log_2 e\} = \text{the fractional part of } x \log_2 e,$$

where $0 < \beta < 1$. It follows that

$$e^x = 2^{\alpha+\beta} = 2^\alpha e^{\beta \log_e 2}.$$

We evaluate $e^{\beta \log_e 2}$ from (1.4), curtailing the series at the term $x^{10}/10!$, say. Since

$$0<\beta<1, \qquad \log_e 2<1, \qquad 1<e^{\beta \log_e 2}<2,$$

it follows that our truncation error is small because of the rapid convergence of the series (1.4). Further, our representation and computational errors are small because of the small number of terms we are considering. We finally make an adjustment to obtain our result for e^x in the form

$$e^x = a_1 . 2^{b_1},$$

with

$$b_1 = \alpha + 1, \qquad a_1 = \tfrac{1}{2} e^{\beta \log_e 2}.$$

It should be pointed out that errors are also incurred in this example in the representation of $\log_2 e$, $\log_e 2$, $\{x \log_2 e\}$, etc., the cumulative effects of which are not considered here, the error analysis involved being outside the scope of the present section (see section 1.9 and exercise 24).

In the next two sections we consider the effects of errors in other procedures These errors arise mainly as a result of multiplication.

1.5 Linear equations

Consider the linear equations discussed by Fox [1],

$$0{\cdot}0003x + 0{\cdot}1000y = 1, \qquad (1.7)$$

$$-0{\cdot}8000x + 0{\cdot}1249y = 0{\cdot}6. \qquad (1.8)$$

The mathematician solves these simultaneous equations by dividing the first equation by 0·0003 (the first non-zero pivot) thus obtaining

$$x + \left(\frac{0{\cdot}1000}{0{\cdot}0003}\right)y = \frac{1}{0{\cdot}0003} \qquad (1.9)$$

and adding 0·8000 times this to (1.8), obtaining

$$0x + \left[0{\cdot}1249 + 0{\cdot}8000\left(\frac{0{\cdot}1000}{0{\cdot}0003}\right)\right]y = 0{\cdot}6 + \frac{0{\cdot}8000}{0{\cdot}0003} \qquad (1.10)$$

From (1.10) we can compute y as the ratio of two numbers and substitute back in (1.9) to produce x. Let us now see what happens when we use our four decimal digit floating point computer. Equations (1.9) and (1.10) become respectively

$$x + 10^3(0{\cdot}3333)y = 10^4(0{\cdot}3333) \qquad (1.11)$$

$$0x + 10^3(0{\cdot}2668)y = 10^4(0{\cdot}2668) \qquad (1.12)$$

yielding $\bar{y} = 10$, $\bar{x} = 0$ as our computed solution.* Clearly (1.7) is satisfied by this

*We shall use the bar notation frequently to denote computed solutions.

solution, but substitution in (1.8) reveals a discrepancy. Moreover, the same solution would have been obtained if the elements 0·1249 and 0·6 had been replaced by any values in the ranges 0·0501 to 0·1499 and 0·5001 to 1·499 respectively.

Now suppose that we take the numerically largest coefficient of x or y as the pivot, in this case $-0{\cdot}8000$. Our process of elimination now gives

$$x - 0{\cdot}1561y = -0{\cdot}7500 \tag{1.13}$$

$$0x + 0{\cdot}1000y = 1{\cdot}000 \tag{1.14}$$

which leads to the computed values $\bar{y} = 10{\cdot}00$ and $\bar{x} = 0{\cdot}8110$, which is very close to the exact solution ($y = 9{\cdot}998$, $x = 0{\cdot}8109$ to four significant figures).

1.6 Recurrence relations

Computers are particularly well-suited to iterative methods so it is natural to restate problems in this form. Consider, for example,

$$I_n = e^{-1}\int_0^1 x^n e^x \, dx. \tag{1.15}$$

I_n satisfies the recurrence relation

$$I_n = 1 - nI_{n-1} \tag{1.16}$$

and on our four decimal digit floating point computer with $\bar{I}_0 = 0{\cdot}6321$, the correctly rounded value of I_0, the computer produces the following results

$\bar{I}_0$	$\bar{I}_1$	$\bar{I}_2$	$\bar{I}_3$	$\bar{I}_4$	$\bar{I}_5$	$\bar{I}_6$	$\bar{I}_7$
0·6321	0·3680	0·2640	0·2080	0·1680	0·1600	0·0400	0·7200.

The computed result for I_7 is clearly wrong since

$$I_n < e^{-1} \max_{0 \leqslant x \leqslant 1} (e^x) \int_0^1 x^n \, dx < \frac{1}{n+1}$$

i.e.,

$$I_7 < \tfrac{1}{8} = 0{\cdot}1250,$$

and, further, the sequence $\{I_n\}$ is a decreasing sequence. The discrepancy can be accounted for by rapid accumulation of errors originating with a small representation error in $\bar{I}_0$, and a computational error in obtaining $\bar{I}_1$, being multiplied successively by the factors 2, 3, . . . , 7.

If, however, we use (1.16) in the form

$$I_{n-1} = (1 - I_n)/n, \tag{1.17}$$

and start with the correctly rounded value of I_7 our computer obtains

$\bar{I}_7$	$\bar{I}_6$	$\bar{I}_5$	$\bar{I}_4$	$\bar{I}_3$	$\bar{I}_2$	$\bar{I}_1$	$\bar{I}_0$
0·1124	0·1269	0·1455	0·1708	0·2073	0·2643	0·3680	0·6320
(0·1124)	(0·1268)	(0·1456)	(0·1709)	(0·2073)	(0·2642)	(0·3679)	(0·6321)

and every result is correct to within one unit in the last digit (the correctly rounded results appear in brackets). It is interesting to note that even if we start with $\bar{I}_7 = 0$ we obtain $\bar{I}_0 = 0{\cdot}6320$. This is because the error is being reduced substantially at each stage as we are multiplying it successively by $\frac{1}{7}, \frac{1}{6}, \ldots, 1$.

1.7 Good and bad methods

Besides seeking methods which produce the least errors in any given circumstance we also seek methods which reduce the amount of computation. These methods often require some mathematical rearrangement of the problem in the first instance, but this is usually worthwhile. Take as an example

$$I = \int_0^1 \int_0^1 \int_0^1 \int_0^1 \int_0^1 e^{-x_1 x_2 x_3 x_4 x_5}\, dx_1\, dx_2\, dx_3\, dx_4\, dx_5 \tag{1.18}$$

This has been used by many authors to test their methods of integration.* Haselgrove, [2], found that he required 12 000 function evaluations with the (unmodified) Monte Carlo technique to obtain a value of I correct to 4 decimal places. With his own method of Cesàro sums he evaluated the integral correctly to 6 decimal places with 12 000 function evaluations. Lyness and McHugh [3], using Richardson's extrapolation to the limit, required 4425 function evaluations to obtain 8 decimal place accuracy. However, by straightforward integration we can show that

$$I = 1 - \frac{1}{2^4 2!} + \frac{1}{3^4 3!} - \frac{1}{4^4 4!} + \ldots . \tag{1.19}$$

The truncated series terminating with $-1/(8^4 8!)$ is equivalent, in machine time, to one function evaluation and gives

$$\bar{I} = 0{\cdot}970\,657\,191,$$

which is the correct value to nine decimal places. It is worth noting that one can evaluate I on a desk calculator by means of the series (1.19) far quicker, and more accurately, than an electronic digital computer using the three earlier methods.

*It should be stressed that these authors knew result (1.19).

It has been the intention, in this introductory chapter, to deter the reader from tackling problems by the first approach that comes to mind. From the examples it should be clear that this path is fraught with difficulties. We hope, however, that the reader has not been deterred from reading the remainder of this book. This is devoted to giving methods for the solution of a wide variety of problems together with an appraisal of the methods. The majority of these methods are standard equipment in most computer installations and so the reader will not be required to do any elaborate programming. It may still happen that, even with a tested programme, difficulties are encountered. In this connection, one final point cannot be stressed often enough: 'If in doubt consult a competent numerical analyst'.

We conclude this chapter with a more detailed examination, and an extension, of some of the ideas contained earlier in the chapter. This will form the basis for the later error analyses in this book.

1.8 Interval analysis

The subject known as Interval Analysis has been developed in order to estimate the worthwhile precision in physical problems. To illustrate the basic ideas of this topic we require some definitions. Firstly, we define an *interval number* to be an ordered pair of real numbers, $[a, b]$, with $a \leqslant b$. An interval number is also a set of real numbers. The interval number, $[a, b]$, is the set of real numbers x such that $a \leqslant x \leqslant b$. Degenerate intervals of the form $[a, a]$ are equivalent to real numbers and we denote an interval of this type by a.

We define four arithmetical operations on intervals (+, −, ., /) by the formulae

$$[a, b] + [c, d] = [a + c, b + d] \tag{1.20}$$

$$[a, b] - [c, d] = [a - d, b - c] \tag{1.21}$$

$$[a, b] \,.\, [c, d] = [\min(ac, ad, bc, bd), \max(ac, ad, bc, bd)] \tag{1.22}$$

and, if 0 does not belong to $[c, d]$,

$$[a, b] \,/\, [c, d] = [a, b] \,.\left[\frac{1}{d}, \frac{1}{c}\right]. \tag{1.23}$$

Thus, for example,

$$[1, 2] + [2, 3] = [3, 5]$$
$$[1, 2] - [2, 3] = [-2, 0]$$
$$[1, 2] \,.\, [2, 3] = [2, 6]$$
$$[1, 2] \,/\, [2, 3] = [1, 2] \,.\, [\tfrac{1}{3}, \tfrac{1}{2}] = [\tfrac{1}{3}, 1].$$

We now consider a simple application of interval arithmetic to computing bounds for the range of values taken by a function. We take, for example,

$$f(x) = x(1 - x), \qquad 0 \leqslant x \leqslant 1.$$

Then

$$x \in [0, 1],^* \qquad (1 - x) \in [0, 1],$$

so that

$$f(x) \in [0, 1].[0, 1] = [0, 1].$$

Alternatively, we can write

$$f(x) = x - x^2,$$

and
$$x \in [0, 1], \qquad x^2 \in [0, 1],$$

so that
$$f(x) \in [0, 1] - [0, 1] = [-1, 1].$$

Again,
$$f(x) = \tfrac{1}{4} - (x - \tfrac{1}{2})^2,$$

and
$$(x - \tfrac{1}{2}) \in [-\tfrac{1}{2}, \tfrac{1}{2}], \qquad (x - \tfrac{1}{2})^2 \in [-\tfrac{1}{4}, \tfrac{1}{4}]$$

so that this time we have

$$f(x) \in \tfrac{1}{4} - [-\tfrac{1}{4}, \tfrac{1}{4}] = [0, \tfrac{1}{2}].$$

The actual range of values of $f(x)$ for x in the interval $[0, 1]$ is $[0, \frac{1}{4}]$, as can be deduced from the differential calculus. The reason for the intervals being unequal, although the three formulae for $f(x)$ are algebraically equivalent, is the failure of the distributive law for intervals. By this we mean that if I, J, K, are intervals,

$$I.(J + K) \subset I.J + I.K.^\dagger$$

This is in contrast to most algebraic operations where, as with real numbers a, b, c,

$$a.(b + c) = a.b + a.c.$$

The failure to obtain the exact range is due to no allowance having been made for a special property of x^2. Namely, that if $x \in [-1, 1]$, say, then $x^2 \in [0, 1]$ although, from our definition of interval multiplication, we can only obtain $x^2 \in [-1, 1]$. If we use this special property then, for

$$f(x) = \tfrac{1}{4} - (x - \tfrac{1}{2})^2,$$

$$(x - \tfrac{1}{2}) \in [-\tfrac{1}{2}, \tfrac{1}{2}], \qquad (x - \tfrac{1}{2})^2 \in [0, \tfrac{1}{4}],$$

and we now have

$$f(x) \in \tfrac{1}{4} - [0, \tfrac{1}{4}] = [0, \tfrac{1}{4}],$$

– the optimum result.

The interval arithmetic described is exact-interval arithmetic. Modifications are necessary to cover the operations involved in computer arithmetic. For instance, if we consider the multiplication of two interval numbers $[a, b]$ and $[c, d]$, where

*By $x \in [0, 1]$ we mean 'x belongs to the interval $[0, 1]$'.

† $\subset$ means 'is contained in'.

a, b, c, d are stored exactly in our four decimal digit floating point machine, the product $I = [a, b].[c, d]$ involves the calculation of the terms ac, ad, bc, bd, which we know can only be obtained approximately. Allowances have to be made as it is possible that the rounding operations will give us a value of I which is smaller than the true value of the interval. Since we can keep track of the remainder in the computer multiplication we can overcome this snag by computing an interval I_1 which, while not exact, just contains the interval I.

1.9 Error bounds for computer arithmetic

In earlier sections we describe how errors arose in the course of a calculation using a digital computer. In later chapters we shall be looking in greater detail at the errors incurred in certain numerical procedures and, in order to estimate these errors, we specify here bounds for the errors involved in fixed and floating point arithmetical operations. It is useful to adopt a notation to indicate whether our calculations are fixed point or floating point so we write, for example,

$$\bar{d} = fi\,(ab + c) \qquad \text{or} \qquad \bar{d} = fl\,(ab + c)$$

to denote that $\bar{d}$ is the value obtained by computing $ab + c$ in fixed or floating point form respectively.

1.10 Rounding errors in four digit floating point computation

(*a*) *Multiplication* satisfies

$$fl(x_1 x_2) = x_1 x_2 (1 + \epsilon), \qquad |\epsilon| \leqslant \tfrac{1}{2} 10^{-3}, \tag{1.24}$$

where ϵ is the *relative* error in the computation. It is sometimes convenient to think of the computed product as being the exact product of $x_1(1 + \epsilon)$ and x_2, or of x_1 and $x_2(1 + \epsilon)$ or any combination $x_1(1 + \epsilon)^p, x_2(1 + \epsilon)^{1-p}, 0 < p < 1$.

(*b*) *Division* satisfies

$$fl(x_1/x_2) = (x_1/x_2)(1 + \epsilon), \qquad |\epsilon| \leqslant \tfrac{1}{2} 10^{-3}. \tag{1.25}$$

The computed quotient is the exact value of $x_1(1 + \epsilon)$ divided by x_2. ϵ is again a relative error.

(*c*) *Addition* is slightly more complicated than division and multiplication since we have the possibility of two rounding errors arising, one when the smaller number is shifted down and the other when the mantissa goes out of bounds. We can express floating point addition in the form

$$fl(x_1 + x_2) = x_1(1 + \epsilon_1) + x_2(1 + \epsilon_2) \tag{1.26}$$

where ϵ_1, ϵ_2 are relative errors which satisfy

$$|\epsilon_1|, |\epsilon_2| \leqslant \tfrac{1}{2}10^{-3}$$

but are in general different. The above result can be expressed more conveniently as

$$fl(x_1 + x_2) = (x_1 + x_2)(1 + \epsilon), \qquad |\epsilon| \leqslant 0{\cdot}55 \times 10^{-3} \tag{1.27}$$

A similar result holds for subtraction.

We now examine the errors involved in some common floating point operations. Firstly, we consider the extended product p_n,

$$p_n = fl(x_1 x_2 \ldots x_n),$$

where p_n is obtained by evaluating recursively the relations

$$p_1 = x_1,$$
$$p_r = fl(p_{r-1}x_r) = p_{r-1}x_r(1 + \epsilon_r), \qquad r = 2, \ldots, n$$

where $$|\epsilon_r| \leqslant \tfrac{1}{2}10^{-3}.$$

We have

$$p_n = x_1 x_2 \ldots x_n (1 + \epsilon_2)(1 + \epsilon_3) \ldots (1 + \epsilon_n),$$

which implies that

$$fl(x_1 x_2 \ldots x_n) = x_1 x_2 \ldots x_n (1 + E),$$

where

$$(1 - \tfrac{1}{2}10^{-3})^{n-1} \leqslant 1 + E \leqslant (1 + \tfrac{1}{2}10^{-3})^{n-1}. \tag{1.28}$$

If $n - 1$ is much smaller than 2×10^3 the computed product has a low relative error. It should be noted that the bounds (1.28) are independent of the order in which the multiplication is performed.

Secondly, we consider the extended addition

$$s_n = fl(x_1 + x_2 + \ldots + x_n).$$

We define the quantities s_r recursively by the relations

$$s_1 = x_1,$$
$$s_r = fl(s_{r-1} + x_r) = (s_{r-1} + x_r)(1 + \epsilon_r), \qquad r = 2, \ldots, n$$

where $$|\epsilon_r| \leqslant 0{\cdot}55 \times 10^{-3}$$

Combining these equations we have

$$s_n = x_1(1+\eta_1) + x_2(1+\eta_2) + \ldots + x_n(1+\eta_n)$$

$$1+\eta_1 = (1+\epsilon_2)(1+\epsilon_3)\ldots(1+\epsilon_n)$$

$$1+\eta_r = (1+\epsilon_r)(1+\epsilon_{r+1})\ldots(1+\epsilon_n), \qquad (r = 2, \ldots, n)$$

$$(1 - 0{\cdot}55 \times 10^{-3})^{n-1} \leqslant 1+\eta_1 \leqslant (1 + 0{\cdot}55 \times 10^{-3})^{n-1} \tag{1.29}$$

$$(1 - 0{\cdot}55 \times 10^{-3})^{n+1-r} \leqslant 1+\eta_r \leqslant (1 + 0{\cdot}55 \times 10^{-3})^{n+1-r} \tag{1.30}$$

where $r = 2, \ldots, n$. In this instance the bounds are dependent on the order of summation. The upper bound for the error is smallest if the terms are added in order of increasing absolute magnitude since then the largest factor $(1 + \eta_1)$ is associated with the smallest x_r.

Finally, we consider the error involved in computing the inner product

$$s_n = fl(a_1b_1 + a_2b_2 + \ldots + a_nb_n),$$

where s_n is calculated recursively by the relations

$$\left.\begin{aligned} &t_r = fl(a_rb_r), \qquad r = 1, 2, \ldots, n \\ &s_1 = t_1, \qquad s_r = fl(s_{r-1} + t_r), \qquad r = 2, \ldots, n. \end{aligned}\right] \tag{1.31}$$

From these relations we deduce

$$t_r = a_rb_r(1+\xi_r), \qquad |\xi_r| \leqslant 0{\cdot}5 \times 10^{-3}$$

$$s_r = (s_{r-1} + t_r)(1+\eta_r), \qquad |\eta_r| \leqslant 0{\cdot}55 \times 10^{-3}.$$

Hence

$$s_n = a_1b_1(1+\epsilon_1) + a_2b_2(1+\epsilon_2) + \ldots + a_nb_n(1+\epsilon_n),$$

$$1+\epsilon_1 = (1+\xi_1)(1+\eta_2)\ldots(1+\eta_n),$$

$$1+\epsilon_r = (1+\xi_r)(1+\eta_r)\ldots(1+\eta_n), \qquad r = 2, \ldots, n.$$

Hence

$$(1 - 0{\cdot}5 \times 10^{-3})(1 - 0{\cdot}55 \times 10^{-3})^{n-1} \leqslant 1+\epsilon_1 \leqslant (1 + 0{\cdot}5 \times 10^{-3})(1 + 0{\cdot}55 \times 10^{-3})^{n-1},$$

$$(1 - 0{\cdot}5 \times 10^{-3})(1 - 0{\cdot}55 \times 10^{-3})^{n-r+1} \leqslant 1+\epsilon_r \leqslant (1 + 0{\cdot}5 \times 10^{-3})(1 + 0{\cdot}55 \times 10^{-3})^{n-r+1},$$

where $\qquad r = 2, \ldots, n.$

Since

$$1-0{\cdot}55\times10^{-3} \leqslant 1-0{\cdot}5\times10^{-3}, \qquad 1+0{\cdot}5\times10^{-3} \leqslant 1+0{\cdot}55\times10^{-3},$$
$$(1-0{\cdot}55\times10^{-3})^{n+1} \leqslant (1-0{\cdot}55\times10^{-3})^{n}, \qquad (1+0{\cdot}55\times10^{-3})^{n} \leqslant (1+0{\cdot}55\times10^{-3})^{n+1},$$

we can simplify the above expressions and avoid having a special formula for the bounds on ϵ_1. We can now write

$$(1-0{\cdot}55\times10^{-3})^{n-r+2} \leqslant 1+\epsilon_r \leqslant (1+0{\cdot}55\times10^{-3})^{n-r+2}, \qquad (1.32)$$

where $r = 1, \ldots, n$.

1.11 Rounding errors in four digit fixed point computation

Addition and subtraction involve no round-off error in this mode of computation although we have to bear in mind the possibility that our sum or difference is outside the permitted range. For multiplication we have

$$c = fi\,(ab) = ab + \epsilon, \qquad |\epsilon| \leqslant \tfrac{1}{2}10^{-4}, \qquad (1.33)$$

and the exact and rounded products always lie in the permitted range. The quotient a/b will not lie in the permitted range unless $|b| \geqslant |a|$ and this operation will thus require special consideration in practice. If $|b| \geqslant |a|$, then

$$c = fi\,(a/b) = \frac{a}{b} + \epsilon, \qquad |\epsilon| \leqslant \tfrac{1}{2}\times10^{-4}. \qquad (1.34)$$

Note that for fixed point arithmetic the error incurred is an *absolute* error.

For floating point computation we gave examples of error analyses for extended multiplication, extended addition and inner product accumulation. We content ourselves here with a consideration of the fixed point computation of an inner product s,

$$s = fi\left(\sum_{1}^{n} a_r b_r\right).$$

Clearly (neglecting the possibility of going out of range), the only error incurred is the rounding error in each multiplication. Thus

$$s = \sum_{1}^{n} a_r b_r + \epsilon, \qquad |\epsilon| \leqslant \tfrac{1}{2}n\times10^{-4},$$

1.12 Notes

The problem of disastrous error growth in the computation of recurrence relations has received a great deal of attention. One of the most common types of recurrence relations is of the form

$$y_{n+1} + a_n y_n + b_n y_{n-1} = 0, \qquad n = 1, 2, \ldots \tag{1.35}$$

where a_n, b_n are given sequences of real (or complex) numbers and $b_n \neq 0$. This problem is treated in great detail in Gautschi [4], where convergent algorithms are given for evaluating the 'minimal' solution of (1.35). See also section 10.3.

The reader is referred to Moore [5] and Rall [6] for further information about interval analysis.

Further examples of error analyses will be introduced in later chapters. The reader is referred to Wilkinson [7], [8], and Rall [6] for a fuller discussion of the subject. Wilkinson also considers computational errors in machines which work on a slightly different principle to our hypothetical computer.

REFERENCES

[1] Fox, L., 'The Proper Use of Recurrence Relations', *Math. Gazette*, Vol. XL1X, **370** (1965) 371–87.

[2] Haselgrove, C. B., 'A Method for Numerical Integration', *Math. Comp.*, **15** (1961) 323–37.

[3] Lyness, J. N. and McHugh, B. J. J., 'Integration over Multidimensional Hypercubes. I. A Progressive Procedure', *Comp. J.*, **6** (1963) 264–70.

[4] Gautschi, W., 'Computational Aspects of Three-term Recurrence Relations', *SIAM Review*, **9** (1967) 24–82.

[5] Moore, R. E., *Interval Analysis*, Prentice-Hall Inc., New Jersey (1966).

[6] Rall, L. B. (ed.), *Error in Digital Computation*, Vols I and II, J. Wiley & Sons, New York (1965).

[7] Wilkinson, J. H., Rounding Errors in Algebraic Processes, *N.P.L. Notes on Applied Science* No. 32, H.M.S.O., London (1963).

[8] Wilkinson, J. H., *The Algebraic Eigenvalue Problem*, Oxford University Press, London (1965).

Exercises

1. We denote a positive number having the form

$$a_r N^r + a_{r-1} N^{r-1} + \ldots + a_1 N + a_0 + \frac{a_{-1}}{N} + \frac{a_{-2}}{N^2} + \ldots + \frac{a_{-s}}{N^s}$$

by $a_r a_{r-1} \ldots a_1 a_0 \cdot a_{-1} a_{-2} \ldots a_{-s}$ (scale N), where the digits a_i

$(i = -s, -s+1, \ldots, 0, \ldots, r)$ satisfy $0 \leqslant a_i < N$. Show that the number 12·5 (scale 10/decimal) corresponds to 1100·1 (scale 2/binary) and 14·4 (scale 8/octal).

2. Find the numbers in the binary and octal scales corresponding to the decimal numbers 136·7, 4093, 0·3333.

3. Find the numbers in the octal and decimal scales corresponding to the binary numbers 11 111 000 111, 0·111 000 110 01.

4. Express the numbers 136·7, 4093, 0·3333 as floating point binary numbers of the form

$$a \,.\, 2^b$$

where b is an integer, $\frac{1}{2} \leqslant a < 1$, and the number a possesses (i) 10 digits, (ii) 20 digits. In each case find a bound for the representational error.

5. Express the binary number 11 111 000·111 011 1 as a floating point decimal number of the form

$$a \,.\, 10^b$$

where b is an integer, $0{\cdot}1000 \leqslant a < 1$, and the number a possesses (i) 4 digits, (ii) 8 digits.

6. Given the numerical value of $\cosh x$ and $\sinh x$ for a large positive value of x, explain why the direct evaluation of $\cosh x - \sinh x$ produces a very unsatisfactory numerical result. How could the result be improved?

7. Evaluate $\sum_1^{20} \frac{1}{n^2}$ on a four decimal digit floating point computer by summing terms

(*a*) in descending order of magnitude,

(*b*) in ascending order of magnitude.

Why do the answers to (*a*) and (*b*) differ, and which one is nearest to the exact value?

8. Show that the series $\sum_1^{\infty} \frac{1}{n}$ diverges. Why is this series convergent when the terms are summed in descending order of magnitude on a floating point computer? What value would you expect your computer to obtain for the sum (approximately)?

9. Given that I_0 lies in the interval [0·6321, 0·6322], obtain an interval estimate for I_7 where I_n satisfies the recurrence relation (1.16).

10. Suppose there is an error ϵ_n in I_n, where I_n satisfies the recurrence relation (1.16). Show that, if no further errors occur in the computation, the error ϵ_{n+k} in I_{n+k} is given by

$$\epsilon_{n+k} = (n+k)(n+k-1) \ldots (n+1)\epsilon_n.$$

11. The Bessel functions $J_n(x)$ satisfy the recurrence relation

$$J_{n+1}(x) - (2n/x)J_n(x) + J_{n-1}(x) = 0.$$

If $x = 1$ and $J_0(1) = 0{\cdot}7652$, $J_1(1) = 0{\cdot}4401$, compute $J_n(1)$ for $n = 2, 3, \ldots, 10$.

Repeat this calculation starting with the more accurate values $J_0(1) = 0{\cdot}765\ 197\ 69$, $J_1(1) = 0{\cdot}440\ 050\ 59$. How do you reconcile your results with the known fact that $J_n(1) \to 0$ as n increases?

12. Consider the system of simultaneous linear equations

$$a_{11}x_1 + a_{12}x_2 = b_1,$$
$$a_{21}x_1 + a_{22}x_2 = b_2, \qquad a_{11} \neq 0, \qquad a_{11}a_{22} - a_{12}a_{21} \neq 0.$$

The solution of these equations is given by

$$x_2 = \frac{a_{11}b_2 - a_{21}b_1}{a_{11}a_{22} - a_{12}a_{21}},$$

$$x_1 = (b_1 - a_{12}x_2)/a_{11}.$$

Obtain interval estimates for x_1 and x_2 when $a_{11} \in [1{\cdot}25, 1{\cdot}26]$, $a_{12} \in [1{\cdot}25, 1{\cdot}26]$, $a_{21} \in [2{\cdot}60, 2{\cdot}61]$, $a_{22} \in [10{\cdot}1, 10{\cdot}2]$, $b_1 \in [2{\cdot}5, 2{\cdot}51]$, $b_2 \in [12{\cdot}7, 12{\cdot}8]$. Obtain an interval estimate for x_1 using the formula

$$x_1 = \frac{b_1a_{22} - b_2a_{12}}{a_{11}a_{22} - a_{12}a_{21}}.$$

13. Given that $x \in [10, 12]$ find an interval estimate for y, where

$$y = x/(x - 1).$$

Obtain the exact range of values of y. Explain why the two estimates differ.

14. Construct numerical examples to show that the bounds given for the error in the floating point arithmetic operations (1.24)–(1.26) cannot be improved.

15. On a floating point computer using a word length of t digits and operating in the number scale N ($N = 2, 8$, or 10) the relative errors involved in multiplication, division, and addition are given by

$$fl(x_1x_2) = x_1x_2(1 + \epsilon), \qquad |\epsilon| \leqslant \tfrac{1}{2}N^{1-t}$$
$$fl(x_1/x_2) = (x_1/x_2)(1 + \epsilon), \qquad |\epsilon| \leqslant \tfrac{1}{2}N^{1-t}$$
$$fl(x_1 + x_2) = (x_1 + x_2)(1 + \epsilon), \qquad |\epsilon| \leqslant \tfrac{1}{2}(N + 1)N^{-t}.$$

For such a machine, determine bounds for the error incurred in evaluating extended products, extended sums, and inner products.

16. If $r \times 0{\cdot}55 \times 10^{-3} < 0{\cdot}1$, show that

$$1 - 0{\cdot}6083r \times 10^{-3} < (1 - 0{\cdot}55 \times 10^{-3})^r,$$
$$(1 + 0{\cdot}55 \times 10^{-3})^r < 1 + 0{\cdot}6083r \times 10^{-3},$$

and hence, in (1.32),

$$|\epsilon_r| < 0{\cdot}6083\,(n - r + 2)10^{-3}.$$

17. Suppose the number x is evaluated on a t-binary digit floating point computer. x is 1 initially and is increased by steps of $1/10$ in a loop of instructions, exit from the loop being by means of the Algol instruction

$$\textbf{if } x = 2 \textbf{ then go to } \text{LABEL}.$$

Explain why this instruction is ineffective.

18. The quadratic equation $x^2 - 18x + 1 = 0$ in section 1.3 may be rewritten in the form

$$x = 18 - 1/x.$$

Consider the iterative process (for given x_1)

$$x_{n+1} = 18 - 1/x_n, \qquad (n = 1, 2, \ldots).$$

Show that under certain conditions the sequence $\{x_n\}$ tends to a limit and this limit is the largest root of the original equation. Hence determine the largest root. Suggest an iterative procedure for evaluating the smallest root.

19. Put $J_{10}(1) = 0$, $J_9(1) = \tau$. Use the formula of exercise 11, in the form

$$J_{n-1}(1) = 2nJ_n(1) - J_{n+1}(1),$$

to evaluate $J_8(1)$, $J_7(1)$, ... Find τ from the identity

$$J_0(x) + 2J_2(x) + 2J_4(x) + 2J_6(x) + \ldots = 1,$$

and hence evaluate $J_9(1), \ldots, J_0(1)$. How do these results compare with the exact values of $J_9(1), \ldots, J_0(1)$?

20. The area of a triangle, Δ, is measured in terms of two adjacent sides and the included angle by the formula

$$\Delta = \tfrac{1}{2}ab \sin C \qquad (0 < C < \tfrac{1}{2}\pi, C \text{ in radians}).$$

If errors δa, δb, and δC are made in the measurements of a, b, and C show that the corresponding error $\delta\Delta$ in Δ (as found from the formula) satisfies

$$\left|\frac{\delta\Delta}{\Delta}\right| \leqslant \left|\frac{\delta a}{a}\right| + \left|\frac{\delta b}{b}\right| + \left|\frac{\delta C}{C}\right|.$$

21. If g is measured from the formula for the period of oscillation of a simple pendulum

$$T = 2\pi\sqrt{(l/g)},$$

and δT, δl are errors in the recorded values of T and l show that the error δg in the value of g found from the formula satisfies

$$\frac{\delta g}{g} = \frac{\delta l}{l} - 2\frac{\delta T}{T}.$$

22. If $a^{\frac{1}{2}}$ $(a > 0)$ is calculated on a four decimal digit floating point computer using the iterative formula

$$x_{n+1} = \tfrac{1}{2}(x_n + a/x_n),$$

show that

$$fl(a^{\frac{1}{2}}) = a^{\frac{1}{2}}(1 + \epsilon) \qquad \text{where} \qquad |\epsilon| < 1{\cdot}3 \times 10^{-3}.$$

23. If x and y are numbers represented exactly in a four decimal digit floating point computer show that

$$\frac{x}{\sqrt{(x^2 + y^2)}}(1 - \epsilon) < fl\frac{x}{\sqrt{(x^2 + y^2)}} < \frac{x}{\sqrt{(x^2 + y^2)}}(1 + \epsilon),$$

where $|\epsilon| < 2{\cdot}4 \times 10^{-3}$.

24. Carry out an error analysis of the method of calculating e^x given at the end of section 1.4. [Hint: Show that the calculated value of e^x is $e^x(1 + \epsilon)$, where ϵ is a small quantity.]

25. If $x \leqslant 0$ show that

$$e^x \in 1 + x + \ldots + \frac{x^{k-1}}{(k-1)!} + \frac{x^k}{k!}[0, 1].$$

2 Polynomials and their zeros

2.1 Introduction

The computation of a polynomial function is usually done by means of 'nested multiplication'. The procedure is as follows. Let our polynomial be

$$p_n(x) = a_0x^n + a_1x^{n-1} + \ldots + a_{n-1}x + a_n,$$

where the coefficients $a_r (r = 0, 1, \ldots, n)$ are real. Then we define a recurrence relation

$$p_0 = a_0, \qquad p_r = xp_{r-1} + a_r, \qquad r = 1, \ldots, n \tag{2.1}$$

giving $p_n = p_n(x)$. To illustrate this method, suppose we are required to find the value of

$$p_3(x) = 0{\cdot}1296x^3 + 0{\cdot}3424x^2 - 0{\cdot}2134x + 0{\cdot}7892,$$

when $x = 0{\cdot}1$, using the four decimal digit floating point computer described in chapter 1. We obtain the approximations (where the bars indicate computed values)

$$\begin{aligned}\bar{p}_1 &= 0{\cdot}1296 \times 0{\cdot}1 + 0{\cdot}3424 = 0{\cdot}3554,\\ \bar{p}_2 &= 0{\cdot}3554 \times 0{\cdot}1 - 0{\cdot}2134 = -0{\cdot}1779,\\ \bar{p}_3 &= -0{\cdot}1779 \times 0{\cdot}1 + 0{\cdot}7892 = 0{\cdot}7714.\end{aligned}$$

The correct value of $p_3(0{\cdot}1)$ is $0{\cdot}771\,413\,6$. Since, from (2.1),

$$a_r = p_r - xp_{r-1}, \qquad r = 1, \ldots, n$$

we see, for example, that $\bar{p}_3$ would be obtained exactly from $\bar{p}_2$ if a_3 were $0{\cdot}789\,19$. In fact, $\bar{p}_3$ is the exact value for $x = 0{\cdot}1$ of the 'perturbed polynomial'

$$0{\cdot}1296x^3 + 0{\cdot}342\,44x^2 - 0{\cdot}213\,44x + 0{\cdot}789\,19.$$

An analysis of the errors in nested multiplication proceeds as follows. We have

$$\begin{aligned}\bar{p}_0 &= a_0,\\ \bar{p}_r &= fl\,(x\bar{p}_{r-1} + a_r),\\ &= x\bar{p}_{r-1}(1 + \xi_r)(1 + \eta_r) + a_r(1 + \eta_r),\end{aligned}$$

where, for all r,

$$|\xi_r| \leqslant \tfrac{1}{2} \times 10^{-3}, \qquad |\eta_r| \leqslant 0{\cdot}55 \times 10^{-3}.$$

Combining these equations we have

$$\bar{p}(x) = a_0(1 + \epsilon_0)x^n + a_1(1 + \epsilon_1)x^{n-1} + \ldots + a_n(1 + \epsilon_n),$$

where

$$(1 - 0{\cdot}55 \times 10^{-3})^{2n+1-2r} \leqslant 1 + \epsilon_r \leqslant (1 + 0{\cdot}55 \times 10^{-3})^{2n+1-2r}.$$

Now
$$|\epsilon_r| \leqslant 0{\cdot}6083\,(2n - 2r + 1) \times 10^{-3},$$

and a very rough estimate for the relative error is given by

$$|p(x) - \bar{p}(x)| \leqslant \max_r |\epsilon_r| \sum_0^n |a_r x^r|.$$

If each $a_r \geqslant 0, x \geqslant 0$ then

$$|p(x) - \bar{p}(x)| \leqslant p(x) \max_r |\epsilon_r|$$

and we have a low relative error.

It is worth pointing out at this juncture that the upper bounds that we have determined for the relative error in polynomial evaluation are much larger than those actually obtained in practice. This is particularly so in the case of polynomials derived by truncating power series where the terms, for appropriate x, decrease rapidly in size.

We now turn our attention to the classical problem of computing the zeros or roots of a polynomial equation. We consider first a question which arises naturally from the previous discussion.

2.2 Condition of a polynomial with respect to the computation of its zeros

Consider the polynomial equation

$$p(z) = a_0 z^n + a_1 z^{n-1} + \ldots + a_{n-1} z + a_n = 0, \tag{2.2}$$

having as zeros $z_1, z_2, \ldots, z_n$ and assume that z_r is an isolated zero of the polynomial. As we saw in the previous section, we obtain a perturbed polynomial $\bar{p}(z)$,

$$\bar{p}(z) = p(z) + \epsilon q(z), \tag{2.3}$$

where

$$q(z) = b_0 z^n + b_1 z^{n-1} + \ldots + b_{n-1} z + b_n, \tag{2.4}$$

when we compute $p(z)$. Now, from the theory of functions we know that a zero $z_r(\epsilon)$ of $\bar{p}(z) = 0$ is given by

$$z_r(\epsilon) = z_r + \sum_1^\infty c_k \epsilon^k = z_r + h, \tag{2.5}$$

where the series converges for sufficiently small ϵ. Again, since $z_r(\epsilon)$ satisfies (2.3), we have

$$\sum_0^n a_{n-i}(z_r + h)^i + \epsilon \sum_0^n b_{n-i}(z_r + h)^i = 0, \tag{2.6}$$

which may be rewritten in the form

$$\sum_0^n \alpha_k h^k + \epsilon \sum_0^n \beta_k h^k = 0, \tag{2.7}$$

where

$$\alpha_k = \frac{1}{k!} p^{(k)}(z), \qquad \beta_k = \frac{1}{k!} q^{(k)}(z), \tag{2.8}$$

and $p^{(k)}(z)$, $q^{(k)}(z)$ denote the kth derivatives of $p(z)$ and $q(z)$ respectively. If, in (2.7), we now substitute for h its power series in ϵ and equate the coefficients of ϵ^0, ϵ^1 to zero we obtain

$$\alpha_0 = p(z_r) = 0,$$
$$\alpha_1 c_1 + \beta_0 = 0,$$

giving

$$c_1 = -\beta_0/\alpha_1 = -q(z_r)/p'(z_r).$$

From (2.5) it now follows that, for sufficiently small ϵ, we have

$$\left| z_r(\epsilon) - z_r + \epsilon \frac{q(z_r)}{p'(z_r)} \right| < K\epsilon^2 \tag{2.9}$$

where K is a constant. Provided we have some knowledge of $q(z)$, (2.9) is a particularly useful result for determining how close the perturbed zero $z_r(\epsilon)$ is to the actual zero z_r.

We give an immediate application of (2.9) to the case of a polynomial having a linear distribution of zeros which is discussed in slightly greater detail in Wilkinson [1]. Let

$$p(z) = (z-1)(z-2)\ldots(z-20) = 0$$

be the polynomial equation of degree 20 having as its zeros the natural numbers from 1 to 20. Let

$$q(z) = z^{19}, \qquad \bar{p}(z) = p(z) + \epsilon q(z).$$

Then, from (2.9), the perturbation $\delta z_r = z_r(\epsilon) - z_r$ in the rth zero which corresponds to a perturbation of ϵ in the coefficient a_1, satisfies the relation

$$\delta z_r \mp -\epsilon z_r^{19}/p'(z_r) = \pm \frac{\epsilon r^{19}}{(r-1)!\,(20-r)!}. \tag{2.10}$$

For given ϵ, δz_r has maximum magnitude when $r = 16$, its value then being approximately $0{\cdot}24 \times 10^{10}\epsilon$. If $\epsilon = -2^{-23}$, i.e., the relative error in the coefficient a_1 (= 210) is about $\frac{1}{2}$ in 10^9, we find that the exact roots of the perturbed polynomial $\bar{p}(z)$ are quite different from those of $p(z)$. For instance, the exact root of $\bar{p}(z)$ corresponding to the root 20 of $p(z)$ is now 20·8469 (to four decimal places), while the roots of $\bar{p}(z)$ corresponding to the real pairs 10, 11; 12, 13; 14, 15; 16, 17; 18, 19; of $p(z)$ are the complex pairs $10{\cdot}0953 \pm 0{\cdot}6435i$; $11{\cdot}7936 \pm 1{\cdot}6523i$; $13{\cdot}9924 \pm 2{\cdot}5188i$; $16{\cdot}7307 \pm 2{\cdot}8126i$; $19{\cdot}5024 \pm 1{\cdot}9403i$. The changes here are so great that linearized perturbation theory no longer applies! If $\epsilon = -2^{-55}$ then the largest perturbation occurs in the zero 16 which now takes the value 16·000 000 067 a result which is now accurately predicted by linearized perturbation theory. On the other hand when r is small δz_r is extremely small for both the values of ϵ which we have considered, i.e., the smallest zeros are insensitive to changes in the coefficient of z^{19}, or 'well-conditioned'. This is in contrast to the larger zeros which, as we have seen, are very sensitive to changes in the coefficient of z^{19}, or 'ill-conditioned'.

The above example might be considered a little artificial as, in practice, errors will occur in all the coefficients. We can assume in this case that the coefficients all have the same relative error and, assuming that we also know the magnitude of the coefficients, carry out our perturbation analysis, obtaining similar startling results. Further examples of the effects of small errors on the condition of the zeros are given in the exercises.

We now consider four methods for determining the zeros of a polynomial – the *method of bisection*, the *method of Newton*, *Bairstow's method*, and *Lehmer's method*.

2.3 The method of bisection

Suppose, as before, $p(z)$ is a polynomial of degree n in z and z_r is an isolated real zero which is well-conditioned. Further, suppose we are given a and b such that $a < z_r < b$, z_r is the only zero in (a, b), and

$$p(a) < 0, \qquad p(b) > 0.$$

Let $x_0 = a$, $x'_0 = b$. We define a sequence $\{x_k, x'_k\}$ such that

$$p(x_k) \leqslant 0, \qquad p(x'_k) > 0,$$

and new terms in the sequence are defined by (see Fig. 2.1)

$$\left.\begin{array}{llll} x_{k+1} = \frac{1}{2}(x_k + x'_k), & x'_{k+1} = x'_k & \text{if} & p\{\frac{1}{2}(x_k + x'_k)\} \leqslant 0 \\ x_{k+1} = x_k, & x'_{k+1} = \frac{1}{2}(x_k + x'_k) & \text{if} & p\{\frac{1}{2}(x_k + x'_k)\} > 0. \end{array}\right] \tag{2.11}$$

Clearly, if all our computations are exact, z_r lies in each of the intervals (x_k, x'_k). Since the width of the kth interval is $2^{-k}(b-a)$ we can, by increasing k, locate our zero z_r to any required precision. In practice, since it is not possible to compute

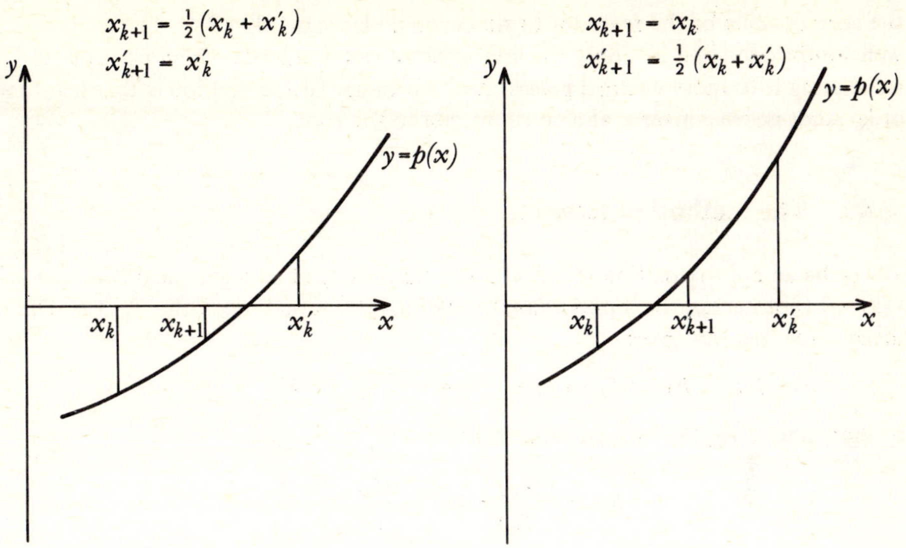

Fig. 2.1 Geometrical interpretation of method of bisection

our polynomial exactly, our previous error analysis shows that the best we can hope for is an estimate $\bar{z}_r$ for z_r where

$$|\bar{z}_r - z_r| < \epsilon_r, \text{ say,}$$

and ϵ_r depends on the condition of the root z_r. We are also limited to some extent by the number of digits we are using in our computation. To illustrate the method of bisection, we carry out on our four decimal digit floating point machine (the calculations are recorded in Table 2.1) the first few stages in the calculation of the real root of the polynomial equation

$$p(z) = z^3 - z - 1 = z(z^2 - 1) - 1 = 0,$$

given that it lies between 1 and 2.

Table 2.1 The method of bisection for $p(z) = z^3 - z - 1 = 0$

k	$\bar{x}_k$	$\bar{x}'_k$	$\bar{p}(\bar{x}_k)$	$\bar{p}(\bar{x}'_k)$	$\bar{p}\{\frac{1}{2}(\bar{x}_k + \bar{x}'_k)\}$
0	1·000	2·000	−1·000	5·000	0·875
1	1·000	1·500	−1·000	0·875	−0·296
2	1·250	1·500	−0·296	0·875	0·225
3	1·250	1·375	−0·296	0·225	−0·049
4	1·313	1·375	−0·049	0·225	0·083
5	1·313	1·344	−0·049	0·083	0·018

After five further iterations we find $\bar{x}_{10} = 1{\cdot}324$, $\bar{p}(\bar{x}_{10}) < 0$, $\bar{x}'_{10} = 1{\cdot}325$, $\bar{p}(\bar{x}'_{10}) > 0$ so we would expect our root to lie between 1·324 and 1·325. In fact, the correct value of the real root to six decimal places is 1·324 718. The root is well conditioned and it is only the limitation of our computer which prevents us calculating it to more decimal places. A disadvantage of the method is that it is first order, that is, we converge at a constant rate to the root.

2.4 The method of Newton

Let ξ_k be an approximation to an isolated real root z_r of an algebraical equation $f(z) = 0$ (not necessarily a polynomial equation) the actual root being $\xi_k + h$. The mean value theorem gives

$$0 = f(\xi_k + h) = f(\xi_k) + hf'(\xi_k) + \tfrac{1}{2}h^2 f''(\xi_k + \theta_2 h), \qquad 0 < \theta_2 < 1,$$

so that, since $f'(\xi_k) \neq 0$ for an isolated root,

$$h = -\frac{f(\xi_k)}{f'(\xi_k)} - \tfrac{1}{2}h^2 \frac{f''(\xi_k + \theta_2 h)}{f'(\xi_k)}, \tag{2.12}$$

and, if h is sufficiently small, a better approximation ξ_{k+1} to z_r is given by

$$\xi_{k+1} = \xi_k - \frac{f(\xi_k)}{f'(\xi_k)}. \tag{2.13}$$

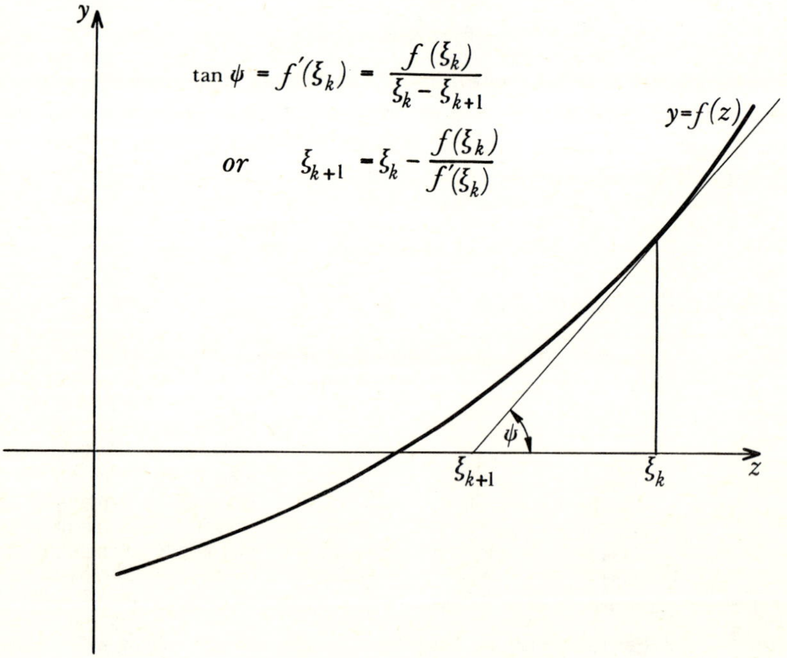

Fig. 2.2 Geometrical interpretation of Newton's method

It follows from (2.12) and (2.13) that the original small error, of magnitude $|h|$, has now been replaced by an error which is a multiple of h^2. A method which possesses this property is termed a quadratically convergent method. This is clearly an improvement from a computational point of view over the method of bisection where the error (which is approximately proportional to the range) can be shown to have been replaced by a multiple of the range – the method of bisection is consequently termed a linearly convergent method, or a first order method. However, the price paid for quadratic convergence is the evaluation of the derivative at each stage.

We now illustrate Newton's method with our polynomial example $p(z) = z^3 - z - 1$, taking $\xi_0 = 1$ as our first approximation. Our results are tabulated in Table 2.2. The value 1·325 is the closest we can get to our root with our four decimal digit floating point computer and, once again, it is the limitation of our hypothetical computer which prevents us achieving greater accuracy.

Table 2.2 Newton's method for the real root of $p(z) = z^3 - z - 1 = 0$

k	ξ_k	$p(\xi_k) = \xi_k^3 - \xi_k - 1$	$p'(\xi_k) = 3\xi_k^2 - 1$
0	1·000	−1·000	2·000
1	1·500	0·875	5·750
2	1·348	0·101	4·451
3	1·325	0·002	4·268
4	1·325	..	..

Newton's method can also be used to obtain isolated complex roots. Thus, suppose we take $\xi_0 = -0{\cdot}5 + 0{\cdot}5i$ as a first approximation to one of the complex roots of $p(z) = z^3 - z - 1 = 0$. Then our four decimal digit floating point computer obtains the results tabulated in Table 2.3.

Table 2.3 Newton's method for a complex root of $p(z) = z^3 - z - 1 = 0$

k	ξ_k	$p(\xi_k) = \xi_k^3 - \xi_k - 1$	$p'(\xi_k) = 3\xi_k^2 - 1$
0	$-0{\cdot}5000 + 0{\cdot}5000i$	$-0{\cdot}250 - 0{\cdot}2500i$	$-1{\cdot}000 - 1{\cdot}500i$
1	$-0{\cdot}6923 + 0{\cdot}5385i$	$-0{\cdot}037 + 0{\cdot}0795i$	$-0{\cdot}432 - 2{\cdot}237i$
2	$-0{\cdot}6611 + 0{\cdot}5611i$	$-0{\cdot}003 - 0{\cdot}0022i$	$-0{\cdot}633 - 2{\cdot}225i$
3	$-0{\cdot}6624 + 0{\cdot}5621i$		

In practice, the complex arithmetic can be minimized by noting that if $\xi_k = x_k + iy_k$ is an approximate root of $p(z) = 0$ so is $\bar{\xi}_k = x_k - iy_k$ and, in particular, $(z - \xi_k)(z - \bar{\xi}_k) = z^2 - \alpha z - \beta$, where $\alpha = 2x_k$, $\beta = -(x_k^2 + y_k^2)$ is an approximate quadratic factor of $p(z)$. We can then use the results (2.20)–(2.24), which we shall develop in connection with Bairstow's method, to obtain our Newton algorithm for complex roots in the form

$$\xi_{k+1} = \xi_k - (q_{n-1}\xi_k + q_n)/\{\xi_k(\alpha t_{n-3} + 2t_{n-2}) + 2\beta t_{n-3} - \alpha t_{n-2} + q_{n-1}\},$$

although it is preferable to use Bairstow's method for finding complex roots, as the Bairstow algorithm is easier to apply.

We now consider the effect of rounding errors on the determination of an isolated root z_r of a polynomial by Newton's method. We assume that we start from an approximation, ξ_0, for which Newton's process with exact arithmetic converges to z_r. Further, we assume the relative error in the computed $p'(z)$ is much smaller than the relative error in $p(z)$. This will certainly be true for values of z close to z_r. Let

$$\xi_k = z_r + h_k.$$

Then a rearrangement of equation (2.12) shows that

$$h_{k+1} \doteqdot \frac{p''(z_k)}{2p'(z_k)} h_k^2 \qquad (h_k \to 0). \tag{2.14}$$

Let $\bar{\xi}_k = z_r + \bar{h}_k$ be the computed approximation to ξ_k and let $z_r + h_{k+1}$ be the exact approximation obtained by using Newton's method with $z_r + \bar{h}_k$ as the previous approximation. Let the true values of $p(z_r + \bar{h}_k)$ and $p'(z_r + \bar{h}_k)$ be p_k and p'_k and the computed values be $\bar{p}_k$ and $\bar{p}'_k$, respectively. From (2.14) we have, for sufficiently small $\bar{h}_k$,

$$|h_{k+1}| < A|\bar{h}_k|^2, \tag{2.15}$$

and if we assume that $\bar{h}_k$ satisfies

$$A|\bar{h}_k| < 0{\cdot}1, \tag{2.16}$$

then the exact Newton process converges rapidly. Since the computed value of $p(z_r + \bar{h}_k)$ is the exact value of a polynomial with coefficients $a_i(1 + \epsilon_i)$ having zeros $z_i(\epsilon)$ we have

$$\bar{p}_k = a_0(1 + \epsilon_0) \prod_{i=1}^{n} [z_r + \bar{h}_k - z_i(\epsilon)], \tag{2.17}$$

while

$$p_k = a_0 \prod_{i=1}^{n} [z_r + \bar{h}_k - z_i], \tag{2.18}$$

so that division of (2.17) by (2.18) gives

$$\frac{\bar{p}_k}{p_k} = (1 + \epsilon_0) \frac{z_r + \bar{h}_k - z_r(\epsilon)}{\bar{h}_k} \prod_{i \neq r} \left(\frac{z_r + \bar{h}_k - z_i(\epsilon)}{z_r + h_k - z_i} \right). \tag{2.19}$$

In (2.19), the final product is close to unity for all $\bar{h}_k$ in which we are interested. It follows that $|\bar{p}_k / p_k|$ is close to 1 as long as $|\bar{h}_k|$ is appreciably greater than $|z_r - z_r(\epsilon)|$.

Hence we may write, since $\bar{p}'_k/p'_k$ has a low relative error,

$$\frac{\bar{p}_k}{\bar{p}'_k} = (1+\eta_k)\frac{p_k}{p'_k}, \qquad (\eta_k \text{ possibly complex}).$$

We now have

$$z_r + \bar{h}_{k+1} = z_r + \bar{h}_k - \bar{p}_k/\bar{p}'_k,$$

or

$$\bar{h}_{k+1} = \bar{h}_k - (1+\eta_k)\frac{p_k}{p'_k} = h_{k+1} - \eta_k \frac{p_k}{p'_k}.$$

Hence

$$\begin{aligned} |\bar{h}_{k+1}| &\leqslant |h_{k+1}| + |\eta_k||p_k/p'_k|, \\ &= |h_{k+1}| + |\eta_k||h_{k+1} - \bar{h}_k|, && \text{from (2.13)} \\ &< (1+|\eta_k|)A|\bar{h}_k|^2 + |\eta_k||\bar{h}_k|, && \text{from (2.15)} \\ &< [0{\cdot}1(1+|\eta_k|) + |\eta_k|]\,|\bar{h}_k|, && \text{from (2.16)} \end{aligned}$$

If $|\eta_k| < 0{\cdot}1$, we have

$$|\bar{h}_{k+1}| < 0{\cdot}21\,|\bar{h}_k|,$$

and convergence is quite satisfactory. In fact, the error diminishes in size even if $|\eta_k|$ is as high as 0·8. Moreover, from our argument, we see that convergence ceases in practice when the error made in computing p_k is of the same order of magnitude as its true value and, consequently, the computed value of the polynomial has no correct significant figure. In particular, the error made in evaluating the function $\sum_{i=0}^{n} a_i \epsilon_i z_r^{\,n-i}$ is of the same order as $p'(z_r)\bar{h}_k$ for limiting accuracy.

2.5 The method of Bairstow

Let

$$p(z) = a_0 z^n + a_1 z^{n-1} + \ldots + a_{n-1} z + a_n, \qquad (n \geqslant 2)$$

and suppose that

$$p(z) = (z^2 - \alpha z - \beta)q(z) + q_{n-1}z + q_n, \tag{2.20}$$

where

$$q(z) = q_0 z^{n-2} + q_1 z^{n-3} + \ldots + q_{n-3}z + q_{n-2}.$$

Comparison of the coefficients on both sides of (2.20) gives

$$\left.\begin{aligned} q_0 &= a_0, \qquad q_1 = a_1 + \alpha q_0, \\ q_k &= a_k + \alpha q_{k-1} + \beta q_{k-2}, \qquad (k = 2, \ldots, n-1) \\ q_n &= a_n + \beta q_{n-2}. \end{aligned}\right] \tag{2.21}$$

Similarly, we may write

$$q(z) = (z^2 - \alpha z - \beta)t(z) + t_{n-3}z + t_{n-2}, \tag{2.22}$$

where

$$t(z) = t_0 z^{n-4} + t_1 z^{n-3} + \ldots + t_{n-5}z + t_{n-4}.$$

Comparison of the coefficients on both sides of (2.22) gives

$$\left.\begin{aligned} &\left.\begin{aligned} t_0 &= q_0, \qquad t_1 = q_1 + \alpha t_0, \\ t_k &= q_k + \alpha t_{k-1} + \beta t_{k-2}, \qquad (k = 2, \ldots, n-3) \\ t_{n-2} &= q_{n-2} + \beta t_{n-4} \end{aligned}\right] (n \geqslant 4) \\ &t_0 = q_0 (n = 2, 3), \qquad t_1 = q_1 (n = 3). \end{aligned}\right] \tag{2.23}$$

Further, from (2.20), we have

$$\begin{aligned} p'(z) &= (z^2 - \alpha z - \beta)q'(z) + (2z - \alpha)q(z) + q_{n-1} \\ &= (z^2 - \alpha z - \beta)[q'(z) + (2z - \alpha)t(z) + 2t_{n-3}] \\ &\qquad + z(\alpha t_{n-3} + 2t_{n-2}) + 2\beta t_{n-3} - \alpha t_{n-2} + q_{n-1} \end{aligned} \tag{2.24}$$

The result (2.24) is useful in connection with finding complex roots by Newton's method but does not enter into the derivation of Bairstow's method.

Now if $z^2 - \alpha z - \beta$ is an exact quadratic factor of $p(z)$ then we know that q_{n-1} and q_n are zero. In Bairstow's method we assume that $z^2 - \alpha z - \beta$ is a good approximation to a quadratic factor. A better approximation is obtained as follows. We consider the function $q_{n-1}z + q_n$ as a function of the two variables α and β and suppose that when α and β are increased to $\alpha + \delta\alpha$ and $\beta + \delta\beta$ respectively, the function $q_{n-1}z + q_n$ is zero. We then have, by Taylor's theorem for two variables (neglecting second order terms),

$$0 = (q_{n-1}z + q_n)_{(\alpha+\delta\alpha,\ \beta+\delta\beta)}$$

or

$$0 = (q_{n-1}z + q_n) + \left[\frac{\partial}{\partial\alpha}\left(q_{n-1}z + q_n\right)\right]\delta\alpha + \left[\frac{\partial}{\partial\beta}\left(q_{n-1}z + q_n\right)\right]\delta\beta, \tag{2.25}$$

where the functions on the right-hand side of (2.25) are evaluated at (α, β). Again, partial differentiation of (2.20) gives, since $p(z)$ is independent of α and β,

$$0 = -\,q(z) + (z^2 - \alpha z - \beta)\frac{\partial q(z)}{\partial\beta} + \frac{\partial}{\partial\beta}(q_{n-1}z + q_n), \tag{2.26}$$

$$0 = -zq(z) + (z^2 - \alpha z - \beta)\frac{\partial q(z)}{\partial\alpha} + \frac{\partial}{\partial\alpha}(q_{n-1}z + q_n), \tag{2.27}$$

where, here again, all functions are evaluated at (α, β). Thus, from (2.22), (2.26), and (2.27) we have

$$\frac{\partial}{\partial \beta}(q_{n-1}z + q_n) = (z^2 - \alpha z - \beta)\mathcal{R}_1 + t_{n-3}z + t_{n-2}, \tag{2.28}$$

$$\frac{\partial}{\partial \alpha}(q_{n-1}z + q_n) = (z^2 - \alpha z - \beta)\mathcal{R}_2 + (\alpha t_{n-3} + t_{n-2})z + \beta t_{n-3}, \tag{2.29}$$

where
$$\mathcal{R}_1 = t(z) - \frac{\partial q(z)}{\partial \beta},$$

$$\mathcal{R}_2 = zt(z) - \frac{\partial q(z)}{\partial \alpha} + t_{n-3}.$$

Now let z_1, z_2 (assumed distinct) be the roots of $z^2 - \alpha z - \beta = 0$. Then substitution of (2.28) and (2.29) in (2.25) gives for each root in turn

$$q_{n-1}z_1 + q_n + [(\alpha t_{n-3} + t_{n-2})z_1 + \beta t_{n-3}]\,\delta\alpha + [t_{n-3}z_1 + t_{n-2}]\,\delta\beta = 0, \tag{2.30}$$

and

$$q_{n-1}z_2 + q_n + [(\alpha t_{n-3} + t_{n-2})z_2 + \beta t_{n-3}]\,\delta\alpha + [t_{n-3}z_2 + t_{n-2}]\,\delta\beta = 0, \tag{2.31}$$

and since $z_1 \neq z_2$ it follows that

$$q_{n-1} + (\alpha t_{n-3} + t_{n-2})\delta\alpha + t_{n-3}\delta\beta = 0, \tag{2.32}$$

and
$$q_n + \beta t_{n-3}\delta\alpha + t_{n-2}\delta\beta = 0. \tag{2.33}$$

Equations (2.32) and (2.33) are linear simultaneous equations in $\delta\alpha$ and $\delta\beta$ whose solution can be expressed in the form

$$\left.\begin{aligned} \delta\alpha &= (t_{n-3}q_n - t_{n-2}q_{n-1})/D; \qquad & \delta\beta &= (Mq_{n-1} - t_{n-2}q_n)/D \\ M &= \beta t_{n-3} + \alpha t_{n-2}; & D &= t_{n-2}^2 - Mt_{n-3}. \end{aligned}\right] \tag{2.34}$$

An improved quadratic factor of our polynomial $p(z)$ is now $z^2 - (\alpha + \delta\alpha)z - (\beta + \delta\beta)$. Note that the formulae (2.34) also hold in the case $z_1 = z_2$ since (2.30) is valid and we can still arrive at formula (2.32) by differentiating (2.25) with respect to z and evaluating the resulting derivative at $z = z_1$.

In practice, it should be noted that t_k can be calculated simultaneously with q_k so that it is not necessary to store in our computer intermediate values of q_k. We now carry out on our four decimal digit floating point computer the first stage of the Bairstow process for finding a quadratic factor of the polynomial $p(z) = z^6 + 4z^5 + 9z^4 + 15z^3 + 12z^2 + 10z + 6$, given the approximate quadratic factor $z^2 - 0{\cdot}28z + 0{\cdot}91$.

It follows that an improved quadratic factor is $z^2 - 0{\cdot}2141\,z + 0{\cdot}8840$. Repetition of the process will lead to a value for the quadratic factor which is as near as possible (on our computer) to the exact quadratic factor. Good approximations to two roots

Table 2.4 One iteration of the Bairstow process

$\alpha = 0{\cdot}28, \qquad \beta = -0{\cdot}91$

	k	a_k	q_k	t_k
	0	1	1·000	1·000
	1	4	4·280	4·560
	2	9	9·288	9·655
$n-3$	3	15	13·710	12·260
$n-2$	4	12	7·390	−1·396
$n-1$	5	10	−0·411	
n	6	6	−0·725	

$M = -11{\cdot}55, \quad D = 143{\cdot}5, \quad \delta\alpha = -0{\cdot}065\,94, \quad \delta\beta = 0{\cdot}026\,03$

of the original polynomial equation are then the roots of the 'updated' equation $z^2 - \alpha z - \beta = 0$. The roots of this quadratic equation are obtained, as in section 1.3. by first finding the root of largest modulus and then calculating the smaller one from the product of the roots. Subject to the provisions of the next section, and the coefficients of the original polynomial being well-conditioned, the remaining roots of $p(z) = 0$ will be well approximated by those of $q(z) = 0$.

2.6 Deflation (synthetic division)

The polynomial $q(z)$ finally determined in the last section by dividing $p(z)$ by a factor, which is acceptable as an exact quadratic factor, is termed a deflated polynomial. If we know an approximate quadratic factor of $q(z)$ we can again use the Bairstow process to improve on this approximation to find two more roots of $p(z) = 0$, and so on. Eventually, we reach a stage where factorization is complete (when $p(z)$ is a polynomial of even order) or we remain with a linear factor ($p(z)$ of odd order). In either case all the roots will be determined. This process we have just described, in which our original polynomial is reduced in degree at each stage, is called deflation. The technique of deflation can also be applied to Newton's method and we now proceed to analyse this application in some detail.

Suppose that an acceptable approximation to the isolated root z_1 of the polynomial equation $p(z) = 0$, call it α for convenience, has been found on our computer using Newton's process. To obtain the coefficients $q_k (k = 0, \ldots, n-1)$ of the deflated polynomial $q(z)$ we compare the coefficients in $p(z)$ and $(z - \alpha)q(z)$. If q_n is the remainder, i.e., $p(z) = (z - \alpha)q(z) + q_n$, we have

$$\left.\begin{aligned} q_0 &= a_0 \\ q_k &= fl(\alpha q_{k-1} + a_k). \qquad k = 1, \ldots, n. \end{aligned}\right] \tag{2.35}$$

Thus the q_k are exactly the partial sums which are obtained when the polynomial $p(z)$ is evaluated for $z = \alpha$ by 'nesting'. Let

$$a'_0 = q'_0, \qquad a'_k = q_k - \alpha q_{k-1} \qquad (k = 1, \ldots, n),$$

and consider the connection between the zeros of $q(z)$ and the zeros of

$$a'_0 z^n + a'_1 z^{n-1} + \ldots + a'_n.$$

We have

$$\begin{aligned}(z-\alpha)q(z) &= q_0 z^n + (q_1 - \alpha q_0)z^{n-1} + \ldots + (q_{n-1} - \alpha q_{n-2})z - \alpha q_{n-1},\\ &= a'_0 z^n + a'_1 z^{n-1} + \ldots + a'_{n-1}z - \alpha q_{n-1},\\ &= a'_0 z^n + a'_1 z^{n-1} + \ldots + a'_{n-1}z + a'_n - q_n = \bar{p}(z), \text{ say.}\end{aligned}$$

The zeros of $q(z)$ are thus the zeros of the polynomial $\bar{p}(z)$ except for $z = \alpha$. If α were exact and $p(\alpha)$ could be computed exactly then its computed value, q_n, would be zero and the roots of $q(z)$ would be exactly those of $p(z)$ (except for $z = \alpha$). Now, for limiting accuracy, the error in the computed value of $p(\alpha)$ is of the same order of magnitude as its true value. Hence the computed value q_n is of the same magnitude as the error which, from the error analysis of section 2.1, is of the form

$$a_0\epsilon_0\alpha^n + a_1\epsilon_1\alpha^{n-1} + \ldots + a_n\epsilon_n.$$

Now suppose $z'_2, \ldots, z'_n$ are the zeros of $\bar{p}(z)$ which correspond to the zeros $z_2, \ldots, z_n$ of $p(z)$. We are interested in comparing the errors $z_i - z'_i$ which can be attributed to the deflation process with the errors which would arise if we were to use the original polynomial. If z''_i are the zeros computed from the original polynomial we have, from section 2.4, that

$$|z''_i - z_i| \leqslant K_2 |a_0\epsilon'_0 z_i^n + \ldots + a_n\epsilon'_n| / |p'(z_i)|, \tag{2.36}$$

where K_2 is of order unity and the ϵ'_i have the same bounds as the ϵ_i.

We consider the error arising in z'_i in two parts:

(a) That coming from the error $-q_n$ in the constant term

This gives rise to an error whose order of magnitude is

$$K|a_0\epsilon_0\alpha^n + \ldots + a_n\epsilon_n| / |p'(z_i)|, \; K \text{ of order unity.} \tag{2.37}$$

If $|\alpha| < |z_i|$ then one would expect this error (in the statistical sense) to be less than the right-hand side of (2.36) and if $|\alpha| \ll |z_i|$ we could be fairly confident that (2.37) is much the smaller of the two estimates. However, if $|\alpha| \gg |z_i|$, e.g., $|\alpha| = 1$, $|z_i| = 10^{-2}$ then the element $a_k\epsilon_k\alpha^{n-k}$ may be larger than $a_k\epsilon'_k z_i^{n-k}$ by the factor $10^{2(n-k)}$ and thus even for polynomials of small degree the error introduced by deflation may be of a much greater order of magnitude than the error inherent in obtaining our root z_i from the original polynomial.

(b) That coming from having coefficients a'_k instead of a_k

We have, for arbitrary α, q_{k-1}, a_k,

$$fl(\alpha q_{k-1} + a_k) = [\alpha q_{k-1}(1 + \eta_1) + a_k](1 + \eta_2).$$

Hence

$$fl(\alpha q_{k-1} + a_k) - (\alpha q_{k-1} + a_k) = \alpha q_{k-1}[(1 + \eta_1)(1 + \eta_2) - 1] + a_k \eta_2,$$

$$|fl(\alpha q_{k-1} + a_k) - (\alpha q_{k-1} + a_k)| \leqslant (2|\alpha q_{k-1}| + |a_k|) 0{\cdot}56 \times 10^{-3},$$

on our four decimal digit machine. If

$$|\alpha q_{k-1}| < |a_k|, \tag{2.38}$$

we have

$$|fl(\alpha q_{k-1} + a_k) - (\alpha q_{k-1} + a_k)| < 1{\cdot}68 \times 10^{-3} |a_k|,$$

or

$$fl(\alpha q_{k-1} + a_k) = \alpha q_{k-1} + a_k(1 + \epsilon_k),$$
$$= \alpha q_{k-1} + a'_k.$$

Thus

$$a'_k - a_k = a_k \epsilon_k,$$

where

$$|\epsilon_k| < 1{\cdot}68 \times 10^{-3}.$$

When (2.38) holds for all k, the errors arising from having coefficients a'_k instead of a_k are bounded by those arising from perturbations of up to $1{\cdot}68$ in 10^3 in the original coefficients. We cannot, however, ensure (2.38) holds at all stages particularly as some of the coefficients a_k might well be zero. By choosing α to be the root of smallest modulus, however, we will keep $a'_k - a_k$ as small as possible so that it is then less likely that $|\alpha q_{k-1}|$ is greater than $|a_k|$.

To summarize, we see that the error in deflation is a minimum when we deflate by dividing by a factor $z - \alpha$ where α is an approximation to the zero of smallest modulus. Similarly, it can be shown that deflation applied to the Bairstow process is best carried out by dividing at each stage by the quadratic factors corresponding to the smallest roots.

We illustrate the danger of dividing out by the root of largest modulus first, by considering the example of Wilkinson [1], in which

$$p(z) = (z - \tfrac{1}{2})(z - \tfrac{1}{4}) \ldots \left(z - \frac{1}{2^{20}}\right),$$

and we accept $2^{-1} + 2^{-40}$ (on a 12 decimal digit floating point computer) as a good approximation to the root of largest modulus. We have

$$p(z) = (z - 2^{-1} - 2^{-40})q(z) + q_n,$$

and, if we assume for convenience that there are no rounding errors in the deflation, we find

$$
\begin{aligned}
q_n &= p(2^{-1} + 2^{-40}),\\
&> 2^{-40}(2^{-1} - 2^{-2})(2^{-1} - 2^{-3}) \ldots (2^{-1} - 2^{-20}),\\
&> 2^{-78},
\end{aligned}
$$

since for $k = 2, 3, \ldots, 20$,

$$2^{-1} - 2^{-k} \geqslant 2^{-2}.$$

The inequality for q_n implies that the constant term in the polynomial equation $p(z) - q_n = 0$, is greater in magnitude than $2^{-78} - 2^{-210}$ (2^{-210} being the constant term of $p(z)$). This means that the constant term in $q(z)$ is greater in magnitude than

$$(2^{-78} - 2^{-210})/(2^{-1} + 2^{-40}) > 2^{-78}.$$

Had we obtained the deflated polynomial exactly by dividing out by the factor $(z - 2^{-1})$, the magnitude of its constant term would have been the product of the remaining roots $\frac{1}{4}, \frac{1}{8}, \ldots, 2^{-20}$ of $p(z) = 0$, that is 2^{-209}. In view of the large relative difference between 2^{-78} and 2^{-209} we can infer that the roots of the deflated polynomial are vastly different from the remaining roots of $p(z) = 0$, and the deflation process is unsound in this case. However, if we deflate by dividing by the factor $(z - 2^{-20})$ first, then the roots are obtained exactly to the degree of accuracy of the computer.

2.7 The method of D. H. Lehmer

The methods already discussed have one feature in common, namely, they all require a fairly good starting approximation to a zero. It often happens that this information is not available and then we have to devise some technique for estimating the approximate whereabouts of the zeros in order to set up a good starting value. Our task is often eased if, for example, we know that all our roots are real and satisfy a Stürm sequence property (see chapter 9) but again this information might not be known in advance and all we may know about our polynomial is its coefficients and degree. It is exactly in this situation that the method of D. H. Lehmer [2], which *systematically* finds the zeros in more or less increasing order of magnitude, is invaluable.

Before describing Lehmer's method it will be necessary to recall some results from classical analysis. We let

$$p(z) = a_0 z^n + a_1 z^{n-1} + \ldots + a_{n-1} z + a_n,$$

but now we allow the coefficients to be complex also. We define

$$p^*(z) = z^n \bar{p}(1/\bar{z}) = \bar{a}_0 + \bar{a}_1 z + \ldots + \bar{a}_n z^n \tag{2.39}$$

where bars denote, in this section, complex conjugates. Further, let

$$T[p(z)] = \bar{a}_n p(z) - a_0 p^*(z) \tag{2.40}$$

so that $T[p(z)]$ is a polynomial of degree $n-1$ (or less) and, in particular,

$$T[p(0)] = \bar{a}_n a_n - a_0 \bar{a}_0 = |a_n|^2 - |a_0|^2,$$

is real. By defining

$$T^j[p(z)] = T\{T^{j-1}[p(z)]\}, \qquad j = 1, 2, \ldots \tag{2.41}$$

we obtain a sequence of polynomials of decreasing degree for which we shall assume that m is the smallest integer for which

$$T^m[p(0)] = 0.$$

We are now in a position to quote the theorem which is the basis of Lehmer's method. This theorem is due principally to Schur and we sketch its proof in the exercises.

THEOREM 2-1 *Let $p(z)$ have no root on the unit circle Γ and suppose $p(0) \neq 0$. If for some k, $1 \leqslant k \leqslant m$, $T^k[p(0)] < 0$, then $p(z)$ has at least one root inside Γ. If, $T^k[p(0)] > 0$ for $1 \leqslant k < m$ and $T^{m-1}[p(z)]$ is a constant, then there are no roots of $p(z)$ inside Γ.*

The restriction that $p(z)$ should have no root on Γ is no real handicap and, in fact, we can show the stronger theorem.

THEOREM 2-2 *Theorem 2-1 holds even when $p(z)$ has roots on Γ.*

Theorem 2-2 provides us with the necessary material to conduct an investigation to decide whether or not a polynomial has a root in the unit circle, and hence any circular region in the complex plane. This is the case because, if $p(z)$ has a zero in the disc $|z-c| = \rho$, where ρ and c are arbitrary, $p(\rho z + c)$ has a zero in the unit circle. The investigation proceeds as follows:

Step 1 Is $p(0) = 0$? If answer is 'yes', then $z = 0$ is a root. If 'no' we carry on with step 2.

Step 2 Calculate $T[p(z)]$. Is $T[p(0)] < 0$? If answer is 'yes' there is a root inside the unit circle and we carry on with step 4. If 'no' we carry on with step 3.

Step 3 Calculate $T^k[p(z)]$, $k = 1, 2, \ldots$, until we reach a stage where $T^k[p(0)] < 0$, $k < m$ or $T^m[p(0)] = 0$. If the former is true then a root occurs inside the unit circle and we move on to step 4. If the latter and $T^{m-1}[p(z)]$ is a constant there is no root inside the unit circle and we move on to step 5. If $T^{m-1}[p(z)]$ is not a constant we move on to step 8.

Step 4 We consider in turn the circles $|z| = 1/2^j, j = 1, 2, \ldots$. $p(z)$ will have a root inside the circle $|z| = 1/2^j$ if $p_j(z) = p(2^{-j}z)$ has a root inside the unit circle. We repeat steps 2 and 3 for the polynomial $p_j(z)$ and continue to increase j until we find that $p(z)$ has no roots inside the circle of radius $R = 2^{-j}$ but has a root inside $|z| = 2R = 2^{-j+1}$, i.e., the annulus $R < |z| \leqslant 2R$ contains a root of $p(z)$. We now proceed to step 6.

Step 5 We consider in turn the circles $|z| = 2^j, j = 1, 2, \ldots$. $p(z)$ will have a root in the circle $|z| = 2^j$ if $p_j(z) = p(2^j z)$ has a root in the unit circle. We repeat steps 2 and 3 for this polynomial $p_j(z)$ and continue to increase j until the circle of radius $R = 2^j$ contains no root of $p(z)$, but the circle of radius $2R$ does contain a root, i.e., the annulus $R < |z| \leqslant 2R$ contains at least one zero of $p(z)$. We continue with Step 6.

Step 6 The annulus $R < |z| \leqslant 2R$, found in steps 4 or 5, can be completely covered by eight overlapping circles each of radius $4R/5$ and with centres at

$$c_h = \frac{3R}{2\cos(\pi/8)} e^{\pi i h/4}, \qquad h = 0, 1, \ldots, 7; \quad i = \sqrt{-1}.$$

Now if $p(z)$ has a zero inside the circle $|z - c_h| = \frac{4}{5}R$, $p(\frac{4}{5}Rz + c_h)$ has a zero inside the unit circle. We can thus test each circle in turn by resorting to steps 2 and 3 and we shall find one at least, call it Γ_1, containing a root of $p(z)$. Our work is simplified if the coefficients a_r are real, as we have only to consider $h = 0, 1, 2, 3, 4$ since we know that any roots with $I(z) < 0$ are just complex conjugates of those already found. We now move on to step 7.

Step 7 If γ_1 is the centre of the circle Γ_1 we proceed as in step 4 halving the radius at each stage until we find an annulus

$$R_1 < |z - \gamma_1| \leqslant 2R_1,$$

containing a zero of $p(z)$. We now repeat steps 6 and 7 until we get as close to our root as we desire or, alternatively, we get a sufficiently good starting value for Newton's or Bairstow's methods.

Step 8 If, in performing step 3, $T^m[p(z)] = 0$ but $T^{m-1}[p(z)]$ is not constant – this happens, for example, when $a_0 = a_n$ – we choose a new radius which is β times our current radius. If we are at step 3 for the first time, or we entered step 3 from step 4, we take $1 < \beta < 2$, say $\beta = 3/2$. If we entered step 3 from step 5 we take $\frac{1}{2} < \beta < 1$, say $\beta = \frac{3}{4}$. We now go back to step 2 noting that our circle has radius βR.

This completes all the different stages of our systematic investigation. The method is illustrated pictorially in Fig. 2.3.

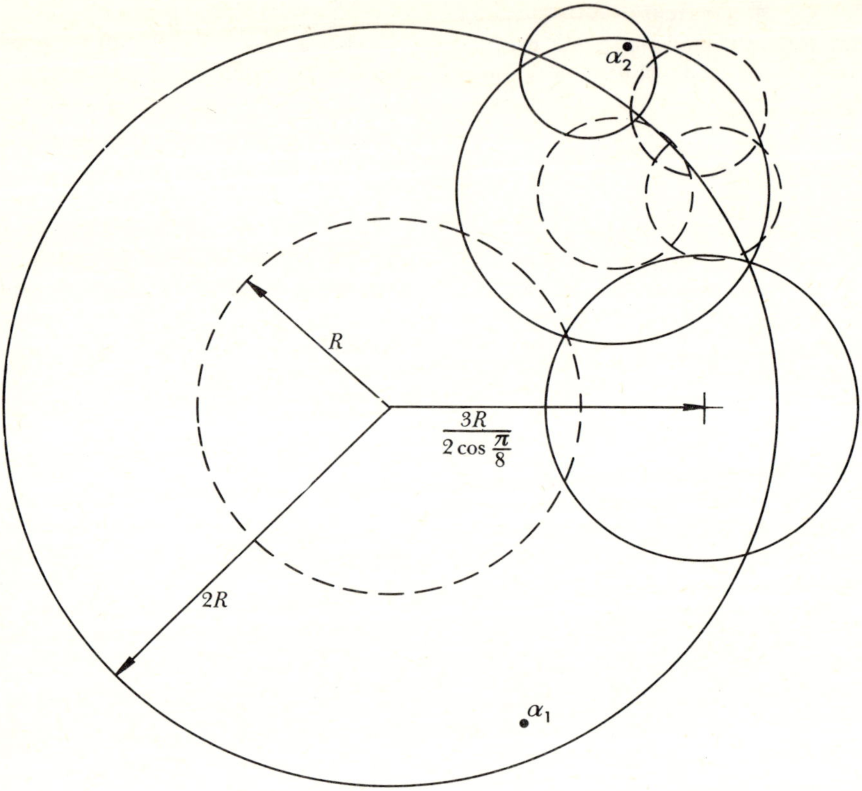

Fig. 2.3 Pictorial representation of D. H. Lehmer's method

As can be seen from this figure we might not converge to the root α_1, which determines our original annulus $R < |z| \leqslant 2R$, but we might converge instead to a root α_2. We can show that α_2 does not exceed $\frac{5}{2}R \cos(\pi/8)$ in magnitude so that α_2, while not necessarily being the smallest root, will in general be one of the smaller roots of the polynomial equation (in modulus). We can remove this root by deflation once we have calculated it sufficiently accurately and we then proceed to find the others taking $2R$ as the starting radius for the next root. Note that convergence is not affected by multiplicity or clustering of the roots which is a drawback of many other methods. However, the method does have the disadvantage of being first order.

The following example was given by Lehmer in his original paper. Show that the smallest root of $p(z) = 9 + 3z - 14z^2 - 8z^3 = 0$ lies in the annulus $\frac{1}{2} < |z| \leqslant 1$. The following calculations are performed:

R	$p(Rz)$	$T[p(Rz)]$ $\propto$	$T[p(0)]$	$T^2[p(Rz)]$ $\propto$	$T^2[p(0)]$	$T^3[p(Rz)]$ $\propto$
1	$9+3z-14z^2-8z^3$	$1-5z-6z^2$	>0	$-1-z$	<0	24
$\frac{1}{2}$	$9+\frac{3}{2}z-\frac{7}{2}z^2-z^3$	$8+z-3z^2$	>0	$5+z$	>0	24

From Theorem 2-2 it is clear that the assertion is true. The roots of the given polynomial equation are $\frac{3}{4}$, -1, and $-\frac{3}{2}$.

2.8 Application of the methods of sections 2.3 – 2.7

The method of bisection proves useful for finding eigenvalues of symmetric matrices and will be brought in again, in that context, in chapter 9. Newton's method is extremely satisfactory for finding the real roots of polynomial equations when a fairly good approximation is available, while Bairstow's method is profitably used for finding quadratic factors of polynomials, leading to conjugate complex roots, when approximate factors are known. In both these methods difficulties arise when multiple roots and pathologically close roots occur, but these difficulties can often be overcome by using a slightly different computational formula (see exercises 6 and 9). When no approximations are available to the roots of the polynomial equation, or the coefficients of the polynomial are complex, Lehmer's method can be used to advantage. In the former situation we determine a small circle enclosing the root and apply Newton's or Bairstow's methods, as appropriate. Should either of these methods fail to converge, that is our successive approximations by these methods falls outside the given small circle, our starting value is not sufficiently good and we must return to Lehmer's method to obtain an improved initial approximation. Alternatively, it might indicate that we have multiple or pathologically close roots which means that $p'(z)$ is approximately zero in the region of the small circle and it is then worthwhile to continue with Lehmer's method. In the latter situation, too, it is best to remain with Lehmer's method. Note that for extremely ill-conditioned polynomials we cannot expect a very high degree of accuracy in our resulting zeros although, if we resort to multiple length arithmetic for our calculations, we can achieve greater accuracy at the expense of time.

Finally, in all methods, we require some criterion for deciding whether at any stage of the process our current approximation is acceptable as a zero and the iteration process can be terminated for that particular zero. If our zero is $\alpha \neq 0$ then a satisfactory criterion for accepting the current approximation ξ_{k+1} as being sufficiently close to α is

$$|(\xi_{k+1}-\xi_k)/\xi_k| < \epsilon,$$

where ϵ is a pre-assigned 'small' quantity. If $\alpha = 0$, then it is not necessary to use this criterion as this zero can be detected by a straightforward substitution and the polynomial deflated. Great care must be taken in choosing ϵ since, if ϵ is chosen too small, the criterion will never be satisfied and, if too large, we might accept a value which is less accurate than the one which can ultimately be obtained on the available machine. In practice, ϵ is determined by the condition of the root although this information might not be known in advance.

2.9 Notes

A large number of methods have been devised for the determination of the zeros of a polynomial equation. Besides the methods discussed in this chapter there are the following: The secant method (the method of false positions), Lin's method, Graeffe's method, Bernoulli's method, Laguerre's method, Müller's method, to name just a few. Some of these methods are sketched in the exercises. Fuller details of the methods in the text and other methods can be found in Wilkinson [1], [3], Ralston [4], Hilderbrand [5], and Traub [6]. The reader is referred to Wilkinson's books for extremely good detailed error analyses of a large selection of the methods given.

REFERENCES

[1] Wilkinson, J. H., Rounding Errors in Algebraic Processes, *N.P.L. Notes on Applied Science* No. 32, H.M.S.O., London (1963).

[2] Lehmer, D. H., 'A Machine Method for Solving Polynomial Equations', *J.A.C.M.*, (1961) 151–62.

[3] Wilkinson, J. H., *The Algebraic Eigenvalue Problem*, Oxford University Press, London (1965).

[4] Ralston, A., *A First Course in Numerical Analysis*, McGraw-Hill Book Co. Ltd., New York (1965).

[5] Hilderbrand, F. B., *Introduction to Numerical Analysis*, McGraw-Hill Book Co. Ltd., New York (1956).

[6] Traub, J. F., *Iterative Methods for the Solution of Equations*, Prentice-Hall Inc. New Jersey (1965).

Exercises

1. Let z_r be a zero of multiplicity m of the polynomial equation $p(z) = 0$. Then there are m zeros of $p(z) + \epsilon q(z) = 0$ defined by the series (convergent for small ϵ)

$$z_r(\epsilon) = \sum_1^\infty p_k \epsilon^{k/m},$$

one zero corresponding to each interpretation of $\epsilon^{1/m}$. Show that

$$p_1 = \left[\frac{-m!q(z_r)}{p^m(z_r)}\right]^{1/m}$$

and

$$|z_r(\epsilon) - (z_r + p_1\epsilon^{1/m})| < K\epsilon^{2/m} \qquad (K \text{ a constant}).$$

2. Show that the zeros of $z^{15} - 1 = 0$ are well conditioned.
3. Let z_r be a root of the polynomial of degree 20 with zeros $2^{-1}, 2^{-2}, \ldots, 2^{-20}$. If an error δa_k is made in the coefficient a_k show that the error δz_r in z_r satisfies

$$\left|\frac{\delta z_r}{z_r}\right| < 64 \left|\frac{\delta a_k}{a_k}\right| \qquad \text{(Wilkinson [1])}$$

4. Consider the polynomial equation

$$p_0(z) = p(z) = z^{20} - 2z + 3 = 0.$$

Compute the roots $z_i^{(0)}$ $(i = 1, 2, \ldots, 20)$ on a four decimal digit floating point computer. Next, evaluate the polynomials

$$p_{r+1}(z) = \prod_i (z - z_i^{(r)}),$$

where $z_i^{(r)}$ $(i = 1, 2, \ldots, 20)$ are the computed roots of $p_r(z) = 0$. Show that eventually the polynomials $p_r(z)$ will bear no resemblance to the original polynomial $p(z)$ and account for this phenomena.
5. Show that Newton's iteration formula (2.13) applied to the function $f(x) = x^2 - a$, $(a > 0)$ leads to the iteration formula

$$x_{n+1} = \tfrac{1}{2}(x_n + a/x_n), \qquad (x_0 > 0),$$

for evaluating $\sqrt{a}$. By similarly considering the function $f(x) = x^p - a$ show that the sequence given by

$$x_{n+1} = [(p-1)x_n + (a/x_n^{p-1})]/p, \qquad (x_0 > 0),$$

can be used to evaluate $a^{1/p}$. Use this last result to tabulate

$$g(\alpha) = (2^\alpha - 1)/\alpha, \qquad \alpha = 0{\cdot}1\,(0{\cdot}1)\,0{\cdot}5$$

and deduce $g(0)$ from your table (see section 3.7). To what does $g(0)$ correspond?
6. If the equation $f(x) = 0$ has a root of multiplicity r show that

$$x_{n+1} = x_n - rf(x_n)/f'(x_n)$$

provides a closer approximation to the multiple root than does Newton's formula. Hence find the double root near $x = 1{\cdot}1$ of $x^3 - 4x^2 + 5x - 2 = 0$.
7. Let X_n denote an interval and let x_n be the midpoint of X_n. Show that the iterative procedure

$$X_{n+1} = x_n - f(X_n)/f'(X_n)$$

is the extension of Newton's method to interval arithmetic.
8. Use the iterative procedure of exercise 7 to find an interval estimate for the root of $e^x - 3x = 0$ given that the root lies in $[0, 1]$.

9. If $f'(a) = 0$ and $f(a) \neq 0$ show that if there exist close roots near $x = a$ they are given approximately by

$$x - a = \mp \sqrt{[2f(a)/f''(a)]}.$$

Hence find the roots near $x = 4$ of

$$x^3 - 7x^2 + 8x + 15{\cdot}95 = 0,$$

by using the above formula and then iterating with (2.13).

10. Show that if $z^2 + z + 1$ is an approximate quadratic factor of the polynomial $p(z) = z^3 - z - 1$ then one iteration of Bairstow's process gives the improved approximation $z^2 + 1{\cdot}333z + 0{\cdot}6667$. By continuing the process further, estimate the complex zeros of the polynomial equation $p(z) = 0$.

11. Let

$$p(z) = z^3 - z^2 + 1 = (z - \alpha)(z^2 + bz + c)$$

where b, c are calculated from the formulae

$$b = -1 + \alpha, \qquad c = b\alpha.$$

If α lies in the interval $[-0{\cdot}7548, -0{\cdot}7550]$ find the intervals in which the numbers b and c lie. Hence give interval estimates for the complex zeros of $z^3 - z^2 + 1 = 0$. Find the estimates when c is calculated from the alternative formula $c\alpha = -1$.

12. Find, correct to four significant digits, the roots of the polynomial equation

$$x^4 - 6{\cdot}798x^3 + 2{\cdot}995x^2 - 0{\cdot}043\,69x + 0{\cdot}000\,089\,25 = 0.$$

If the largest root is determined first and the polynomial is deflated, show that the zeros of the deflated polynomial equation differ from those of the original polynomial and find their exact values. (Assume your computer is a four decimal digit floating point computer.)

13. Locate the zeros of $p(z) = z^3 - z^2 - 1 = 0$ by Lehmer's method.

14. Let Γ denote the unit circle. Show that

$$\int_\Gamma \frac{dz}{z - a} = \begin{cases} 2\pi i, & |a| < 1 \\ 0; & |a| > 1. \end{cases}$$

15. Let $P(z)$ be a polynomial with no root on the unit circle Γ. Show that the number of roots of $P(z)$ inside Γ, multiple roots being counted according to their multiplicities, is given by

$$\frac{1}{2\pi i}\int_\Gamma \frac{P'(z)}{P(z)}\,dz$$

(Hint: Use the result of exercise 14.)

16. Let P and Q be two polynomials such that

$$|P(z)| < |Q(z)| \quad \text{on} \quad |z| = 1.$$

Prove that Q and $P + Q$ have the same number of roots inside the unit circle. (Hint: Use the result of exercise 15.)

17. Let ϕ be a polynomial of degree d with no root on the unit circle Γ and m roots inside Γ. Let $T(\phi(0)) \neq 0$, where T is defined as in (2.40). Show that $T(\phi)$ has no root on Γ and has m or $d-m$ roots inside Γ according as $T(\phi(0))$ is positive or negative. Hence deduce Theorems 2-1 and 2-2 of Lehmer's method. (Hint: Use the result of exercise 16.)

18. Let $\alpha_1, \alpha_2, \ldots, \alpha_n$ be the roots of the polynomial equation

$$p(x) = x^n + a_1x^{n-1} + a_2x^{n-2} + \ldots + a_{n-1}x + a_n = 0,$$

where $0 < |\alpha_1| < |\alpha_2| < \ldots < |\alpha_n|$. Let

$$s_m = \alpha_1{}^m + \alpha_2{}^m + \ldots + \alpha_n{}^m, \qquad m = 1, 2, \ldots.$$

Prove the Newton–Girard formulae

$$s_1 + a_1 = 0$$

$$s_m + a_1s_{m-1} + \ldots + a_{m-1}s_1 + ma_m = 0 \qquad (1 < m \leqslant n)$$

$$s_m + a_1s_{m-1} + \ldots + a_{n-1}s_{m-n+1} + a_ns_{m-n} = 0 \qquad (m > n)$$

and show that

$$\lim_{m\to\infty} (s_m/s_{m-1}) = \alpha_n.$$

Use this result of Bernoulli to find the largest root of the equation

$$x^3 - 4x^2 + 5x - 2 = 0.$$

19. Rearrange the polynomial $p(x)$ of exercise 18 in the form

$$b_0 + b_1x + b_2x^2 + \ldots + b_nx^n = \prod_{m=1}^{n} (1 - x/\alpha_m), \qquad (b_0 = 1).$$

By logarithmically differentiating the above equation show that

$$-mb_m = b_{m-1}\sigma_1 + b_{m-2}\sigma_2 + \ldots + b_0\sigma_m, \qquad 1 \leqslant m \leqslant n$$

$$0 = b_n\sigma_{m-n} + b_{n-1}\sigma_{m-n+1} + \ldots + b_0\sigma_m, \qquad m > n$$

where

$$\sigma_m = \sum_1^n \alpha_i^{-m}.$$

Hence deduce that

$$\lim_{m\to\infty} (\sigma_m/\sigma_{m+1}) = \alpha_1$$

and use this result to find the smallest root of

$$x^3 - 4x^2 + 5x - 2 = 0.$$

20. (Secant method) Let x_1 and x'_1 be two values of x such that (x_1, x'_1) contains exactly one root α of the polynomial equation $p(x) = 0$ and $p(x_1)p(x'_1) < 0$. Let

$$x_{i+1} = \frac{x_i p(x'_i) - x'_i p(x_i)}{p(x'_i) - p(x_i)}, \qquad i = 1, 2, \ldots$$

and let x'_{i+1} be the value of x_i or x'_i such that $p(x_{i+1})p(x'_{i+1}) < 0$. Show that the sequence $\frac{1}{2}(x_i + x'_i)$ converges to the root α.

21. Show that the iteration formula

$$z_{k+1} = z_k - \frac{p(z_k)}{p'(z_k)} - \frac{\{p(z_k)\}^2 \{p''(z_k)\}}{2\{p'(z_k)\}^3},$$

is a cubically convergent formula for finding the roots of the polynomial equation $p(z) = 0$.

22. (Müller's method) Let z_{k-2}, z_{k-1}, z_k be approximations to a root of a polynomial equation $p(z) = 0$ and let $p_i = p(z_i)$. We write

$$h_i = z_i - z_{i-1}, \qquad \lambda_i = h_i/h_{i-1}, \qquad \delta_i = 1 + \lambda_i.$$

Then the next approximation z_{k+1} to a zero is taken to be the zero of the quadratic through (z_{k-2}, p_{k-2}), (z_{k-1}, p_{k-1}), (z_k, p_k) which is closest to z_k. Show that in this case λ_{k+1} satisfies the quadratic equation

$$\lambda_{k+1}^2 \lambda_k \delta_k^{-1} (p_{k-2}\lambda_k - p_{k-1}\delta_k + p_k) + \lambda_{k+1}\delta_k^{-1} g_k + p_k = 0,$$

where

$$g_k = p_{k-2}\lambda_k^2 - p_{k-1}\delta_k^2 + p_k(\lambda_k + \delta_k).$$

Thus

$$\lambda_{k+1} = -2p_k\delta_k / \{g_k \pm [g_k^2 - 4p_k\delta_k\lambda_k(p_{k-2}\lambda_k - p_{k-1}\delta_k + p_k)]^{\frac{1}{2}}\}$$

where we take the sign in the denominator which makes λ_{k+1} smallest in absolute magnitude. Hence find z_{k+1}.

23. If, in exercise 5, $\epsilon_n = x_n a^{-\frac{1}{2}} - 1$ show that

$$\epsilon_{n+1} = \tfrac{1}{2}\epsilon_n^2/(1 + \epsilon_n) < \tfrac{1}{2}\epsilon_n^2.$$

Use this result to show that for Eve's starting approximations (*Computer J.*, 1964) to $\sqrt{a}$

$$x_0 = 1{\cdot}797\,210 - 1{\cdot}710\,324\,(a + 1{\cdot}068\,628)^{-1}, \qquad \tfrac{1}{4} \leqslant a < \tfrac{1}{2}$$

$$x_0 = 2{\cdot}541\,639 - 4{\cdot}837\,528\,(a + 2{\cdot}137\,255)^{-1}, \qquad \tfrac{1}{2} \leqslant a < 1$$

ϵ_0 satisfies $|\epsilon_0| < 0{\cdot}33 \times 2^{-10}$ and two iterations of the formula in exercise 5 yield an approximation satisfying $0 < \epsilon_2 < 0{\cdot}2 \times 2^{-47}$. Estimate ϵ_0 when taking the result of exercise 21 (chapter 13) as a first approximation to $\sqrt{a}$.

24. Let $F(z)$ be a function of the complex variable z which is regular inside and on the circle $|z| = \rho$; let $F(0) \neq 0$, and let $\xi_1, \xi_2, \ldots, \xi_N$ be all the zeros of $F(z)$ satisfying $|\xi_\upsilon| \leqslant \rho$. Jensen's integral formula states that

$$\frac{1}{2\pi}\int_0^{2\pi} \ln|F(\rho e^{i\theta})|\,d\theta = \ln|F(0)| + \sum_{\upsilon=1}^{N} \ln\frac{\rho}{|\xi_\upsilon|}.$$

Suggest how this formula can be used for determining the zeros of $F(z)$, and state one advantage of your method.

25. Find the roots of the Legendre polynomial equation $P_n(x) = 0$ where $P_n(x)$ is defined by (4.15). Using the result (5.28) evaluate the coefficients α_k. Verify your answers are correct by comparison with Table 5.3. Carry out similar exercises for Laguerre and Hermite polynomials to verify Tables 5.4 and 5.5.

26. If

$$p(x) = x^n + a_1x^{n-1} + \ldots + a_{n-1}x + a_n$$

and α is the largest root in modulus of $p(x) = 0$ show that

$$|\alpha| < \max(1, |a_1| + |a_2| + \ldots + |a_n|).$$

Give a practical use for this result.

3 Interpolation and differentiation

3.1 Introduction

It frequently happens, particularly as a result of experimental work, that information about a function $f(x)$ is given in tabular form with the numerical values of $f(x)$ given for specified values of x, particularly equally spaced values of x. For a function specified in this way we are often required to find the value of $f(x)$ at some value of x in between two tabular values – this is the problem of direct interpolation. Alternatively, we may be given a value of $f(x)$ between two tabular values of $f(x)$ and our problem is to find the corresponding value of x – this is the problem of inverse interpolation (see section 3.8). Another problem is to determine the value of $f(x)$ at some point x outside the range of tabular values – this is the problem of extrapolation. Again, it might be desired to differentiate the function at the tabular points, or at some value in between tabular points – this is the problem of numerical differentiation. Other problems which arise are the numerical integration of data and the approximation to data by specified functions – these will be discussed in later chapters.

However, a table does not define a function uniquely and does not establish that a function is differentiable (one or more times) anywhere within the range of the table. It will be necessary to assume therefore, in order to derive and apply the methods of this chapter, that the given function $f(x)$ specified by the table is continuous and differentiable. On this assumption, interpolation is usually a satisfactory numerical process but, unless some a priori information is available, attempts at extrapolation can be disastrous, since little is known about the function outside the range of tabular values. Numerical differentiation, too, can be a dubious process for it tends to exaggerate the effect of rounding errors present in the tabular values (in contrast to integration which smoothes them).

3.2 Lagrangian interpolation

It is instructive to first look at the problem of interpolation from a geometrical standpoint. Let $x_0, x_1, \ldots, x_n$ be $n + 1$ given tabular values of x and let $f_0, f_1, \ldots, f_n$ be the corresponding function values of $f(x)$. We construct a polynomial of degree n which passes through the tabular points $(x_0, f_0), \ldots, (x_n, f_n)$ and we assume that this polynomial approximates reasonably well to the function $f(x)$

over the tabular range. We then take the value of the polynomial at any non-tabul point x to be the value of the function $f(x)$. Thus, in the simplest case where we have two tabular values, illustrated in Fig. 3.1, we approximate the function $f(x)$ between P and Q by the chord PQ. The value of the function at C will then be approximately RC. By similar triangles,

$$\frac{RC - f_0}{f_1 - f_0} = \frac{x - x_0}{x_1 - x_0}$$

i.e.,

$$RC = f_0 + \frac{x - x_0}{x_1 - x_0}(f_1 - f_0)$$

$$= \frac{x - x_1}{x_0 - x_1} f_0 + \frac{x - x_0}{x_1 - x_0} f_1$$

Thus

$$f(x) \doteqdot \frac{x - x_1}{x_0 - x_1} f_0 + \frac{x - x_0}{x_1 - x_0} f_1. \tag{3.1}$$

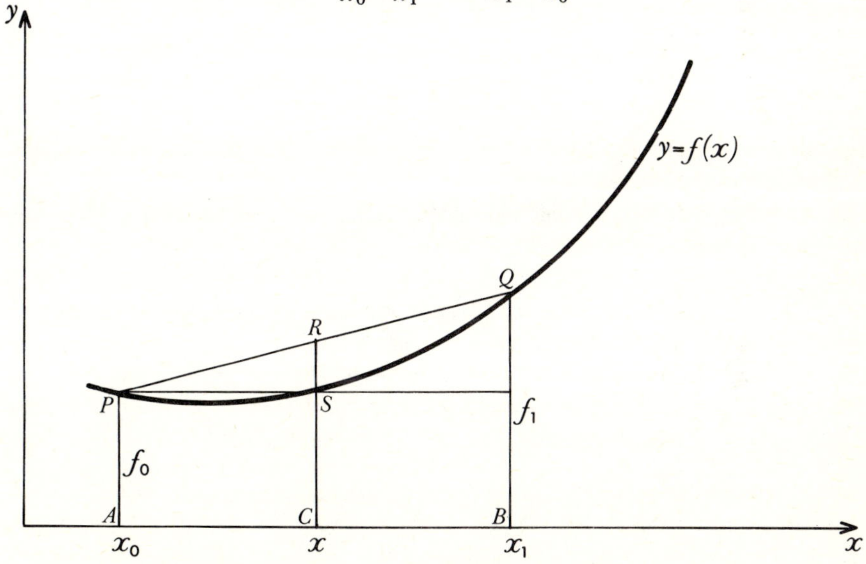

Fig. 3.1 Diagrammatic representation of linear interpolation

Equation (3.1) is known as the linear interpolation polynomial. It is exact at the tabular points x_0 and x_1 and, provided x_0 and x_1 are close enough, it gives fairly accurate results.

Consider, as an example, the evaluation of $\sin 20^0 18'$ given $\sin 20^0 = 0{\cdot}342\,02$, $\sin 21^0 = 0{\cdot}358\,37$. The formula gives

$$\sin 20^0 18' \doteqdot \left(\frac{20{\cdot}3 - 21}{20 - 21}\right) 0{\cdot}342\,02 + \left(\frac{20{\cdot}3 - 20}{21 - 20}\right) 0{\cdot}358\,37,$$

$$= 0{\cdot}7 \times 0{\cdot}342\,02 + 0{\cdot}3 \times 0{\cdot}358\,37 = 0{\cdot}346\,92_5,$$

where the number 5 at a lower level indicates the next digit in the computed answer. Thus, to five decimal places, sin $20^{0}18' = 0{\cdot}346\,93$. The correct answer to five decimal places is 0·346 94.

In the general case the polynomial $P(x)$ of degree n passing through the tabular points is given by

$$P(x) = \sum_{k=0}^{n} L_k(x) f_k, \tag{3.2}$$

where

$$L_k(x) = \frac{(x-x_0)(x-x_1)\ldots(x-x_{k-1})(x-x_{k+1})\ldots(x-x_n)}{(x_k-x_0)(x_k-x_1)\ldots(x_k-x_{k-1})(x_k-x_{k+1})\ldots(x_k-x_n)} \tag{3.3}$$

$$= \prod_{r \neq k} (x-x_r) / \prod_{r \neq k} (x_k - x_r).$$

$P(x)$ is known as the Lagrangian interpolation polynomial of degree n. Since

$$L_k(x_i) = \begin{cases} 0, & i \neq k \\ 1, & i = k \end{cases} \tag{3.4}$$

it follows that $P(x_k) = f_k$, that is, the polynomial $P(x)$ has the same values as the function $f(x)$ at the tabular points.

Before giving methods for calculating the Lagrangian polynomial $P(x)$ we consider how well $P(x)$ approximates to the function $f(x)$. We prove the following theorem.

THEOREM 3-1 *Suppose the function $f(x)$ is $n + 1$ times differentiable in a given interval $[a, b]$ and assume that it takes the values $f_0, f_1, \ldots, f_n$ at the points $x_0, x_1, \ldots, x_n$ where $a \leqslant x_0 < x_1 < \ldots < x_n \leqslant b$. Then to each x in (a, b) there exists a point ξ (dependent on x), located in the smallest interval containing x and the points $x_0, \ldots, x_n$, such that*

$$f(x) = P(x) + \{\prod_{i=0}^{n} (x - x_i)\} f^{(n+1)}(\xi)/(n+1)! \tag{3.5}$$

Proof If x is identical with one of the tabular points x_i then (3.5) is clearly satisfied and we have nothing to prove. If x is not a tabular point we construct the function

$$F(t) = f(t) - P(t) - [f(x) - P(x)] \frac{(t-x_0)(t-x_1)\ldots(t-x_n)}{(x-x_0)(x-x_1)\ldots(x-x_n)}. \tag{3.6}$$

Then $\quad F(x_i) = f(x_i) - P(x_i), \quad i = 0, 1, \ldots, n$
$\qquad\qquad = 0, \quad$ by the construction of $P(x_i)$.

Further,

$$F(x) = f(x) - P(x) - [f(x) - P(x)].1 = 0.$$

Thus $F(t)$ is zero at $n+2$ distinct points. Now if a real continuous differentiable function $g(t)$ is zero at two distinct points t_1 and t_2, Rolle's theorem asserts that there exists a value of t, such that $t_1 < t < t_2$, for which $g'(t) = 0$. Hence, we may infer that $F'(t)$ is zero at $n+1$ distinct points located in the smallest interval containing x and $x_0, \ldots, x_n$. Another application of Rolle's theorem shows that $F''(t)$ is zero at n distinct points in the region containing x and $x_0, \ldots, x_n$. Repeated application shows that there exists some value of t, call it ξ, located in the smallest interval containing x and $x_0, \ldots, x_n$ for which

$$F^{(n+1)}(\xi) = 0.$$

But differentiating (3.6) $n+1$ times shows that

$$F^{(n+1)}(t) = f^{(n+1)}(t) - P^{(n+1)}(t) - \frac{[f(x) - P(x)](n+1)!}{(x-x_0)(x-x_1)\ldots(x-x_n)}.$$

Since $P(t)$ is a polynomial of degree n, $P^{(n+1)}(t) = 0$. Thus for $t = \xi$

$$f^{(n+1)}(\xi) = \frac{[f(x) - P(x)](n+1)!}{(x-x_0)\ldots(x-x_n)},$$

which, on rearrangement, gives (3.5).

It may be remarked here that a slight extension of the arguments used in the proof enables us to show that a function $f(x)$, which is continuous and $2n$ times differentiable, can be expressed in the form

$$f(x) = y(x) + \{\prod_{k=1}^{n} (x - x_k)^2\} f^{(2n)}(\xi)/(2n)! \tag{3.7}$$

where $y(x)$ is a polynomial of degree $2n-1$, $x_1 < x_2 < \ldots < x_n$ and ξ lies in (x_1, x_n) and depends on $x (\in (x_1, x_n))$ (see exercise 28 and Ralston [1]). This result will be required in section 5.5.

We now apply Theorem 3-1 to determine the truncation error in our example on linear interpolation. From (3.5), the error in formula (3.1) is

$$\tfrac{1}{2}(x-x_0)(x-x_1)f''(\xi).$$

When $f(x) = \sin x$, $f''(x) = -\sin x$ and, since $\sin x$ is an increasing function, the maximum value of $f''(x)$ in the range $\left(\frac{\pi}{9}, \frac{7\pi}{60}\right)$, which corresponds to $(20^\circ, 21^\circ)$, is 0·358 37.

Thus the truncation error is less in modulus than

$$\tfrac{1}{2}(0{\cdot}3)\frac{\pi}{180} \times (0{\cdot}7)\frac{\pi}{180} \times 0{\cdot}358\,37$$

$$\doteqdot 1{\cdot}1 \times 10^{-5}.$$

3.3 Neville's method of interpolation

This is an iterative method for evaluating the Lagrangian polynomial $P(x)$ for given x. We form the scheme

$$
\begin{array}{lllllll}
x_i & x_i - x & f(x_i) & & & & \\
x_0 & x_0 - x & f_0 & & & & \\
 & & & f_{0,1} & & & \\
x_1 & x_1 - x & f_1 & & f_{0,1,2} & & \\
 & & & f_{1,2} & & & \\
x_2 & x_2 - x & f_2 & & f_{1,2,3} & & f_{0,1,\ldots,n} \\
 & & & f_{2,3} & & & \\
x_3 & x_3 - x & f_3 & & f_{2,3,4} & & \\
\cdot & \cdot & \cdot & \cdot & \cdot & & \\
x_{n-1} & x_{n-1} - x & f_{n-1} & & f_{n-2,\,n-1,\,n} & & \\
 & & & f_{n-1,\,n} & & & \\
x_n & x_n - x & f_n & & & &
\end{array}
$$

where

$$f_{r,r+1} = \frac{1}{(x_{r+1} - x_r)} \begin{vmatrix} f_r & x_r - x \\ f_{r+1} & x_{r+1} - x \end{vmatrix} \quad (r = 0, 1, \ldots, n-1) \tag{3.8}$$

$$f_{r,r+1,r+2} = \frac{1}{(x_{r+2} - x_r)} \begin{vmatrix} f_{r,r+1} & x_r - x \\ f_{r+1,r+2} & x_{r+2} - x \end{vmatrix} \quad (r = 0, 1, \ldots, n-2) \tag{3.9}$$

$$f_{r,r+1,r+2,r+3} = \frac{1}{(x_{r+3} - x_r)} \begin{vmatrix} f_{r,r+1,r+2} & x_r - x \\ f_{r+1,r+2,r+3} & x_{r+3} - x \end{vmatrix} \quad (r = 0, 1, \ldots, n-3) \tag{3.10}$$

and so on until

$$f_{0,1,\ldots,n} = \frac{1}{(x_n - x_0)} \begin{vmatrix} f_{0,1,\ldots,n-1} & x_0 - \\ f_{1,2,\ldots,n} & x_n - x \end{vmatrix}. \tag{3.11}$$

The quantity $f_{r,r+1,\ldots,r+k}$ is easily shown to be the interpolation polynomial through $x_r, \ldots, x_{r+k}$ and $f_{0,1,\ldots,n}$ at the apex of the table is $P(x)$. A similar scheme has been devised by Aitken (see Kopal [**5**]).

In Table 3.1 we give the numerical results obtained in the evaluation of sin (0·44) from a table of given values of sin x for $x = 0{\cdot}3(0{\cdot}1)0{\cdot}6$. The numbers along the

Table 3.1 Evaluation of sin (0·44) by Neville's method

x_i	$x_i - 0{\cdot}44$	$f(x_i)$			
0·3	−0·14	0·2955			
			0·426 96		
0·4	−0·04	0·3894		0·425 87	
			0·425 40		0·425 91
0·5	0·06	0·4794		0·425 98	
			0·428 28		
0·6	0·16	0·5646			

sloping lines exhibit a consistent trend and this leads us to assert with confidence that 0·4259 is the value of sin (0·44). Consider, however, the evaluation of $f(1{\cdot}5)$ from the data given in Table 3.2. The function values are the values of $1/x$ so that the correct value to four decimal places should be 0·6667. The explanation of the failure of the interpolation process can be found by recourse to (3·5). The modulus of the truncation error is found to be about $13{\cdot}54/\xi^8$, where ξ is some number between 0·1 and 10. It follows that the truncation error is large if ξ is at the lower end of the range, as in this example. Note that a substantial reduction in the truncation error is obtained by having tabular points close together, for then $\prod_i (x - x_i)$ is smaller.

Table 3.2 Example showing the failure of Neville's method

x_i	$x_i - 1{\cdot}5$	$f(x_i) = 1/x_i$						
0·1	−1·4	10·0						
			−60					
0·2	−1·3	5·0		122				
			−8		−6			
0·5	−1·0	2·0		5		−0·2895		
			0		1·75		0·2004	
1·0	−0·5	1·0		0·5		1·425		0·3573
			0·75		0·55		1·310	
2·0	0·5	0·5		0·725		0·5609		
			0·55		0·6532			
5·0	3·5	0·2		0·5675				
			0·27					
10·0	8·5	0·1						

3.4 Finite differences

It frequently happens in practice that the tabular values of x in interpolation problems are equally spaced. In such a case, if x_0 is an arbitrary tabular value of x, the remaining tabular values are denoted by

$$x_r = x_0 + rh, \qquad r = \pm 1, \pm 2, \ldots .$$

The corresponding function values are denoted by

$$f_r = f(x_r) = f(x_0 + rh), \qquad r = 0, \pm 1, \ldots .$$

If we write the values of x_r and f_r in vertical columns, starting with the smallest value of x_r and leaving a blank line between successive entries (as in Table 3.3), we are in a position to construct a difference table. We start with first differences which are obtained by subtracting f_r from f_{r+1}, for each r, and placing the result on the line between f_r and f_{r+1}. A column of second differences is obtained in exactly the same way by subtraction of adjacent first differences. And so on for third, fourth,

Table 3.3 Difference table for $f(x) = x^4 - 2x + 1$ for $x = -0{\cdot}3(0{\cdot}1)0{\cdot}3$

x	$f(x)$	*Difference* First	Second	Third	Fourth
−0·3	1·6081				
		−2065			
−0·2	1·4016		50		
		−2015		−36	
−0·1	1·2001		14		24
		−2001		−12	
0·0	1·0000		2		24
		−1999		12	
0·1	0·8001		14		24
		−1985		36	
0·2	0·6016		50		
		−1935			
0·3	0·4081				

and higher differences. It is common practice to express all the differences in terms of units of the last decimal place of the tabulated values of $f(x)$, that is, the decimal point and leading zeros are omitted. Note that in Table 3.3 fourth differences are constant and consequently fifth and subsequent differences will be zero. This is a special case of a general result for polynomial functions (exercise 1) and is only true if the function values are exact. If just one function value, say f_0, contains an error ϵ (due to rounding or incorrect recording) then the error influences differences spreading out from f_0 in the shape of a fan (Table 3.4). In the table $f_{i,\,i+1}$ denotes the first difference associated with f_i and f_{i+1}, $f_{i,\,i+1,\,i+2}$ denotes the second difference associated with $f_{i,\,i+1}$ and $f_{i+1,\,i+2}$ and so on. Table 3.4 is very useful for correcting mistakes in tables. In the practical application of difference tables it is convenient to adopt different notations for identifying the elements in the difference table. We

Table 3.4 Effect of an isolated error

x_{-2}	f_{-2}			
				$f_{-3,-2,-1,0}+\epsilon$
x_{-1}	f_{-1}		$f_{-2,-1,0}+\epsilon$	
		$f_{-1,0}+\epsilon$		$f_{-2,-1,0,1}-3\epsilon$
x_0	$f_0+\epsilon$		$f_{-1,0,1}+2\epsilon$	
		$f_{0,1}-\epsilon$		$f_{-1,0,1,2}+3\epsilon$
x_1	f_1		$f_{0,1,2}+\epsilon$	
				$f_{0,1,2,3}-\epsilon$
x_2	f_2			

therefore introduce the symbols Δ, ∇, δ (called the forward, backward, and central difference operators respectively) which are defined by

$$\Delta_r = \Delta f_r = f_{r+1} - f_r \tag{3.12}$$

$$\nabla_{r+1} = \nabla f_{r+1} = f_{r+1} - f_r \tag{3.13}$$

$$\delta_{r+\frac{1}{2}} = \delta f_{r+\frac{1}{2}} = f_{r+1} - f_r. \tag{3.14}$$

Since $f_{r+1} - f_r$ represents the first difference associated with the function values f_r and f_{r+1} it is clear that Δ_r, ∇_{r+1}, and $\delta_{r+\frac{1}{2}}$ can be used to identify the elements in the column of first differences. Likewise, denoting kth order differences by the symbols Δ^k, ∇^k, or $\delta^k (k \geqslant 1)$ we have

$$\Delta_r^{\,k} = \Delta_{r+1}^{k-1} - \Delta_r^{\,k-1} \tag{3.15}$$

$$\nabla_{r+1}^{\,k} = \nabla_{r+1}^{k-1} - \nabla_r^{\,k-1} \tag{3.16}$$

and

$$\left.\begin{aligned} \delta_r^k &= \delta_{r+\frac{1}{2}}^{k-1} - \delta_{r-\frac{1}{2}}^{k-1}, \qquad && k \text{ even} \\ \delta_{r+\frac{1}{2}}^k &= \delta_{r+1}^{k-1} - \delta_r^{k-1}, \qquad && k \text{ odd.} \end{aligned}\right\} \tag{3.17}$$

Other symbols we shall use are E the shift operator, μ the averaging operator, and D the differentiation operator. These are defined by

$$Ef_r = f_{r+1} \tag{3.18}$$

$$\mu f_r = \tfrac{1}{2}(f_{r+\frac{1}{2}} + f_{r-\frac{1}{2}}) \tag{3.19}$$

$$Df_r = \left[\frac{\mathrm{d}}{\mathrm{d}x} f(x)\right]_{x=x_r}. \tag{3.20}$$

We shall write f_{r+p} to denote the function value at $x_r + ph$, that is,

$$f_{r+p} = f(x_r + ph). \tag{3.21}$$

In particular, we have the following relationships holding between operators

$$\Delta \equiv E - 1 \tag{3.22}$$

$$\begin{aligned} E^n &\equiv (1 + \Delta)^n \\ &\equiv 1 + \binom{n}{1}\Delta + \ldots + \binom{n}{k}\Delta^k + \ldots + \Delta^n \end{aligned} \tag{3.23}$$

(we understand $E^n f_r$ to be f_{r+n})

$$E \equiv \exp(hD) \tag{3.24}$$

and

$$\mu\delta_r = \mu\delta f_r = \tfrac{1}{2}(f_{r+1} - f_r). \tag{3.25}$$

These results are easily deduced from the definitions.

3.5 Finite difference interpolation formulae

We are now in a position to establish interpolation formulae in terms of the entries in the finite difference tables. The object of all such finite difference formulae is to express the required function value in terms of finite differences in some systematic way. For instance, we may wish to express $f_p = f(x_0 + ph)$ in terms of some forward differences (and a remainder), e.g., in the form

$$f_p = f_0 + A_1(p)\Delta_0 + A_2(p)\Delta_0^2 + A_3(p)\Delta_0^3 + R,$$

or, alternatively, in terms of some backward differences (and a remainder), e.g., in the form

$$f_p = f_0 + A'_1(p)\nabla_0 + A'_2(p)\nabla_0^2 + A'_3(p)\nabla_0^3 + R'.$$

Again, we might wish to represent f_p in terms of central differences. However, whatever form we eventually choose, the formula must be identical with some Lagrangian interpolation formula, and this enables us to determine the coefficients of the difference terms and the remainder. Take, for example, the proposed forward difference formula given above. In terms of function values it can be written as

$$f_p = f_0 + A_1(p)(f_1 - f_0) + A_2(p)(f_2 - 2f_1 + f_0) + A_3(p)(f_3 - 3f_2 + 3f_1 - f_0) + R.$$

The third order Lagrangian interpolation formula, which is exact at (x_0, f_0) (x_1, f_1), (x_2, f_2), (x_3, f_3), is, in terms of $p = (x - x_0)/h$, (by (3.5))

$$f_p = -\frac{(p-1)(p-2)(p-3)}{3!}f_0 + \frac{p(p-2)(p-3)}{2!}f_1 - \frac{p(p-1)(p-3)}{2!}f_2$$
$$+ \frac{p(p-1)(p-2)}{3!}f_3 + \frac{p(p-1)(p-2)(p-3)}{4!}h^4 f^{(4)}(\xi), \qquad x_0 < \xi < x_3.$$

Comparison of these two formulae shows

$$R = \frac{p(p-1)(p-2)(p-3)}{4!}h^4 f^{(4)}(\xi)$$

$$A_3(p) = \frac{p(p-1)(p-2)}{3!}$$

$$A_2(p) = 3A_3(p) - \frac{p(p-1)(p-3)}{2!} = \frac{p(p-1)}{2!}$$

$$A_1(p) = A_2(p) - 3A_3(p) + \frac{p(p-2)(p-3)}{2!} = p$$

and we have thus established a particular case of Newton's forward difference interpolation formula

$$f_p = f_0 + p\Delta_0 + \frac{p(p-1)}{2!}\Delta_0^2 + \frac{p(p-1)(p-2)}{3!}\Delta_0^3 + \frac{p(p-1)(p-2)(p-3)}{4!}h^4 f^{(4)}(\xi),$$

$$x_0 < \xi < x_3. \qquad (3.26)$$

Similarly we may deduce the special case of Newton's backward difference interpolation formula

$$f_p = f_0 + p\nabla_0 + \frac{1}{2!}p(p+1)\nabla_0^2 + \frac{1}{3!}p(p+1)(p+2)\nabla_0^3$$
$$+\frac{1}{4!}p(p+1)(p+2)(p+3)h^4 f^{(4)}(\xi), \qquad x_{-3} < \xi < x_0. \qquad (3.27)$$

Again, we could establish central difference formulae such as Bessel's and Everett's, which we shall quote in the respective forms

$$f_p = f_0 + p\delta_{\frac{1}{2}} + B_2(p)\mu\delta_{\frac{1}{2}}^2 + B_3(p)\delta_{\frac{1}{2}}^3 + B_4(p)\mu\delta_{\frac{1}{2}}^4$$
$$+ B_5(p)\delta_{\frac{1}{2}}^5 + \frac{1}{6!}(p+2)(p+1)p(p-1)(p-2)(p-3)h^6 f^{(\mathrm{VI})}(\xi), \qquad (3.28)$$

where $x_{-2} < \xi < x_3$, $B_2(p) = p(p-1)/2!$, $B_3(p) = p(p-1)(p-\frac{1}{2})/3!$,

$B_4(p) = (p+1)p(p-1)(p-2)/4!$, $\qquad B_5(p) = (p+1)p(p-1)(p-2)(p-\frac{1}{2})/5!$;

and

$$f_p = (1-p)f_0 + pf_1 + E_2(p)\delta_0^2 + F_2(p)\delta_1^2 + E_4(p)\delta_0^4 + F_4(p)\delta_1^4 +$$
$$+\frac{1}{6!}(p+2)(p+1)p(p-1)(p-2)(p-3)h^6 f^{(\mathrm{VI})}(\xi), \qquad x_{-2} < \xi < x_3, \qquad (3.29)$$

where $E_2(p) = -p(p-1)(p-2)/3!$, $F_2(p) = (p+1)p(p-1)/3!$,

$E_4(p) = -(p+1)p(p-1)(p-2)(p-3)/5!$, $\qquad F_4(p) = (p+2)(p+1)p(p-1)(p-2)/5!$.

The coefficients $B_2(p), B_3(p), \ldots$; $E_2(p), F_2(p), \ldots$ are tabulated for a large range of values of p in *Interpolation* and *Allied Tables* [2].* Note that the coefficients E_{2k} and F_{2k} are connected by the formula

$$E_{2k}(p) = F_{2k}(1-p), \qquad k = 1, 2. \qquad (3.30)$$

This means that it is only necessary to tabulate the coefficient E_{2k} in the range $(0, 1)$ or E_{2k} and F_{2k} in $(0, \frac{1}{2})$ in order to use formula (3.29). We illustrate the use of formula (3.29) to evaluate $e^{2\cdot82}$ using the data in Table 3.5. In this case $p = (2\cdot82 - 2\cdot7)/0\cdot2 = 0\cdot6$ and, with this value of p,

$$E_2 = -0\cdot056, \qquad F_2 = -0\cdot064,$$
$$E_4 = 0\cdot010\,752, \qquad F_4 = 0\cdot011\,648.$$

Hence, from (3.29),

$$\exp(2\cdot82) \doteqdot 0\cdot4 \times 14\cdot88 + 0\cdot6 \times 18\cdot17 - 0\cdot056 \times 0\cdot59 - 0\cdot064 \times 0\cdot74$$
$$+ 0\cdot010\,752 \times 0\cdot05 + 0\cdot011\,648 \times -0\cdot01$$
$$\doteqdot 16\cdot77_4.$$

*Our B_2 and B_4 are twice those tabulated in *Interpolation and Allied Tables.*

Table 3.5 Difference table for exp x for $x = 2{\cdot}3(0{\cdot}2)3{\cdot}3$

x	exp x				
2·3	9·97				
		221			
2·5	12·18		49		
		270		10	
2·7 (x_0)	14·88 (f_0)		59 (δ_0^2)		5 (δ_0^4)
		329		15	
2·9 (x_1)	18·17 (f_1)		74 (δ_1^2)		−1 (δ_1^4)
		403		14	
3·1	22·20		88		
		491			
3·3	27·11				

The truncation error is less than 10^{-5} and it follows that in the above answer, quoted to two decimal places, any error is due to rounding in the original data.

In practice when fifth (and higher) differences do not contribute significantly to the value of f_p we can modify the second differences in the Bessel and Everett formulae so that the effect of omitting the terms B_4, or E_4 and F_4, is very small. Consider, for example, Bessel's formula

$$f_p = f_0 + p\delta_{\frac{1}{2}} + B_2(p)\mu\delta_{\frac{1}{2}}^{2} + B_3(p)\delta_{\frac{1}{2}}^{3} + B_4(p)\mu\delta_{\frac{1}{2}}^{4} \; \ldots .$$

The modified formula is

$$f_p = f_0 + p\delta_{\frac{1}{2}} + B_2\mu\delta_{\frac{1}{2}\,m}^{2} + B_3\delta_{\frac{1}{2}}^{3} \qquad (3.31)$$

where the modified second difference $\mu\delta_{\frac{1}{2}\,m}^{2} = \mu\delta_{\frac{1}{2}}^{2} + C\mu\delta_{\frac{1}{2}}^{4}$. The error is approximately (why not exactly?)

$$(B_4 - CB_2)\mu\delta_{\frac{1}{2}}^{4}.$$

We choose C to make the maximum positive and negative values of $B_4(p) - CB_2(p)$ over the range $0 \leqslant p \leqslant 1$ equal in magnitude. To satisfy this condition we can show that $C \doteqdot -0{\cdot}184$ (exercise 13).

A similar argument can be used for Everett's formula and yields the same result. Values of $0{\cdot}184\delta^4$ are given for various values of δ^4 in *Interpolation and Allied Tables* [2]. These are called throwbacks and are to be subtracted from the second differences for formula (3.31) to be applied. The reader is referred to Fox [3] for the general theory of throwbacks.

We end this section with an example on the use of the modified Bessel formula (3.31). Table 3.6 records the differences necessary for the evaluation of $f(x)$ at $x = 1{\cdot}268$. We have

$$\begin{aligned} f(1{\cdot}268) &\doteqdot 0{\cdot}182\,831 + 0{\cdot}68(0{\cdot}161\,176) - \\ &\quad - 0{\cdot}0544(0{\cdot}011\,819 + 0{\cdot}013\,426) - 0{\cdot}006\,53(0{\cdot}001\,607) \\ &\doteqdot 0{\cdot}291\,047. \end{aligned}$$

Table 3.6 Use of modified Bessel's formula

x	$f(x)$					
$1{\cdot}2(x_0)$	$0{\cdot}182\ 831(f_0)$		$11\ 840(\delta_0^2)$		$116(\delta_0^4)$	$11\ 819(\delta_{0m}^2)$
		$161\ 176$		1607		
$1{\cdot}3(x_1)$	$0{\cdot}344\ 007(f_1)$		$13\ 451(\delta_1^2)$		$135(\delta_1^4)$	$13\ 426(\delta_{1m}^2)$

3.6 Numerical differentiation

One approach to finding the derivative of a function is obtained by considering formula (3.5). Differentiation of this formula gives

$$f'(x) = P'(x) + \frac{\mathrm{d}}{\mathrm{d}x}\left[\left\{\prod_{i=0}^{n}(x - x_i)\right\} f^{(n+1)}(\xi)/(n+1)!\right].$$

The second term on the right-hand side of the equation can be written as

$$\frac{\mathrm{d}}{\mathrm{d}x}\left\{\prod_{i=0}^{n}(x - x_i)\right\}\frac{f^{(n+1)}(\xi)}{(n+1)!} + \prod_{i=0}^{n}(x - x_i)\frac{\mathrm{d}}{\mathrm{d}x}\left\{\frac{f^{(n+1)}(\xi)}{(n+1)!}\right\}.$$

Unfortunately, $\frac{\mathrm{d}}{\mathrm{d}x}\left[f^{(n+1)}(\xi)\right]$ is not easy to determine since ξ is an unknown function of x. However this problem can be overcome if the value of $f'(x)$ is required at one of the tabular points x_k. For then $\prod_i (x - x_i) = 0$, and we obtain after simplification

$$f'(x_k) = P'(x_k) + \prod_{i \neq k}^{n}(x_k - x_i) f^{(n+1)}(\xi)/(n+1)! \tag{3.32}$$

The simplest possible case occurs when we have just two tabular points. Formula (3.32) then reduces to

$$f'(x_0) = \frac{f(x_1) - f(x_0)}{x_1 - x_0} + \frac{1}{2!}(x_0 - x_1)f''(\xi), \qquad x_0 < \xi < x_1 \tag{3.33}$$

which is a restatement of Taylor's theorem. Note that if we omit the error terms in (3.33) we have

$$f'(x_0) \doteqdot \frac{f(x_1) - f(x_0)}{x_1 - x_0},$$

a result which might have been expected since, by definition, $f'(x_0)$ is the limit as $x_1 \to x_0$ of the right-hand side. In the general case with n points available, Neville's iterative method has been developed to evaluate $P'(x_k)$ (see Hunter [4]).

When the data is given at equal intervals, then differentiation formulae can be obtained from the finite difference interpolation formulae which were derived in section 3.5. Thus, starting with Bessel's formula (3.28), and noting that

$$\frac{df(x)}{dx}=\frac{df(x)}{dp}\frac{dp}{dx}=\frac{1}{h}\frac{df_p}{dp}=\frac{1}{h}f'_p$$

we have

$$\frac{df(x)}{dx}=\frac{1}{h}\frac{d}{dp}\left[f_0+p\delta_{\frac{1}{2}}+B_2(p)\mu\delta_{\frac{1}{2}}^2+\ldots+B_5(p)\delta_{\frac{1}{2}}^5+R\right],$$

where R denotes the remainder term in the formula. Thus

$$\frac{df(x)}{dx}=\frac{1}{h}\left[\delta_{\frac{1}{2}}+B'_2\mu\delta_{\frac{1}{2}}^2+B'_3\delta_{\frac{1}{2}}^3+B'_4\mu\delta_{\frac{1}{2}}^4+B'_5\delta_{\frac{1}{2}}^5+\frac{dR}{dp}\right], \tag{3.34}$$

where $B'_2=\frac{1}{2}(2p-1)$, $B'_3=\frac{1}{12}(6p^2-6p+1)$, $B'_4=\frac{1}{12}(2p^3-3p^2-p+1)$, $B'_5=\frac{1}{120}(5p^4-10p^3+5p-1)$.* At the tabular point x_0

$$\left(\frac{dR}{dp}\right)_{p=0}=-\frac{1}{60}h^6f^{(VI)}(\xi),\qquad x_{-2}<\xi<x_3$$

and we obtain the formula

$$hf'_0=\delta_{\frac{1}{2}}-\frac{1}{2}\mu\delta_{\frac{1}{2}}^2+\frac{1}{12}\delta_{\frac{1}{2}}^3+\frac{1}{12}\mu\delta_{\frac{1}{2}}^4-\frac{1}{120}\delta_{\frac{1}{2}}^5-\frac{1}{60}h^6f^{(VI)}(\xi). \tag{3.35}$$

At the mid-point $x_{\frac{1}{2}}$ we have

$$\left(\frac{dR}{dp}\right)_{p=\frac{1}{2}}=\frac{7}{1024}h^6\frac{d}{dp}f^{(VI)}(\xi)=R',\text{ say.}$$

It is not possible to evaluate the derivative on the right-hand side since ξ is an unknown function of p, but the approximate value can be found from formula (3.40). Hence we have the 'mid-point' formula

$$hf'_{\frac{1}{2}}=\delta_{\frac{1}{2}}-\frac{1}{24}\delta_{\frac{1}{2}}^3+\frac{3}{640}\delta_{\frac{1}{2}}^5+R'. \tag{3.36}$$

Differentiation formulae can also be obtained directly from formula (3.32). For instance, if the original interpolation polynomial passed through the points $x_{-3}, x_{-2}, \ldots, x_3$ we find, after simplification, that

$$hf'_0=\frac{1}{60}(f_3-f_{-3})-\frac{1}{20}(f_2-f_{-2})+\frac{3}{4}(f_1-f_{-1})-\frac{1}{140}f^{(7)}(\xi),$$

where $x_{-3}<\xi<x_3$. This formula can be written in the more concise form

$$hf'_0=\mu\delta_0-\frac{1}{6}\mu\delta_0^3+\frac{1}{30}\mu\delta_0^5-\frac{1}{140}f^{(7)}(\xi). \tag{3.37}$$

To illustrate the use of this formula we estimate $\frac{d}{dx}(\exp x)$ at $x=2{\cdot}7$. The required

*Our B'_2 and B'_4 are twice those tabulated in *Interpolation and Allied Tables.*

data necessary for the computation is given in Table 3.5. We get, ignoring the term in $\mu\delta_0^5$ and the error term,

$$0{\cdot}2 f'_0 = \tfrac{1}{2}(2{\cdot}70 + 3{\cdot}29) - \tfrac{1}{6}.\tfrac{1}{2}(0{\cdot}10 + 0{\cdot}15)$$

which yields

$$f'_0 = 14{\cdot}88.$$

This is the correct answer to two decimal places, since we know that the derivative of the function $\exp x$ is $\exp x$, i.e., the calculated result should be equal to the tabulated values at $x = 2{\cdot}7$.

In some instances it is necessary to express a finite difference in terms of derivatives (in contrast to the above formulae which express a derivative in terms of finite differences). Some of these formulae can be obtained quite readily from Taylor's theorem. We have, with the usual notation for equally spaced data,

$$f_1 = f_0 + hf'_0 + \frac{h^2}{2!}f''_0 + \frac{h^3}{3!}f'''_0 + \frac{h^4}{4!}f^{(IV)}_0 + \frac{h^5}{5!}f^{(V)}(\xi_1), \qquad x_0 < \xi_1 < x_1,$$

$$f_{-1} = f_0 - hf'_0 + \frac{h^2}{2!}f''_0 - \frac{h^3}{3!}f'''_0 + \frac{h^4}{4!}f^{(IV)}_0 - \frac{h^5}{5!}f^{(V)}(\xi_1), \qquad x_{-1} < \xi_2 < x_0.$$

Hence

$$\mu\delta_0 = \frac{1}{2}(f_1 - f_{-1}) = hf'_0 + \frac{h^3}{3!}f'''_0 + \frac{h^5}{5!}\,\frac{1}{2}\{f^{(V)}(\xi_1) + f^{(V)}(\xi_2)\}.$$

As $f^{(V)}(x)$ is assumed to be continuous in (x_{-1}, x_1) there exists some value ξ of x between ξ_1 and ξ_2 for which

$$f^{(V)}(\xi) = \tfrac{1}{2}\{f^{(V)}(\xi_1) + f^{(V)}(\xi_2)\}.$$

This leads to the formula

$$\mu\delta_0 = hf'_0 + \tfrac{1}{6}h^3 f'''_0 + \tfrac{1}{120}h^5 f^{(V)}(\xi), \qquad x_{-1} < \xi < x_1. \tag{3.38}$$

Similarly we can obtain

$$\mu\delta_0^3 = h^3 f'''_0 + \tfrac{1}{4}h^5 f^{(V)}(\eta), \qquad x_{-2} < \eta < x_2, \tag{3.39}$$

and the approximations

$$h^n f^{(n)}_0 \doteqdot \begin{cases} \delta_0^{\,n}, & \text{if } n = 2k \\ \mu\delta_0^{\,n}, & \text{if } n = 2k+1. \end{cases} \tag{3.40}$$

A large number of additional formulae are given in *Interpolation and Allied Tables* [2] (see also exercise 22).

Finally, we remark that derivatives at non-tabular points can be obtained by interpolation. Alternatively, we can use Bessel's formula and approximate to the terms in the truncation formula by taking the term $h^n f^{(n)}(\xi)$ to be the largest central difference in the range of ξ (or mean central difference if n is odd).

3.7 Extrapolation

The formulae which were developed in earlier sections for interpolation can also be used for extrapolation. However, the truncation error term can be quite large, since the further we move from the tabular points the larger $\prod_i (x - x_i)$ becomes. Thus, in general, extrapolation can only be guaranteed for arguments very close to the tabulated arguments. The dangers in extrapolation can be demonstrated by very simple examples (see exercise 18).

As an example consider the estimation of sin 21°18′ given sin 20° = 0·342 02 and sin 21° = 0·358 37. We use the interpolation formula

$$f_p \doteqdot (1-p) f_0 + p f_1$$

with

$$p = \frac{21{\cdot}3 - 20}{21 - 20} = 1{\cdot}3.$$

We have

$$\begin{aligned} \sin 21°18' &= (1 - 1{\cdot}3)0{\cdot}342\,02 + 1{\cdot}3(0{\cdot}358\,37) \\ &= 0{\cdot}363\,28. \end{aligned}$$

The truncation error is approximately

$$\tfrac{1}{2}(1{\cdot}3)(0{\cdot}3)(\pi/180)^2 \times 0{\cdot}363\,28 \doteqdot 2 \times 10^{-5}$$

in modulus, which is a fairly favourable result.

There are some cases where extrapolation is a very satisfactory process. For instance, if we have a sequence of function values $\{f_r\}$, which (as r increases) tends to a limit f in such a way that the sequence $\{f - f_r\}$ is a geometric progression. Then, for three consecutive values f_{r-1}, f_r, f_{r+1}, we have

$$(f - f_{r-1})(f - f_{r+1}) = (f - f_r)^2$$

and hence

$$f = f_{r+1} - \frac{(f_{r+1} - f_r)^2}{f_{r+1} - 2f_r + f_{r-1}}. \tag{3.41}$$

This formula is known as Aitken's δ^2-extrapolation formula and it is particularly useful in speeding up the convergence of certain types of iterative methods. Consider, for example, the solution of the equation

$$x = \tfrac{1}{2}\cos x,$$

by means of the iterative formula

$$x_{r+1} = \tfrac{1}{2}\cos x_r, \qquad x_0 = 0, \text{ say.}$$

Let $x_r = x - \xi_r$. Substitution in the iterative formula gives

$$x - \xi_{r+1} = \tfrac{1}{2}\cos(x - \xi_r),$$
$$= \tfrac{1}{2}\cos x \cos \xi_r + \tfrac{1}{2}\sin x \sin \xi_r,$$
$$= \tfrac{1}{2}\cos x + \tfrac{1}{2}\xi_r \sin x + O(\xi_r^2),$$

i.e.,

$$\xi_{r+1} = -\tfrac{1}{2}\xi_r \sin x + O(\xi_r^2).$$

Thus, if ξ_r is sufficiently small, the premise that the sequence $\{x - x_r\}$ forms a geometric progression is valid, and we can apply the Aitken δ^2-extrapolation formula. We have

$$x_1 = \tfrac{1}{2}\cos 0 = 0{\cdot}5$$
$$x_2 = \tfrac{1}{2}\cos 0{\cdot}5 = 0{\cdot}438\,79$$
$$x_3 = \tfrac{1}{2}\cos 0{\cdot}438\,79 = 0{\cdot}452\,63$$
$$x_4 = \tfrac{1}{2}\cos 0{\cdot}452\,63 = 0{\cdot}449\,65$$

and

$$x \doteqdot 0{\cdot}449\,65 - \frac{(0{\cdot}449\,65 - 0{\cdot}452\,63)^2}{0{\cdot}449\,65 - 2 \times 0{\cdot}452\,63 + 0{\cdot}438\,79} = 0{\cdot}450\,18.$$

This is the correct answer to five decimal places.

Another method of extrapolation, which will be applied in chapter 5, is called Richardson's deferred approach to the limit. We suppose that a quantity f is evaluated by means of an approximation $g(h)$ which depends on a parameter h. More precisely, suppose

$$f = g(h) + ch^n + O(h^m), \qquad (3.42)$$

where c, m, n are constants and $n < m$. Then

$$f = g(\alpha h) + c(\alpha h)^n + O(\alpha^m h^m), \qquad \alpha \neq 1$$

and, by elimination of c,

$$f = \frac{\alpha^n g(h) - g(\alpha h)}{\alpha^n - 1} + O(h^m). \qquad (3.43)$$

In general, $[\alpha^n g(h) - g(\alpha h)]/(\alpha^n - 1)$ is a better approximation to the values of f than $g(h)$ and $g(\alpha h)$.

As an example of Richardson's method of extrapolation consider the evaluation of the derivative f'_0. Assuming that higher derivatives are continuous, Taylor's theorem gives, if we combine the expansions for f_1 and f_{-1} and rearrange

$$f'_0 = \frac{f_1 - f_{-1}}{2h} - \frac{h^2}{6} f_0''' - \frac{h^4}{120} f^{(5)}(\xi), \qquad x_{-1} < \xi < x_1.$$

Likewise

$$f'_0 = \frac{f_2 - f_{-2}}{4h} - \frac{4h^2}{6} f_0''' - \frac{16h^4}{120} f^{(5)}(\eta), \qquad x_{-2} < \eta < x_2.$$

Let $A(h) = (f_1 - f_{-1})/2h$, $A(2h) = (f_2 - f_{-2})/4h$. Then we can eliminate f'''_0 between the two formulae with the result that

$$f'_0 = \tfrac{4}{3}A(h) - \tfrac{1}{3}A(2h) + O(h^4). \tag{3.44}$$

We apply this result to find $f'(0{\cdot}2)$ where $f(x) = 1/(1 + \sin x)$. From Table 5.1,

$$A(0{\cdot}1) = (0{\cdot}7719 - 0{\cdot}9092)/0{\cdot}2 = -0{\cdot}6865,$$
$$A(0{\cdot}2) = (0{\cdot}7197 - 1{\cdot}0000)/0{\cdot}4 = -0{\cdot}7008.$$

Formula (3.44) gives approximately

$$f'(0{\cdot}2) \doteqdot \tfrac{4}{3}(-0{\cdot}6865) - \tfrac{1}{3}(-0{\cdot}7008) = -0{\cdot}6817.$$

The correct value of the derivative to four decimal places is $-0{\cdot}6820$.

3.8 Summary and notes

Lagrangian interpolation, performed by the methods of Neville or Aitken, is well suited to automatic computation in view of its iterative nature. It is probably the most widely used method in practice, as it is applicable whatever the distribution of the data. It is advisable, however, to check results by inverse interpolation, that is, taking the function values as the arguments and the arguments as the function values. Finite difference formulae are only applicable to data given for equally spaced values of the argument and are suitable for computation on desk machines and automatic computation. Central difference formulae are generally to be preferred to forward or backward difference formulae and the latter are only used when the function value is required near the beginning or the end of the tabular range – and then if only the first two or three differences make a significant contribution to the interpolated value. Forward and backward differences do however find useful application in chapters 6 and 11 in connection with series summation and predictor-corrector methods for the solution of differential equations. Central difference formulae are also preferable as differentiation formulae (for equally spaced values of the argument).

With regard to extrapolation, Aitken's δ^2-formula has found considerable application to speeding up the convergence of iterative sequences while Richardson's deferred approach to the limit is particularly useful in the evaluation of integrals.

Finally, a problem which occurs frequently is that of interpolation in two dimensions. This can be tackled by constructing a two-dimensional analogue of the Lagrange interpolation polynomial, or by constructing finite difference formulae by recourse to Taylor's expansion in two variables. The basic ideas of the Taylor series method can be grasped from section 5.8.

REFERENCES

[1] Ralston, A., *A First Course in Numerical Analysis*, McGraw-Hill Book Co., New York (1965).

[2] *Interpolation and Allied Tables*, H.M.S.O., London (1956).

[3] Fox, L., The Use and Construction of Mathematical Tables, *N.P.L. Mathematical Tables*, Vol. 1, H.M.S.O., London (1956).

[4] Hunter, D. B., 'An Iterative Method for Numerical Differentiation', *The Computer J.*, 3 (1961) 270–71.

[5] Kopal, Z., *Numerical Analysis*, J. Wiley and Sons, New York (1955).

Exercises

1. If $f(x)$ is a polynomial of degree n prove that $(n+1)$th differences are all zero. What is the constant value of the nth difference? (Hint: Assume true for $f(x) = x^p$, $p \leqslant m$, i.e., $\delta^{p+r}(x^p) = 0$, $r \geqslant 1$. Show

$$\delta^{m+3}(x^{m+2}) = \delta^{m+1}\{(x+h)^{m+2} - 2x^{m+2} + (x-h)^{m+2}\} = 0.$$

The result follows by induction.)

2. Prepare a difference table for

$$f(x) = 10x^3 + x^2, \qquad x = 0(0{\cdot}1)0{\cdot}4.$$

Extend the table as far as $x = 1{\cdot}0$. (Hint: Use the fact that 3rd differences are constant.)

3. Two consecutive entries for $f(x)$ in the following table contain identical errors. Find which entries are wrong and correct them.

x	0	1	2	3	4	5
$f(x)$	4·2	3·1	6·0	12·9	23·8	37·8
x	6	7	8	9	10	
$f(x)$	56·7	80·5	107·4	138·3	173·2	

4. Show the following result between nth central differences and function values

$$\delta^n f_j = \sum_{k=0}^{n} (-1)^k \frac{n!}{k!(n-k)!} f_{j+\frac{1}{2}n-k},$$

where $2j$ is an integer if n odd
j is an integer if n even.

[The coefficients in the above expansion are the coefficients in the binomial expansion for $(1-z)^n$.]

5. Prove the following relationships for operators

(i) $\Delta = \mu\delta + \frac{1}{2}\delta^2 = E^{\frac{1}{2}}\delta$;

(ii) $\Delta - \nabla = \Delta\nabla$;

(iii) $\mu(f_n g_n) = \mu f_n \mu g_n + \frac{1}{4}\delta f_n \delta g_n$.

6. Use formulae (3.26) and (3.27) to evaluate $f(0{\cdot}35)$ and $f(9{\cdot}4)$ from the corrected difference table in exercise 3.

7. Derive Stirling's interpolation formula

$$f_p = f_0 + \tfrac{1}{2}p(\delta_{\frac{1}{2}} + \delta_{\frac{1}{2}}) + \tfrac{1}{2}p^2\delta_0{}^2 + \frac{p(p^2-1)}{2\cdot 3!}(\delta_{-\frac{1}{2}}{}^3 + \delta_{\frac{1}{2}}{}^3) + \frac{p^2(p^2-1)}{4!}\delta_0^4 + \ldots .$$

8. Use Stirling's formula to find the values of $f(4{\cdot}3)$ and $f(5{\cdot}8)$ from the corrected data in exercise 3.

9. Bessel's formula in the form

$$p = [(f_p - f_0) - B_2(p)\ \mu\delta_{\frac{1}{2}}{}^2 - B_3(p)\delta_{\frac{1}{2}}{}^3 - \ldots]/\delta_{\frac{1}{2}},$$

is suitable for inverse interpolation. Take $p^{(0)} = (f_p - f_0)/\delta_{\frac{1}{2}}$ and substitute this value in the right-hand side of the above equation to obtain $p^{(1)}$. Repeat with $p^{(1)}$ in the right-hand side of the equation, and so on, until successive iterates are close together. Use this method to find $\sin^{-1}(0{\cdot}35)$ in degrees and decimals from the following data

x^0	0	10	20	30	40	50
$\sin x$	0	0·173 65	0·342 02	0·500 00	0·642 79	0·766 04

10. Use the following data for the sine integral $\mathrm{Si}(x) = \int_0^x \frac{\sin t}{t}\, dt$ to find the value of x when $\mathrm{Si}(x) = 0{\cdot}45$,

(i) by Neville's method,
(ii) by Bessel's formula.

Check your results by direct interpolation.

x	0	0·2	0·4	0·6	0·8	1·0
$\mathrm{Si}(x)$	0	0·199 56	0·396 46	0·588 13	0·772 10	0·946 08

11. Find the greatest value of $|B_3(p)|$, $0 \leqslant p \leqslant 1$. Hence show that the contribution from $\delta_{\frac{1}{2}}{}^3$ to the interpolated value is less than 0·5 in the least significant figure if $|\delta_{\frac{1}{2}}{}^3|$ is less than about 62.

12. Show that the greatest value of $|B_4(p)|$ is 0·0234. What is the greatest value of $|(\delta_0{}^4 + \delta_1{}^4)|$ which contributes an amount less than 0·5 to the least significant figure of the interpolated value?

13. Prove that the exact value of C used for Bessel's modified formula is $-(3+\sqrt{2})/24$. Show that with this value of C the residual contribution from $\frac{1}{2}(B_4 - CB_2)(\delta_0{}^4 + \delta_1{}^4)$ is less than 0·5 in the least significant figure if $|(\delta_0{}^4 + \delta_1{}^4)|$ is less than about 2200. Carry out the corresponding exercise for Everett's formula.

14. Let $\phi_1(\delta)$ be an even function of δ and $\phi_2(\delta)$ be an odd function of δ. Bessel's formula is of the type

$$f_p = \phi_1(\delta)(f_0 + f_1) + \phi_2(\delta)f_{\frac{1}{2}},$$

i.e.,

$$E^p f_0 = \phi_1(\delta)(E + 1)f_0 + \phi_2(\delta)E^{\frac{1}{2}}f_0.$$

Show $\phi_1(\delta)$ and $\phi_2(\delta)$ satisfy

$$\cosh (p - \tfrac{1}{2})U + \sinh (p - \tfrac{1}{2})U = e^{(p-\frac{1}{2})U} = \phi_1(\delta) . 2 \cosh \tfrac{1}{2}U + \phi_2(\delta),$$

where $U = hD$. By comparing odd and even functions on both sides of the equation show that

$$\phi_1(\delta) = \cosh (p - \tfrac{1}{2})U/2 \cosh \tfrac{1}{2}U, \qquad \phi_2(\delta) = \sinh (p - \tfrac{1}{2})U.$$

Hence find $B_2(p), \ldots, B_5(p)$ by expanding these functions in terms of $\delta = 2 \sinh \frac{1}{2}U$ (see Hartree, *Numerical Analysis*, Oxford University Press (1958)).

15. By assuming an interpolation formula has the form

$$f_p = \phi_1(\delta)f_0 + \phi_2(\delta)f_1$$

where $\phi_1(\delta)$ and $\phi_2(\delta)$ are even functions of δ, show that

$$\phi_1(\delta) = \sinh (1 - p)U/\sinh U,$$
$$\phi_2(\delta) = \sinh pU/\sinh U.$$

Hence find the coefficients E_2, F_2, E_4, F_4 in Everett's formula.

16. Gauss's forward interpolation formula has the form

$$f_p = f_0 + G_1\delta f_{\frac{1}{2}} + G_2\delta^2 f_0 + G_3\delta^3 f_{\frac{1}{2}} + G_4\delta^4 f_0 + \ldots .$$

Using the method of exercises 14 and 15 show

$$G_1 = p, \quad G_2 = \frac{p(p-1)}{2!}, \quad G_3 = \frac{(p+1)p(p-1)}{3!}, \quad G_4 = \frac{(p+1)p(p-1)(p-2)}{4!}, \ldots$$

By a similar approach find the coefficients in Gauss's backward interpolation formula

$$f_p = f_1 + g_1\delta f_{\frac{1}{2}} + g_2\delta^2 f_1 + g_3\delta^3 f_{\frac{1}{2}} + g_4\delta^4 f_1 + \ldots .$$

What formula is obtained by averaging these results for f_p?

17. Show, by induction, that the term $f_{0,1,\ldots,n}$ in Neville's method is equivalent to the nth degree Lagrangian polynomial through the points $x_0, \ldots, x_n$.

18. Find the value of $\mathrm{Si}(x)$ when $x = 0$,

(*a*) by linear extrapolation;

(*b*) by using Lagrange's interpolation formula with $x = 0$.

x	0·2	0·4	0·6	0·8	1·0
Si(x)	0·199 56	0·396 46	0·588 13	0·772 10	0·946 08

Explain why linear extrapolation produces a better result in this instance?

19. Deduce the following forward and backward difference formulae for first derivatives.

$$f'_0 = \frac{1}{h}(\Delta - \tfrac{1}{2}\Delta^2 + \tfrac{1}{3}\Delta^3 - \ldots)f_0,$$

$$f'_n = \frac{1}{h}(\nabla + \tfrac{1}{2}\nabla^2 + \tfrac{1}{3}\nabla^3 + \ldots)f_n.$$

20. Evaluate $f'(0{\cdot}1)$ by formula (3.37) and the formulae in exercise 19, where $f(x)$ is the function given in Table 3.3.

21. For the data given in Table 3.3 find the value of $f(x)$ for which $f'(x) = -1{\cdot}98$.

22. Show

$$h^2 f''_0 = \delta_0^2 - \tfrac{1}{12}\delta_0^4 + \tfrac{1}{90}\delta_0^6 - \ldots$$

and

$$h^4 f_0^{(IV)} = \delta_0^4 - \tfrac{1}{6}\delta_0^6 + \ldots .$$

23. For the data given in Table 3.3, evaluate $f''(0{\cdot}1)$ and $f^{(IV)}(0{\cdot}1)$. What is the correspondence between $f_0^{(n)}$ and δ_0^n for a polynomial of degree n?

24. Show that

$$n \sin \pi/n = \pi - \frac{1}{3!}\frac{\pi^3}{n^2} + \frac{1}{5!}\frac{\pi^5}{n^4} - \ldots$$

$$= \pi + \frac{A}{n^2} + \frac{B}{n^4} + \frac{C}{n^6} + \ldots .$$

Use Richardson's method with $n = 1, 2, 3$ substituted in the above formula, to obtain an extrapolated value for π.

25. By assuming that the data in exercise 18 satisfies

$$\mathrm{Si}(h) = \mathrm{Si}(0) + \alpha h + \beta h^3 + \gamma h^5 + \ldots,$$

use Richardson's deferred approach to the limit, in conjunction with linear extrapolation, to estimate Si(0).

26. Use the method of least squares (chapter 13) to fit a third order polynomial to the data of exercise 10. From this polynomial estimate approximately

$$\mathrm{Si}(0{\cdot}42) \quad \text{and} \quad \mathrm{Si}'(0{\cdot}42).$$

27. Let $\{f_n\}$ be a sequence converging to the limit f and such that $d_n = f_n - f$ satisfies

$$d_{n+1} = (c + \epsilon_n)d_n, \qquad d_n \neq 0,$$

where c is a constant such that $|c| < 1$ and $\epsilon_n \to 0$ as $n \to \infty$. Show that the sequence $\{f'_n\}$ defined by

$$f'_{n+1} = f_{n+1} - \frac{(f_{n+1} - f_n)^2}{f_{n+1} - 2f_n + f_{n-1}}, \qquad n \geqslant 1,$$

converges to f faster than the sequence $\{f_n\}$. (Hint: Show $[(f'_n - f)/(f_n - f)] \to 0$ as $n \to \infty$.)

28. (Hermite Interpolation) Show that the interpolation polynomial $y(x)$ of degree $2n-1$ for which $y(x)$ and $f(x)$ coincide at the points $x_1, x_2, \ldots, x_n$, and for which $y'(x)$ and $f'(x)$ coincide at the same points, is

$$y(x) = \sum_{k=1}^{n} \{1 - 2L'_k(x_k)(x - x_k)\}L_k^2(x)f(x_k) + \sum_{k=1}^{n} (x - x_k)L_k^2(x)f'(x_k).$$

Hence deduce (3.7).

4 Orthogonal polynomials

4.1 Some basic properties of orthogonal functions

Orthogonal polynomials will be encountered throughout this work. For this reason we include here some of the basic properties of the four orthogonal polynomials which, we feel, occur most often in practice; namely, those of Chebyshev, Legendre, Hermite, and Laguerre. We begin with a definition of a general class of functions called orthogonal functions.

DEFINITION Suppose we are given a function $w(x)$ which is non-negative in some closed interval $[a, b]$ together with a sequence of functions $\{f_n(x)\}$, $n = 0, 1, 2, \ldots,$ defined in $[a, b]$. Suppose further, that the following relations are satisfied

$$\int_a^b w(x) f_m(x) f_n(x)\, dx = 0, \qquad m \neq n \tag{4.1}$$

$$\int_a^b w(x) f_n^2(x)\, dx = C_n \tag{4.2}$$

where C_n is, in general, a non-zero constant (dependent on n).

Then the functions $f_n(x)$ are termed orthogonal functions relative to the positive weighting function $w(x)$. If, in addition, $C_n = 1$ for all n the functions $f_n(x)$ are termed orthonormal functions.

If we take $w(x) = 1, a = -\pi, b = \pi$, for example, then $\{\cos nx\}$ is seen from the definition to form an orthogonal sequence of functions, since

$$\int_{-\pi}^{\pi} \cos nx \cos mx\, dx = \begin{bmatrix} 0, & m \neq n \\ \pi, & m = n\ (\neq 0) \\ 2\pi, & m = n = 0. \end{bmatrix}$$

In the next sections we shall establish the orthogonality conditions for the special polynomials already referred to, together with many other properties of these polynomials. We just mention here three results which are applicable to all these types of polynomials. The reader is referred to Szego [1] for proofs of these results.

THEOREM 4-1 *The sequence of orthogonal polynomials* $\{f_n(x)\}$, *defined with respect to a positive weighting function* $w(x)$ *in an interval* $[a, b]$, *forms a linearly independent sequence of functions in* $[a, b]$.

THEOREM 4-2 *The polynomial $f_n(x)$ of degree n has all its zeros in the interval* $[a, b]$.

THEOREM 4-3 *Between any two zeros of $f_{n+1}(x)$ there exists a zero of $f_n(x)$.*

The last two theorems are particularly useful for the numerical calculation of the zeros since, once the zeros of $f_n(x)$ have been located, we know that the zeros of $f_{n+1}(x)$ lie between them or between them and the end-points of the interval. The mid-point of each interval then provides a good starting approximation for determining a zero.

4.2 Chebyshev polynomials

There are two kinds of Chebyshev polynomials of which the most important are Chebyshev polynomials of the first kind. These are defined by*

$$T_n(x) = \cos n\,(\cos^{-1} x), \qquad -1 \leqslant x \leqslant 1. \tag{4.3}$$

Substitution of $n = 0, 1, 2, \ldots$ in (4.3) gives rise to the sequence

$$T_0(x) = 1, \qquad T_1(x) = x, \qquad T_2(x) = 2x^2 - 1,$$
$$T_3(x) = 4x^3 - 3x, \qquad T_4(x) = 8x^4 - 8x^2 + 1, \ldots .$$

If, in (4.3), we let $x = \cos\theta$ we find

$$\begin{aligned} T_{n+1}(x) + T_{n-1}(x) &= \cos(n+1)\theta + \cos(n-1)\theta, \\ &= 2\cos\theta\cos n\theta = 2xT_n(x). \end{aligned}$$

Thus we have obtained the recurrence relation for calculating Chebyshev polynomials of the first kind,

$$T_{n+1}(x) = 2xT_n(x) - T_{n-1}(x), \qquad n \geqslant 1. \tag{4.4}$$

It follows from (4.4) that the coefficient of x^n in $T_n(x)$ is 2^{n-1}. We can also express powers of x in terms of the Chebyshev polynomials and we obtain

$$1 = T_0(x), \qquad x = T_1(x),$$
$$x^2 = \tfrac{1}{2}(T_0 + T_2), \qquad x^3 = \tfrac{1}{4}(T_3 + 3T_1)$$
$$x^4 = \tfrac{1}{8}(T_4 + 4T_2 + 3T_0), \ldots .$$

*An alternative definition is

$$T_n(x) = \frac{(-2)^n n!}{(2n)!}(1 - x^2)^{\frac{1}{2}} \frac{d^n}{dx^n}(1 - x^2)^{n-\frac{1}{2}}$$

Higher powers of x can be deduced from the formulae

$$\left.\begin{aligned} x^{2m} &= \frac{1}{2^{2m-1}} \sum_{k=0}^{m}{}' \binom{2m}{m-k} T_{2k}(x) \\ x^{2m+1} &= \frac{1}{2^{2m}} \sum_{k=0}^{m} \binom{2m+1}{m-k} T_{2k+1}(x) \end{aligned}\right] \tag{4.5}$$

where the prime in the first summation indicates that the term in $T_0(x)$ is to be multiplied by the factor $\frac{1}{2}$.

The Chebyshev polynomials are orthogonal with respect to the weighting function $1/\sqrt{(1-x^2)}$ over the interval $[-1, 1]$. More precisely,

$$\int_{-1}^{1} \frac{T_m(x)T_n(x)}{\sqrt{(1-x^2)}}\,dx = \begin{bmatrix} 0, & m \neq n \\ \frac{1}{2}\pi, & m = n\,(\neq 0) \\ \pi, & m = n = 0. \end{bmatrix} \tag{4.6}$$

These results are established immediately on putting $x = \cos\theta$.

Further, it is not difficult to show that $T_n(x)$ satisfies the second order differential equation

$$(1-x^2)y'' - xy' + n^2y = 0. \tag{4.7}$$

Second order differential equations have, in general, two linearly independent solutions. In the case of equation (4.7) it turns out that

$$S_n(x) = \sin n(\cos^{-1} x), \tag{4.8}$$

is the other solution of (4.7). In particular

$$S_0(x) = 0, \qquad S_1(x) = \sqrt{(1-x^2)}, \qquad S_2(x) = 2x\sqrt{(1-x^2)}, \ldots .$$

If we write

$$U_{n-1}(x) = S_n(x)/\sqrt{(1-x^2)}, \qquad n \geqslant 1 \tag{4.9}$$

we obtain the sequence of polynomials known as Chebyshev polynomials of the second kind, viz.:

$$U_0(x) = 1, \qquad U_1(x) = 2x, \qquad U_2(x) = 4x^2 - 1,$$
$$U_3(x) = 8x^3 - 4x, \ldots .$$

These polynomials satisfy the recurrence relation

$$U_{n+1}(x) = 2xU_n(x) - U_{n-1}(x), \quad n \geqslant 1. \tag{4.10}$$

It is easily seen too, that the Chebyshev polynomials of the second kind are orthogonal with respect to the weight function $\sqrt{(1-x^2)}$ in the interval $[-1, 1]$. In particular

$$\int_{-1}^{1} \sqrt{(1-x^2)}U_m(x)U_n(x)\,dx = \begin{bmatrix} 0, & m \neq n \\ \frac{1}{2}\pi, & m = n. \end{bmatrix} \tag{4.11}$$

We remark also on the connection between Chebyshev polynomials of first and second kind, namely

$$\left.\begin{aligned} U_0(x) &= T_0(x), \qquad U_1(x) = 2T_1(x), \\ U_n(x) &= 2T_n(x) + U_{n-2}(x), \qquad n \geqslant 2. \end{aligned}\right] \tag{4.12}$$

One way of proving (4.12) is to compare the generating functions for $T_n(x)$ and $U_n(x)$. These are, respectively

$$\frac{1 - tx}{1 - 2tx + t^2} = \sum_{n=0}^{\infty} T_n(x) t^n, \tag{4.13}$$

$$\frac{1}{1 - 2tx + t^2} = \sum_{n=0}^{\infty} U_n(x) t^n. \tag{4.14}$$

Multiplication of (4.14) by $(1 - tx)$ and subtraction of (4.13) gives

$$U_n(x) - xU_{n-1}(x) = T_n(x), \qquad n \geqslant 1$$

and (4.12) now follows from this and (4.10).

Finally, before leaving Chebyshev polynomials, we prove a remarkable result which was demonstrated by Chebyshev himself. This result leads to the important minimax property (cf. section 7.2) in $(-1, 1)$. Namely, given a polynomial $f(x)$ of degree n with leading coefficient 1, then the polynomial $g(x)$ of lower degree such that $f(x) - g(x)$ has the smallest maximum deviation from zero is given by

$$f(x) - g(x) = T_n(x)/2^{n-1}.$$

Chebyshev's result is

THEOREM 4-4 *Let $f(x)$ be a polynomial of degree n with coefficient of x^n equal to 1. Let*

$$m(f) = \max_{-1 \leqslant x \leqslant 1} |f(x)|.$$

Then
$$m(f) \geqslant 1/2^{n-1}.$$

Proof We consider the polynomial $g(x)$ defined by

$$g(x) = f(x) - 2^{-(n-1)} T_n(x),$$

and we assume

$$|f(x)| < 1/2^{n-1},$$

everywhere in the interval $[-1, 1]$. Further we note that

$$T_n(x) = (-1)^r \quad \text{for} \quad x = x_r = \cos(r\pi/n), \qquad r = 0, 1, \ldots, n.$$

Hence

$$g(x_0) = f(x_0) - 2^{-(n-1)}T_n(x_0) < 0$$
$$g(x_1) = f(x_1) - 2^{-(n-1)}T_n(x_1) > 0$$
$$g(x_2) = f(x_2) - 2^{-(n-1)}T_n(x_2) < 0$$
$$\cdots\cdots\cdots\cdots$$

and it is clear that the polynomial $g(x)$ changes sign $n + 1$ times in the interval $[-1, 1]$ indicating that $g(x)$ has n zeros in the interval. However, since the coefficient of x^n in both $2^{-(n-1)}T_n(x)$ and $f(x)$ is equal to 1, it follows that $g(x)$ is of degree $n - 1$ at most. By the fundamental theorem of algebra $g(x)$ cannot have more than $n - 1$ roots and we have a contradiction. Our premise is therefore incorrect and we must have

$$m(f) \geqslant 1/2^{n-1}.$$

Note that $m(f) = 1/2^{n-1}$ is attained when $f(x) = 2^{-(n-1)}T_n(x)$. The minimax property in $(-1, 1)$ then follows immediately by observing that $f(x) - g(x)$ is a polynomial of degree n with leading coefficient 1, that is, $f(x) - g(x)$ satisfies the conditions of Theorem 4-4. Accordingly the smallest maximum deviation from zero of $f(x) - g(x)$ is $1/2^{n-1}$ and it is attained when

$$f(x) - g(x) = 2^{-(n-1)}T_n(x).$$

Additional properties of Chebyshev polynomials can be found in references [2] and [3].

4.3 Legendre, Laguerre, and Hermite polynomials

In this section we summarize the most important properties of the orthogonal polynomials due to Legendre, Laguerre, and Hermite. The defining relations are

$$\text{Legendre:}\quad P_n(x) = \frac{1}{2^n n!}\frac{\mathrm{d}^n}{\mathrm{d}x^n}(x^2 - 1)^n, \qquad -1 \leqslant x \leqslant 1 \tag{4.15}$$

$$\text{Laguerre:}\quad L_n(x) = \mathrm{e}^x \frac{\mathrm{d}^n}{\mathrm{d}x^n}(x^n \mathrm{e}^{-x}), \qquad 0 \leqslant x \leqslant \infty \tag{4.16}$$

$$\text{Hermite:}\quad H_n(x) = (-1)^n \mathrm{e}^{x^2} \frac{\mathrm{d}^n}{\mathrm{d}x^n}(\mathrm{e}^{-x^2}), \qquad -\infty \leqslant x \leqslant \infty. \tag{4.17}$$

The first few polynomials are given in Table 4.1.

Table 4.1 Legendre, Laguerre, and Hermite polynomials of degree $n(\leqslant 4)$

n	Legendre	Laguerre	Hermite
0	1	1	1
1	x	$1-x$	$2x$
2	$\frac{1}{2}(3x^2-1)$	$2-4x+x^2$	$4x^2-2$
3	$\frac{1}{2}(5x^3-3x)$	$6-18x+9x^2-x^3$	$8x^3-12x$
4	$\frac{1}{8}(35x^4-30x^2+3)$	$24-96x+72x^2-16x^3+x^4$	$16x^4-48x^2+12$

Additional polynomials can be determined from the recurrence relations

$$P_{n+1}(x)=\frac{2n+1}{n+1}\,xP_n(x)-\frac{n}{n+1}P_{n-1}(x), \qquad n\geqslant 1 \tag{4.18}$$

$$L_{n+1}(x)=(1+2n-x)L_n(x)-n^2L_{n-1}(x), \qquad n\geqslant 1 \tag{4.19}$$

$$H_{n+1}(x)=2xH_n(x)-2nH_{n-1}(x), \qquad n\geqslant 1. \tag{4.20}$$

The coefficients of x^n in these polynomials are respectively

$$\frac{(2n)!}{2^n(n!)^2}, \qquad (-1)^n, \qquad 2^n \tag{4.21}$$

and the polynomials satisfy the orthogonality relationships

$$\int_{-1}^{1} P_m(x)P_n(x)\,\mathrm{d}x = \begin{cases} 0, & m\neq n \\ \dfrac{2}{2n+1}, & m=n \end{cases} \tag{4.22}$$

$$\int_{0}^{\infty} \mathrm{e}^{-x}L_m(x)L_n(x)\,\mathrm{d}x = \begin{cases} 0, & m\neq n \\ (n!)^2, & m=n \end{cases} \tag{4.23}$$

$$\int_{-\infty}^{\infty} \mathrm{e}^{-x^2}H_m(x)H_n(x)\,\mathrm{d}x = \begin{cases} 0, & m\neq n \\ 2^n n!\sqrt{\pi}, & m=n. \end{cases} \tag{4.24}$$

The polynomials are also generated by

$$(1-2xt+t^2)^{-\frac{1}{2}}=\sum_{n=0}^{\infty} P_n(x)t^n \tag{4.25}$$

$$(1-t)^{-1}\exp\left[-tx/(1-t)\right]=\sum_{n=0}^{\infty}(-1)^nL_n(x)t^n/n! \tag{4.26}$$

$$\exp(2xt-t^2)=\sum_{n=0}^{\infty}H_n(x)t^n/n!. \tag{4.27}$$

These polynomials arise naturally in practice in such fields as potential theory and wave mechanics to give two instances. Proofs of the above results and additional properties can be found in the references.

REFERENCES

[1] Szego, G., Orthogonal Polynomials, Vol. 23, *Amer. Math. Soc. Colloq. Pubs.* New York (1959).

[2] Fox, L. and Parker, I. B., *Chebyshev Polynomials in Numerical Analysis*, Oxford University Press, London (1968).

[3] Lanczos, C., *Applied Analysis*, Pitman, London (1957).

[4] Ralston, A., *A First Course in Numerical Analysis*, McGraw-Hill Book Co., New York (1965).

[5] Hildebrand, F. B., *Introduction to Numerical Analysis*, McGraw-Hill Book Co., New York (1956).

[6] Kacmarz, S. and Steinhaus, H., *Theorie der Orthogonalreihen*, Chelsea, New York (1951).

Exercises

1. Show that the sequence of functions $\{\cos nx\}$ is orthogonal in $[0, 2\pi]$ with respect to the weight function $w(x) = 1$.

2. Show that the sequence of functions $\{\sin nx\}$ is orthogonal in both $[-\pi, \pi]$ and $[0, 2\pi]$ with respect to the weight function $w(x) = 1$.

3. If $f(\theta)$ is continuous and bounded in $(0, \pi)$ show that the function $f(\theta)$ can be expressed in the form of a Fourier series

$$f(\theta) = \tfrac{1}{2}a_0 + \sum_{1}^{\infty} a_n \cos n\theta,$$

where

$$a_n = \frac{2}{\pi}\int_0^{\pi} f(\theta) \cos n\theta \, d\theta.$$

(compare exercise 7).

4. Show that Chebyshev polynomials of the first kind satisfy

$$T_{n+m}(x) + T_{n-m}(x) = 2T_n(x)T_m(x), \qquad n \geqslant m.$$

5. The shifted Chebyshev polynomials of the first kind are orthogonal polynomials in $[0, 1]$ defined by

$$T_n^*(x) = T_n(2x - 1).$$

Evaluate $T_0^*, T_1^*, \ldots, T_6^*$ and show that

$$T_n^*(x^2) = T_{2n}(x).$$

6. Prove that

$$\int T_n(x) \, dx = \begin{cases} T_1(x), & n = 0 \\ \tfrac{1}{4}T_2(x), & n = 1 \\ \tfrac{1}{2}\left(\dfrac{T_{n+1}(x)}{n+1} - \dfrac{T_{n-1}(x)}{n-1}\right), & n > 1. \end{cases}$$

7. Under certain conditions (known as Dirichlet conditions) the function $f(x)$ can be expanded over the range $[-1, 1]$ in the form of a (Fourier–)Chebyshev series

$$f(x) = \tfrac{1}{2}a_0 + a_1 T_1(x) + \ldots + a_n T_n(x) + \ldots$$

where
$$a_n = \frac{2}{\pi}\int_{-1}^{1} \frac{f(x)T_n(x)\,\mathrm{d}x}{\sqrt{(1-x^2)}}.$$

Assuming that $a_{n+1} = a_{n+2} = \ldots = 0$ show that $f(x)$ may be computed by the recurrence formulae

$$b_n = a_n, \qquad b_{n-1} = a_{n-1} + 2xb_n$$

$$b_i = a_i - b_{i+2} + 2xb_{i+1} \qquad (i = n-2, \ldots, 0)$$

yielding $f(x) = \tfrac{1}{2}(b_0 - b_2)$.

8. With $f(x)$ in the form of exercise 7 verify that

$$\int_{-1}^{x} f(x)\,\mathrm{d}x = \tfrac{1}{2}C_0 + C_1T_1(x) + \ldots + C_{n+1}T_{n+1}(x), \qquad x \leqslant 1$$

where
$$C_i = \frac{1}{2i}(a_{i-1} - a_{i+1}), \qquad (i = 1, 2, \ldots, n-1).$$

$$C_n = a_{n-1}/2n, \qquad C_{n+1} = a_n/2(n+1)$$

$$\tfrac{1}{2}C_0 = C_1 - C_2 + C_3 - \ldots + (-1)^n C_{n+1}.$$

Hence deduce that

$$\int_{-1}^{1} f(x)\,\mathrm{d}x = 2(C_1 + C_3 + C_5 + \ldots).$$

[Hint: Apply the results of exercise 6.]

9. Given that over $[-1, 1]$

$$f(x) = \tfrac{1}{2}a_0T_0 + a_1T_1 + \ldots + a_nT_n$$

show that

$$f'(x) = \tfrac{1}{2}c_0T_0 + c_1T_1 + \ldots + c_{n-1}T_{n-1}$$

where

$$c_{n-1} = 2na_n, \qquad c_{n-2} = 2(n-1)a_{n-1},$$

$$c_{n-3} = 2(n-2)a_{n-2} + 2na_n, \ldots$$

$$c_1 = 4a_2 + 8a_4 + 12a_6 + \ldots$$

$$c_0 = 2a_1 + 6a_3 + 10a_5 + \ldots.$$

10. Obtain similar results to those of examples 8 and 9 for the shifted Chebyshev polynomials.

11. Show that Chebyshev polynomials of the first kind and Legendre polynomials are connected by the relations

$$T_0(x) = P_0(x), \qquad T_1(x) = P_1(x)$$

$$T_n(x) = \sum_{j=0}^{n} P_j(x)P_{n-j}(x) - x \sum_{j=0}^{n-1} P_j(x)P_{n-j-1}(x), \qquad n > 1.$$

Hence show that

$$\int_{-1}^{1} T_n(x)\,dx = \begin{cases} 0, & n = 2m+1 \\ \dfrac{-2}{n^2-1}, & n = 2m. \end{cases}$$

Deduce similar formulae for Chebyshev polynomials of the second kind. [Hint: Compare generating functions.]

12. Let $x_k = \cos\left(\dfrac{2k+1}{N+1}\dfrac{\pi}{2}\right)$. Show that

$$\sum_{k=0}^{N} T_r(x_k)T_s(x_k) = \begin{cases} 0, & \text{if } r \neq s \\ \frac{1}{2}(N+1), & \text{if } r = s\ (\neq 0). \end{cases}$$

How can this result be used to obtain the coefficients $a_0, a_1, \ldots, a_N$ in the Fourier–Chebyshev expansion

$$f(x) = \tfrac{1}{2}a_0 + a_1 T_1(x) + a_2 T_2(x) + \ldots + a_n T_n(x)?$$

(see section 5.6).

5 Numerical integration (quadrature)

5.1 Elementary methods (Newton–Cotes formulae)

As with interpolation (section 3.2) we can derive approximate formulae for the evaluation of the definite integral

$$\int_a^b f(x)\,dx, \tag{5.1}$$

from elementary considerations. For, the integral may be thought of as the area of the region enclosed between the curve $y = f(x)$, the x-axis, and the ordinates $x = a$ and $x = b$. This region will be subdivided into simple geometrical figures, whose areas are easily determined, and by summation of the individual areas we shall obtain an approximation to (5.1). In this section we shall develop three useful formulae in this manner and, at the same time, we shall show how they may be considered a logical development of each other.

(a) The mid-point (mid-ordinate) rule

We shall consider the approximate evaluation of

$$\int_{x_0}^{x_1} f(x)\,dx$$

where $x_1 - x_0$ $(=h)$ is small. It is reasonable to expect that the mean value of $f(x)$ will be at $x_{\frac{1}{2}}$ (using the notation of section 3.4) approximately. Moreover, a fairly good approximation to the integral, which is the area enclosed by $AQRC$ in Fig. 5.1, is the area obtained by taking y as the constant value $f_{\frac{1}{2}}$, i.e., the area of the rectangle $AMNC$. Thus, approximately,

$$\int_{x_0}^{x_1} f(x)\,dx \doteqdot (x_1 - x_0)f_{\frac{1}{2}} = hf_{\frac{1}{2}}. \tag{5.2}$$

(b) The trapezium rule

In this approximation we assume that the area under the curve between Q and R (Fig. 5.1) is of the same order of magnitude as the area of the trapezium $AQRC$. We thus obtain

$$\int_{x_0}^{x_1} f(x)\,dx \doteqdot \tfrac{1}{2}(x_1 - x_0)(f_0 + f_1) = \tfrac{1}{2}h(f_0 + f_1). \tag{5.3}$$

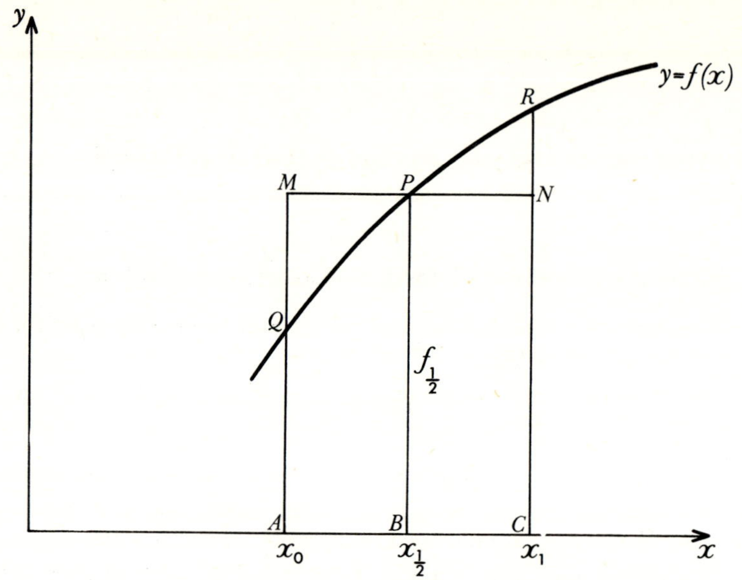

Fig. 5.1 Geometrical representation of the mid-point rule

This can be considered as a logical development of method (*a*) in the sense that we are replacing the curve $y = f(x)$ in the region (x_0, x_1) by the linear interpolation polynomial (section 3.2)

$$y = \frac{x - x_1}{x_0 - x_1} f_0 + \frac{x - x_0}{x_1 - x_0} f_1,$$

and not by a constant.*

(*c*) *Simpson's rule*

This can be considered as the logical development of the previous method. We now take three points x_0, x_1, x_2 on the x-axis, where $x_1 - x_0 = x_2 - x_1 = h$, and replace the curve $y = f(x)$ in the region (x_0, x_2) by the quadratic interpolation polynomial

$$\begin{aligned} y &= \frac{(x - x_1)(x - x_2)}{(x_0 - x_1)(x_0 - x_2)} f_0 + \frac{(x - x_0)(x - x_2)}{(x_1 - x_0)(x_1 - x_2)} f_1 + \frac{(x - x_0)(x - x_1)}{(x_2 - x_0)(x_2 - x_1)} f_2, \\ &= \frac{1}{h^2} [\tfrac{1}{2}(x - x_1)(x - x_2) f_0 - (x - x_0)(x - x_2) f_1 + \tfrac{1}{2}(x - x_0)(x - x_1) f_2]. \end{aligned}$$

We have

$$\int_{x_0}^{x_2} f(x)\,\mathrm{d}x \doteqdot \int_{x_0}^{x_2} y\,\mathrm{d}x = \tfrac{1}{3}h(f_0 + 4f_1 + f_2). \tag{5.4}$$

*For a strict logical development we should have taken the constant value as f_0 and not $f_{\frac{1}{2}}$.

Clearly, we can approximate by higher order interpolation polynomials and extend these results. These formulae are particular examples of a whole class of formulae, known as Newton–Cotes formulae, which approximate to the integral $\int_{x_0}^{x_m} f(x)\,dx$ by means of a weighted sum of function values at the points $x_k = x_0 + kh$ $(k = 0, 1, \ldots, n)$.* A large selection of these formulae are given in *Interpolation and Allied Tables* [1], and a number are derived, with estimates of truncation error, in section 5.3. Note, however, that the x_k need not be equally spaced, and in section 5.5 we shall derive a class of formulae, called Gaussian formulae, where the x_k are the zeros of orthogonal polynomial equations.

Before we proceed to other methods, we demonstrate how the mid-point rule, say, is used in practice (cf. section 5.4). We take as an example the evaluation of $\int_0^{\frac{1}{2}\pi} \sin x\,dx$, whose exact value is 1. First, the range $(0, \frac{1}{2}\pi)$ is divided into a number of equal parts, say 10, and then, to each part, formula (5.2) is applied. This gives

$$\begin{aligned}\int_0^{\frac{1}{2}\pi} \sin x\,dx &\doteqdot \tfrac{1}{20}\pi(\sin 4\tfrac{1}{2}^0 + \sin 13\tfrac{1}{2}^0 + \ldots + \sin 85\tfrac{1}{2}^0),\\ &\doteqdot \tfrac{1}{20}\pi(0{\cdot}078\,459\,1 + 0{\cdot}233\,445\,4 + \ldots + 0{\cdot}996\,917\,3),\\ &\doteqdot 1{\cdot}001\,0.\end{aligned}$$

5.2 The method of series expansion

This method, which is a good way of evaluating certain classes of integrals, consists of expressing the integrand as a convergent power series (see section 7.1) and then integrating the series term by term. The method is best described by considering a simple example, viz.

$$\int_0^{\frac{1}{2}} x \ln(1 + x)\,dx.$$

Since, for $0 < x < \frac{1}{2}$,

$$\ln(1 + x) = x - \tfrac{1}{2}x^2 + \tfrac{1}{3}x^3 - \ldots = \sum_1^{\infty} (-1)^{n+1} \frac{x^n}{n},$$

*This is not strictly true of the mid-point rule as we have stated it, but it is true if we rewrite it in the form

$$\int_{x_0}^{x_2} f(x)\,dx \doteqdot 2hf(x_1), \qquad h = x_{i+1} - x_i \qquad (i = 0, 1).$$

and $$x \ln(1+x) = \sum_1^\infty (-1)^{n+1} x^{n+1}/n.$$

$$\int_0^{\frac{1}{2}} x \ln(1+x)\,\mathrm{d}x = \int_0^{\frac{1}{2}} \left(\sum_1^\infty (-x)^{n+1}/n\right) \mathrm{d}x$$

$$= \sum_1^\infty \int_0^{\frac{1}{2}} \frac{1}{n}(-x)^{n+1}\,\mathrm{d}x$$

the interchange of the orders of summation and integration being justified since the series is absolutely convergent in $(0, \frac{1}{2})$. Hence

$$\int_0^{\frac{1}{2}} x \ln(1+x)\,\mathrm{d}x = \sum_1^\infty \left[(-1)^{n+1} \frac{x^{n+2}}{n(n+2)}\right]_0^{\frac{1}{2}} = \frac{1}{2}\sum_1^\infty \frac{(-\frac{1}{2})^{n+1}}{n(n+2)},$$

$$= \tfrac{1}{3}(0{\cdot}5)^3 - \tfrac{1}{8}(0{\cdot}5)^4 + \tfrac{1}{15}(0{\cdot}5)^5 - \tfrac{1}{24}(0{\cdot}5)^6 + \ldots$$

$$\doteqdot 0{\cdot}035\,45,$$

with a truncation error, when terminated at the fourth term, of at most $\frac{1}{5{\cdot}7}(0{\cdot}5)^7 = 0{\cdot}000\,23$ (see section 6.1). This method is very efficient when the series obtained is rapidly convergent, and another advantage is that the truncation error can often be easily established. Further, where the general term is readily deduced, but convergence is slow, we can employ the methods of chapter 6 to speed up convergence.

5.3 Finite difference methods

Integration formulae, in terms of finite difference expansions, can be obtained in much the same way as the differentiation formulae in section 3.6. If $x = x_0 + ph$, $0 \leqslant p \leqslant 1$, $f_p = f(x)$ then

$$\int_{x_0}^{x_1} f(x)\,\mathrm{d}x = h\int_0^1 f_p\,\mathrm{d}p \tag{5.5}$$

and we can substitute known interpolation formulae for f_p in (5.5) to obtain integration formulae. Thus, recalling Bessel's formula,

$$f_p = f_0 + p\delta_{\frac{1}{2}} + \frac{1}{2}\cdot\frac{1}{2}p(p-1)(\delta_0^2 + \delta_1^2) + \frac{1}{3!}p(p-1)(p-\tfrac{1}{2})\delta_{\frac{1}{2}}^3$$

$$+\frac{1}{2}\cdot\frac{1}{4!}(p+1)p(p-1)(p-2)(\delta_0^4 + \delta_1^4) + \frac{1}{5!}(p+1)p(p-1)(p-2)(p-\tfrac{1}{2})\delta_{\frac{1}{2}}^5$$

$$+\frac{1}{6!}(p+2)(p+1)p(p-1)(p-2)(p-3)h^6 f^{(vi)}(\xi), \qquad x_{-2} < \xi < x_3,$$

$$= f_0 + p\delta_{\frac{1}{2}} + \tfrac{1}{2}(p^2 - p)\mu\delta_{\frac{1}{2}}^2 + \tfrac{1}{6}(p^3 - \tfrac{3}{2}p^2 + \tfrac{1}{2}p)\delta_{\frac{1}{2}}^3$$

$$+ \tfrac{1}{24}(p^4 - 2p^3 - p^2 + 2p)\mu\delta_{\frac{1}{2}}^4 + \tfrac{1}{120}(p^5 - \tfrac{5}{2}p^4 + \tfrac{5}{2}p^2 - p)\delta_{\frac{1}{2}}^5$$

$$+\frac{1}{6!}(p+2)(p+1)p(p-1)(p-2)(p-3)h^6 f^{(vi)}(\xi),$$

where ξ depends on p. We have on substitution in (5.5), and integration with respect to p,

$$\int_{x_0}^{x_1} f(x)\,\mathrm{d}x = h(f_0 + \tfrac{1}{2}\delta_{\frac{1}{2}} - \tfrac{1}{12}\mu\delta_{\frac{1}{2}}^{2} + \tfrac{11}{720}\mu\delta_{\frac{1}{2}}^{4}) + R \tag{5.6}$$

where

$$R = \frac{h^7}{6!}\int_0^1 (p+2)(p+1)p(p-1)(p-2)(p-3)f^{(vi)}(\xi)\,\mathrm{d}p. \tag{5.7}$$

At this stage it is necessary to digress in order to establish a result which will permit the evaluation of error terms such as (5.7). This result is stated as Theorem 5-1.

THEOREM 5-1 *Let $g(x)$ be a function which is not identically zero and which satisfies $g(x) \geqslant 0$ in the interval $[a, b]$. Let $f(x)$ be a continuous function in $[a, b]$. Then*

$$\int_a^b f(x)g(x)\,\mathrm{d}x = f(\xi)\int_a^b g(x)\,\mathrm{d}x \tag{5.8}$$

for some ξ such that $a \leqslant \xi \leqslant b$.

Proof Let m and M be the lower and upper bounds of $f(x)$ in $[a, b]$. Then, for all x in $[a, b]$,

$$f(x) - m \geqslant 0, \qquad M - f(x) \geqslant 0,$$

and thus

$$\int_a^b (f(x) - m)g(x)\,\mathrm{d}x \geqslant 0, \qquad \int_a^b (M - f(x))g(x)\,\mathrm{d}x \geqslant 0,$$

since the integrands are non-negative. Consequently,

$$m\int_a^b g(x)\,\mathrm{d}x \leqslant \int_a^b f(x)g(x)\,\mathrm{d}x \leqslant M\int_a^b g(x)\,\mathrm{d}x.$$

Further, there must be some number c between m and M such that

$$\int_a^b f(x)g(x)\,\mathrm{d}x = c\int_a^b g(x)\,\mathrm{d}x.$$

But, since it is a continuous function, $f(x)$ takes all values between m and M, i.e., there must be one value of $f(x)$ (at least), call it $f(\xi)$, where $a \leqslant \xi \leqslant b$, such that $f(\xi) = c$. Thus we have established (5.8). The same argument shows, too, that the theorem is true if $g(x) \leqslant 0$. That is, the result (5.8) holds if $g(x)$ has the same sign in $[a, b]$.

This enables us to determine R in equation (5.7). For the function $(p+2)(p+1)p(p-1)(p-2)(p-3)$ is negative or zero in $[0, 1]$ and hence, from Theorem 5-1,

$$R = \frac{h^7}{6!} f^{(vi)}(\eta) \int_0^1 (p+2)(p+1)p(p-1)(p-2)(p-3)\, dp, \qquad x_{-2} < \eta < x_3$$

$$= \frac{-191}{60\,480} h^7 f^{(vi)}(\eta). \tag{5.9}$$

Formula (5.6) was obtained by considering a fifth order interpolation polynomial, but we could equally well have obtained formulae of lower or higher order. For instance, by considering Bessel's formula curtailed after the first two terms we obtain

$$\int_{x_0}^{x_1} f(x)\, dx = h(f_0 + \tfrac{1}{2}\delta_{\frac{1}{2}}) - \tfrac{1}{12}h^3 f''(\xi'), \qquad x_0 < \xi' < x_1,$$
$$= \tfrac{1}{2}h(f_0 + f_1) - \tfrac{1}{12}h^3 f''(\xi'). \tag{5.10}$$

This is the trapezium rule with an estimate for the error.

Next we illustrate how (5.6) is used to evaluate an integral, and we choose as our example

$$\int_{0\cdot 1}^{0\cdot 4} \frac{dx}{1 + \sin x}.$$

First, we note that (5.6) can be written as

$$\int_{x_0}^{x_1} f(x)\, dx = h[f_0 + \tfrac{1}{2}(f_1 - f_0) - (\tfrac{1}{12}\mu\delta - \tfrac{11}{720}\mu\delta^3)\delta_{\frac{1}{2}}] + R_1,$$
$$= h[\tfrac{1}{2}(f_0 + f_1) - (\tfrac{1}{12}\mu\delta - \tfrac{11}{720}\mu\delta^3)(f_1 - f_0)] + R_1,$$

where R_1 denotes the truncation error, which has the form (5.9). Thus

$$\int_{x_0}^{x_N} f(x)\, dx = \int_{x_0}^{x_1} f(x)\, dx + \int_{x_1}^{x_2} f(x)\, dx + \ldots + \int_{x_{N-1}}^{x_N} f(x)\, dx,$$
$$= h[\tfrac{1}{2}(f_0 + f_1) + \tfrac{1}{2}(f_1 + f_2) + \ldots + \tfrac{1}{2}(f_{N-1} + f_N)$$
$$- (\tfrac{1}{12}\mu\delta - \tfrac{11}{720}\mu\delta^3)\{(f_1 - f_0) + (f_2 - f_1) + \ldots + (f_N - f_{N-1})\}]$$
$$+ \sum_1^N R_i$$

$$= h[\tfrac{1}{2}\sum_{r=0}^{N-1}(f_r + f_{r+1}) - (\tfrac{1}{12}\mu\delta - \tfrac{11}{720}\mu\delta^3)(f_N - f_0)] + \sum_1^N R_i. \tag{5.11}$$

It now follows from Table 5.1 and (5.11) that

$$\int_{0\cdot 1}^{0\cdot 4} \frac{dx}{1+\sin x} \doteqdot 0\cdot 1\{\tfrac{1}{2} \times 0\cdot 9092 + 0\cdot 8342 + 0\cdot 7719 + \tfrac{1}{2} \times 0\cdot 7197$$
$$- \tfrac{1}{24}[-0\cdot 0522 - 0\cdot 0438 - (-0\cdot 0908 - 0\cdot 0750)]\}$$
$$= 0\cdot 1\,(2\cdot 4205_5 - 0\cdot 0029_1) \doteqdot 0\cdot 2418.$$

In this particular example the terms involving $\mu\delta^3$ were too small to make any significant contribution to the evaluation of the integral (assuming only four decimal place accuracy is required), and thus a lower order formula could have been used. Finally, an estimate is required for the truncation error. Since, from (3.40), $h^6 f_0^{(vi)} \doteqdot \delta_0{}^6$ to a first degree of approximation, the order of magnitude of $h^6 f^{(vi)}(\eta)$ is obtainable from the column of sixth differences of the function. Since these are small and have to be multiplied by a factor $-191h/60\,480$ it is clear that the truncation error is very small and makes no significant contribution to the error. Thus we have virtually only to consider the rounding error of the tabular entries, and the errors involved in the arithmetic operations, to find the error in the quoted answer of 0·2418.

Table 5.1 Difference table for $f(x) = 1/(1 + \sin x)$, $x = -0{\cdot}1(0{\cdot}1)0{\cdot}6$

x	$f(x)$	δ	δ^2	δ^3	δ^4	δ^5	δ^6
−0·1	1·1109						
		−1109					
0·0	1·0000		201				
		−908		−43			
0·1(x_0)	0·9092(f_0)		158		12		
		−750		−31		−7	
0·2	0·8342		127		5		11
		−623		−26		4	
0·3	0·7719		101		9		−10
		−522		−17		−6	
0·4(x_3)	0·7197(f_3)		84		3		
		−438		−14			
0·5	0·6759		70				
		−368					
0·6	0·6391						

The use of Newton's backward difference interpolation formula (3.27) as our starting point in (5.5) will lead to another integration formula. We have

$$f_p = f_0 + p\nabla_0 + \frac{1}{2!}p(p+1)\nabla_0^2 + \frac{1}{3!}p(p+1)(p+2)\nabla_0{}^3$$
$$+ \frac{1}{4!}p(p+1)(p+2)(p+3)h^4 f^{(iv)}(\xi), \qquad x_{-3} < \xi < x_0.$$

Proceeding as above, and using Theorem 5-1, we obtain

$$\int_{x_0}^{x_1} f(x)\,dx = h(f_0 + \tfrac{1}{2}\nabla_0 + \tfrac{5}{12}\nabla_0{}^2 + \tfrac{3}{8}\nabla_0{}^3) + \tfrac{251}{720}h^5 f^{(iv)}(\eta),\ x_{-3} < \eta < x_0. \qquad (5.12)$$

Use of Newton's formula as an extrapolation formula in the form

$$f_p = f_1 + (p-1)\nabla_1 + \frac{1}{2!}(p-1)p\nabla_1^2 + \frac{1}{3!}(p-1)p(p+1)\nabla_1^3$$
$$+ \frac{1}{4!}(p-1)p(p+1)(p+2)\nabla_1^4$$
$$+ \frac{1}{5!}(p-1)p(p+1)(p+2)(p+3)h^5 f^{(v)}(\xi), \qquad x_{-3} < \xi < x_1$$

gives, by the same method, the integration formula

$$\int_{x_0}^{x_1} f(x)\,dx = h(f_1 - \tfrac{1}{2}\nabla_1 - \tfrac{1}{12}\nabla_1^2 - \tfrac{1}{24}\nabla_1^3 - \tfrac{19}{720}\nabla_1^4)$$
$$- \tfrac{3}{160}h^6 f^{(v)}(\eta), \qquad x_{-3} < \eta < x_1. \qquad (5.13)$$

Formulae (5.12) and (5.13) are useful in the predictor-corrector methods of chapter 11. A large number of additional formulae, which are obtainable by the method of this section, are given in *Interpolation and Allied Tables* [1].

5.4 Romberg integration

This technique is based on the method of extrapolation called Richardson's deferred approach to the limit which was explained in chapter 3. If $f(x)$ is continuous and differentiable,

$$I = \int_a^b f(x)\,dx,$$

and we denote by $I_M(N)$ the value of the mid-ordinate approximation (5.2) to the integral using N strips, we can show from Taylor's theorem that (see exercise 14)

$$I = I_M(N) + \beta_2 h_N^2 + \beta_4 h_N^4 + \beta_6 h_N^6 + \ldots, \qquad h_N = (b-a)/N \qquad (5.14)$$
$$(h_N < 1, \quad N > N_0).$$

In particular, we can take $N = 1, 2$, to eliminate β_2 and get a new estimate for I, call it $I_M(1, 2)$, which satisfies

$$I_M(1,2) = I_M(2) + \tfrac{1}{3}\{I_M(2) - I_M(1)\}. \qquad (5.15)$$

Similarly we can take $N = 2, 4$ to eliminate β_2 and get an estimate

$$I_M(2,4) = I_M(4) + \tfrac{1}{3}\{I_M(4) - I_M(2)\}. \qquad (5.16)$$

It is easy to show that

$$I - I_M(1,2) = -\tfrac{1}{4}\beta_4 h_1^4 + \ldots,$$
$$I - I_M(2,4) = -\tfrac{1}{4}\beta_4(\tfrac{1}{2}h_1)^4 + \ldots,$$

and we can eliminate β_4 to find

$$I_M(1,2,4) = I_M(2,4) + \tfrac{1}{15}\{I_M(2,4) - I_M(1,2)\} \tag{5.17}$$

which only differs from I by $O(h_1^6)$ and, more generally,

$$I_M(1,2,\ldots,2^n) = I_M(2,4,\ldots,2^n) + \frac{1}{2^{2n}-1}\{I_M(2,4,\ldots,2^n) - I_M(1,2,\ldots,2^{n-1})\} \tag{5.18}$$

which only differs from I by $O(h_1^{2n+2})$. Exactly the same results hold for the trapezium rule. For Simpson's rule there is no need to eliminate β_2 in (5.14) as this is now zero.

We illustrate the method of Romberg integration in connection with the mid-point rule taking as our example

$$\int_0^{\frac{1}{2}\pi} \sin x \, dx.$$

We have

$$I_M(1) = \frac{\pi}{2} \sin\frac{\pi}{4} = 1{\cdot}110\,720\,7,$$

$$I_M(2) = \frac{\pi}{4}\left(\sin\frac{\pi}{8} + \sin\frac{3\pi}{8}\right) = 1{\cdot}026\,172\,2,$$

$$I_M(4) = \frac{\pi}{8}\left(\sin\frac{\pi}{16} + \sin\frac{3\pi}{16} + \sin\frac{5\pi}{16} + \sin\frac{7\pi}{16}\right) = 1{\cdot}006\,454\,5,$$

$$I_M(8) = \frac{\pi}{16}\left(\sin\frac{\pi}{32} + \sin\frac{3\pi}{32} + \ldots + \sin\frac{15\pi}{32}\right) = 1{\cdot}001\,608\,2.$$

We construct a pyramid of the form

$I_M(1)$			
	$I_M(1,2)$		
$I_M(2)$		$I_M(1,2,4)$	
	$I_M(2,4)$		$I_M(1,2,4,8)$
$I_M(4)$		$I_M(2,4,8)$	
	$I_M(4,8)$		
$I_M(8)$			

Substituting the above numerical values in this pyramid, we obtain

1·110 720 7			
	0·997 989 4		
1·026 172 2		1·000 008 1	
	0·999 881 9		1·000 000 1
1·006 454 5		1·000 000 2	
	0·999 992 8		
1·001 608 2			

The elements along the sloping line show a consistent trend and 1·000 000 1 is clearly a good approximation to the integral. The reader should compare this value (obtained from only 15 function evaluations) and the value 1·000 008 1 (obtained from only 7 function evaluations) with the value of $I(10)$ found at the end of section 5.1.

The method of Romberg integration required the values of $I_M(N)$ for $N = 1, 2, 4, 8, \ldots$. A variation, due to Lyness and McHugh [2] requires the evaluation of $I_M(N)$ for $N = 1, 2, 3, 4, \ldots$. Using these values we can obtain by eliminating $\beta_2, \beta_4, \ldots$, in (5.14) successively better approximations $I_M(1)$, $I_M(1, 2)$, $I_M(1, 2, 3)$, etc., where

$$I_M(1, 2, \ldots, k) = \sum_{r=1}^{k} w_{kr} I_M(r), \tag{5.19}$$

and the coefficients w_{kr} are given in Table 5.2.

Table 5.2 Coefficients w_{kr} in Lyness and McHugh method of integration

r \ k	1	2	3	4	5
1	1·000 000 0	−0·333 333 3	0·041 666 7	−0·002 777 8	0·001 115 7
2		1·333 333 3	−1·066 666 7	0·355 555 5	−0·067 724 9
3			2·025 000 0	−2·603 571 4	1·464 508 9
4				3·250 793 7	−5·779 188 7
5					5·382 288 9

For $\int_0^{\frac{1}{2}\pi} \sin x \, dx$ we have already obtained $I_M(1)$, $I_M(2)$, $I_M(4)$. We require also

$$I_M(3) = \frac{\pi}{6}\left(\sin\frac{\pi}{12} + \sin\frac{\pi}{4} + \sin\frac{5\pi}{12}\right) = 1{\cdot}011\ 515\ 2.$$

From Table 5.2 we have

$$\begin{aligned} I_M(1, 2, 3) &= 0{\cdot}041\ 666\ 7\, I_M(1) - 1{\cdot}066\ 666\ 7\, I_M(2) + 2{\cdot}025\, I_M(3), \\ &= 1{\cdot}000\ 014\ 7 \\ I_M(1, 2, 3, 4) &= -0{\cdot}002\ 777\ 8\, I_M(1) + 0{\cdot}355\ 555\ 5\, I_M(2) \\ &\qquad -2{\cdot}603\ 571\ 4\, I_M(3) + 3{\cdot}250\ 793\ 7\, I_M(4) \\ &= 1{\cdot}000\ 000\ 0 \end{aligned}$$

In practice we proceed with the calculations until

$$|I_M(1, 2, \ldots, N) - I_M(1, 2, \ldots, N-1)| < \epsilon,$$

where ϵ is some pre-assigned small quantity.

5.5 Gaussian quadrature

The quadrature formulae encountered so far have depended on the evaluation of the function $f(x)$ at equally spaced points in x. A number of very powerful formulae – called Gaussian formulae – exist, which depend on the evaluation of $f(x)$ at specially selected points, which are not equally spaced. The basis of these formulae is the following theorem.

THEOREM 5-2 *Let $p_{2n-1}(x)$ be any polynomial of degree $2n-1$ in x, which is well defined in $[a, b]$. Let $\{f_n(x)\}$ be a sequence of orthogonal polynomials in $[a, b]$ relative to a given positive weighting function $w(x)$, such that $f_n(x)$ is a polynomial of degree n. Then*

$$\int_a^b w(x)p_{2n-1}(x)\,dx = \sum_{k=1}^{n} \alpha_k p_{2n-1}(x_k), \tag{5.20}$$

where the α_k $(k = 1, 2, \ldots, n)$ are constants and $x_k (k = 1, 2, \ldots, n)$ are the zeros of $f_n(x) = 0$.

Proof We can express $p_{2n-1}(x)$ in the form

$$p_{2n-1}(x) = f_n(x)q_{n-1}(x) + r_{n-1}(x), \tag{5.21}$$

where $q_{n-1}(x)$ is the quotient and $r_{n-1}(x)$ is the remainder when $p_{2n-1}(x)$ is divided by $f_n(x)$. Clearly $q_{n-1}(x)$ is of degree $n-1$ and $r_{n-1}(x)$ is of degree $n-1$ at most. It follows that

$$\int_a^b w(x)p_{2n-1}(x)\,dx = \int_a^b w(x)f_n(x)q_{n-1}(x)\,dx + \int_a^b w(x)r_{n-1}(x)\,dx.$$

Since $q_{n-1}(x)$ can be represented as a linear sum of orthogonal polynomials $f_m(x)$ with $m \leqslant n-1$ we have, invoking the orthogonality properties of the sequence $\{f_n(x)\}$,

$$\int_a^b w(x)f_n(x)q_{n-1}(x)\,dx = 0,$$

no matter what $q_{n-1}(x)$ happens to be. Hence

$$\int_a^b w(x)p_{2n-1}(x)\,dx = \int_a^b w(x)r_{n-1}(x)\,dx. \tag{5.22}$$

Now we are attempting to find values x_k such that (5.20) holds, i.e., from (5.21),

$$\int_a^b w(x)p_{2n-1}(x)\,dx = \sum_{k=1}^{n} \alpha_k [f_n(x_k)q_{n-1}(x_k) + r_{n-1}(x_k)]. \tag{5.23}$$

Equation (5.22) indicates that the integral $\int_a^b w(x)p_{2n-1}(x)\,dx$ is independent of the polynomial $q_{n-1}(x)$, and so it is natural to arrange that the right-hand side of (5.23) is independent of $q_{n-1}(x)$. This is the case if the x_k are taken to be the roots of the polynomial equation $f_n(x) = 0$. We then have

$$\int_a^b w(x)p_{2n-1}(x)\,dx = \sum_{k=1}^{n} \alpha_k r_{n-1}(x_k).$$

Finally, since $r_{n-1}(x)$ is a polynomial of degree $n-1$ (at most) it is completely specified by its values at the n distinct x_k. In fact, since $r_{n-1}(x_k) = p_{2n-1}(x_k)$, from (5.21), the Lagrange interpolation formula gives

$$r_{n-1}(x) = \sum_{k=1}^{n} L_k(x) p_{2n-1}(x_k), \tag{5.24}$$

where $L_k(x) = \prod_i{}'(x - x_i)/\prod_i{}'(x_k - x_i)$, the prime indicating that $i = k$ is excluded in the product.

Hence

$$\begin{aligned}\int_a^b w(x) p_{2n-1}(x)\,dx &= \int_a^b w(x) r_{n-1}(x)\,dx,\\ &= \int_a^b w(x)\left(\sum_{k=1}^{n} L_k(x) p_{2n-1}(x_k)\right) dx, \qquad \text{from (5.24)}\\ &= \sum_{k=1}^{n}\left(\int_a^b w(x) L_k(x)\,dx\right) p_{2n-1}(x_k).\end{aligned}$$

Thus, by comparison with the right-hand side of (5.20), each α_k is specified by

$$\alpha_k = \int_a^b w(x) L_k(x)\,dx, \tag{5.25}$$

and is constant. This proves the theorem.

We note that a consequence of Theorem 5-2 is that for all polynomials of degree $0, 1, \ldots, 2n-1$ formula (5.20) is exact. This provides us with a simpler way of obtaining the α_k than does (5.25). Take, for example, the case where $(a, b) \equiv (-1, 1)$, $w(x) \equiv 1, \{f_n(x)\} \equiv \{P_n(x)\}$ the Legendre polynomials, and we seek a three-point formula (i.e., $n = 3$). We have, since the roots of $P_3(x) = 0$ are $-\sqrt{(3/5)}, 0, \sqrt{(3/5)}$, that

$$\int_{-1}^{1} 1\,dx = 2 = \alpha_1 . 1 + \alpha_2 . 1 + \alpha_3 . 1$$

$$\int_{-1}^{1} x\,dx = 0 = \alpha_1[-\sqrt{(3/5)}] + \alpha_2[0] + \alpha_3[\sqrt{(3/5)}]$$

$$\int_{-1}^{1} x^2\,dx = \tfrac{2}{3} = \alpha_1[-\sqrt{(3/5)}]^2 + \alpha_2[0]^2 + \alpha_3[\sqrt{(3/5)}]^2.$$

(Note that use of higher order polynomials gives equations dependent on the above.) The solution of these linear equations gives the weight coefficients

$$\alpha_1 = \alpha_3 = \tfrac{5}{9}, \qquad \alpha_2 = \tfrac{8}{9}.$$

This method can be used to find weight coefficients for Gaussian formulae of any order.

In general, of course, the functions we will be required to integrate will not be polynomial functions. From (3.7) it follows that when $f(x)$ is continuous and $2n$ times differentiable we can write $f(x)$ in the form

$$f(x) = y(x) + \prod_{k=1}^{n} (x - x_k)^2 f^{(2n)}(\xi)/(2n)!,$$

where $y(x)$ is a polynomial of degree $2n - 1$, $x_1 < x_2 < \ldots < x_n$ and ξ lies in (x_1, x_n) and depends on x. Further,

$$f(x_k) = y(x_k), \qquad k = 1, \ldots, n. \tag{5.26}$$

Now assign the x_k to be the distinct zeros of the orthogonal polynomial equation $f_n(x) = 0$. It follows that

$$\int_a^b w(x) f(x)\, dx = \int_a^b w(x) y(x)\, dx + \frac{1}{(2n)!} \int_a^b w(x) \prod_{k=1}^{n} (x - x_k)^2 f^{(2n)}(\xi)\, dx.$$

By virtue of Theorem 5-2 and (5.26)

$$\int_a^b w(x) y(x)\, dx = \sum_{k=1}^{n} \alpha_k f(x_k),$$

where the α_k are the constants defined by (5.25). Thus

$$\int_a^b w(x) f(x)\, dx = \sum_{k=1}^{n} \alpha_k f(x_k) + R,$$

where

$$R = \frac{1}{(2n)!} \int_a^b w(x) \prod_{k=1}^{n} (x - x_k)^2 f^{(2n)}(\xi)\, dx.$$

Since $w(x)$ has been assumed to be a positive weighting function, Theorem 5-1 is now applicable and we deduce that

$$R = \frac{f^{(2n)}(\eta)}{(2n)!} \int_a^b w(x) \prod_{k=1}^{n} (x - x_k)^2\, dx, \qquad x_1 < \eta < x_n.$$

If a_0 is the coefficient of the term in x^n in the polynomial function $f_n(x)$ the expression for R can be written as

$$R = \frac{f^{(2n)}(\eta)}{(2n)!\, a_0^2} \int_a^b w(x) f_n^2(x)\, dx. \tag{5.27}$$

The integral in (5.27) has been evaluated in chapter 4 for the four special orthogonal polynomials discussed there. We note also that we can express α_k in terms of the orthogonal polynomial function. For, if a_0 is the coefficient of x^n in $f_n(x)$,

$$\prod_i{}' (x - x_i) = \prod_i (x - x_i)/(x - x_k) = f_n(x)/a_0(x - x_k),$$

and

$$\prod_i{}' (x_k - x_i) = f'_n(x_k)/a_0. \qquad \text{(Why?)}$$

Thus, from (5.25), we have

$$\alpha_k = \int_a^b \frac{w(x)f_n(x)}{(x - x_k)f'_n(x_k)}\, dx. \tag{5.28}$$

A summary of the most important results, for the special orthogonal polynomials of chapter 4, follows.

(a) Gauss–Legendre quadrature, $w(x) = 1$, $[a, b] = [-1, 1]$

$$\int_{-1}^{1} f(x)\, dx = \sum_{k=1}^{n} \alpha_k f(x_k) + R \tag{5.29}$$

where

$$R = \frac{2^{2n+1}(n!)^4}{(2n+1)[(2n)!]^3} f^{(2n)}(\xi), \qquad -1 < \xi < 1.$$

$x_1, x_2, \ldots, x_n$ are the roots of the Legendre polynomial equation $P_n(x) = 0$ and

$$\alpha_k = \frac{1}{P'_n(x_k)} \int_{-1}^{1} \frac{P_n(x)}{x - x_k}\, dx, \qquad k = 1, 2, \ldots, n,$$

$$= \frac{1}{[P'_n(x_k)]^2} \frac{2}{1 - x_k^2}.$$

The values of x_k and α_k ($k = 1, 2, \ldots, n$) are given for several values of n in Table 5.3. More extensive tables of these and later Gaussian quadrature formulae can be found in Abramowitz and Stegun [9].

Table 5.3 Abscissae and weights for Gauss–Legendre quadrature

n	x_k	α_k
1	0·000 000 0	2·000 000 0
2	± 0·577 350 3	1·000 000 0
3	± 0·774 596 7	0·555 555 6
	0·000 000 0	0·888 888 9
4	± 0·861 136 3	0·347 854 8
	± 0·339 981 0	0·652 145 2
5	± 0·906 179 8	0·236 926 9
	0·538 469 3	0·478 628 7
	0·000 000 0	0·568 888 9

(b) Gauss–Laguerre quadrature, $w(x) = e^{-x}$, $[a, b] = [0, \infty]$

$$\int_0^\infty e^{-x} f(x)\, dx = \sum_{k=1}^{n} \alpha_k f(x_k) + R, \tag{5.30}$$

where
$$R = \frac{(n!)^2}{(2n)!} f^{(2n)}(\xi), \qquad 0 < \xi < \infty$$

$x_1, x_2, \ldots, x_n$ are the roots of the Laguerre polynomial equation $L_n(x) = 0$ and

$$\alpha_k = \frac{(n!)^2}{x_k [L_n'(x_k)]^2}.$$

The values of x_k and α_k ($k = 1, 2, \ldots, n$) are given for several values of n in Table 5.4.

Table 5.4 Abscissae and weights for Gauss–Laguerre quadrature

n	x_k	α_k
1	1·000 000 0	1·000 000 0
2	0·585 786 4	0·853 553 4
	3·414 213 6	0·146 446 6
3	0·415 774 6	0·711 093 0
	2·294 280 4	0·278 517 7
	6·289 945 1	0·010 389 3
4	0·322 547 7	0·603 154 1
	1·745 761 1	0·357 418 7
	4·536 620 3	0·038 887 9
	9·395 070 9	0·000 539 3
5	0·263 560 3	0·521 755 6
	1·413 403 1	0·398 666 8
	3·596 425 8	0·075 942 4
	7·085 810 0	0·003 611 8
	12·640 800 8	0·000 023 4

(c) Gauss–Hermite quadrature, $w(x) = e^{-x^2}$, $[a, b] = [-\infty, \infty]$

$$\int_{-\infty}^\infty e^{-x^2} f(x)\, dx = \sum_{k=1}^{n} \alpha_k f(x_k) + R, \tag{5.31}$$

where
$$R = \frac{n!\sqrt{\pi}}{2^n (2n)!} f^{(2n)}(\xi), \qquad -\infty < \xi < \infty.$$

$x_1, x_2, \ldots, x_n$ are the roots of the Hermite polynomial equation $H_n(x) = 0$ and

$$\alpha_k = \frac{2^{n+1} n! \sqrt{\pi}}{[H_n'(x_k)]^2}.$$

The values of x_k and α_k ($k = 1, 2, \ldots, n$) are given for several values of n in Table 5.5.

Table 5.5 Abscissae and weights for Gauss–Hermite quadrature

n	x_k	α_k
1	0·000 000 0	1·772 453 9
2	± 0·707 106 8	0·886 226 9
3	0·000 000 0	1·181 635 9
	± 1·224 744 9	0·295 410 0
4	± 0·524 647 6	0·804 914 1
	± 1·650 680 1	0·081 312 8
5	0·000 000 0	0·945 308 7
	± 0·958 572 5	0·393 619 3
	± 2·020 182 9	0·019 953 2

(d) Gauss–Chebyshev quadrature, $w(x) = 1/\sqrt{(1-x^2)}$, $[a, b] = [-1, 1]$

$$\int_{-1}^{1} \frac{f(x)}{\sqrt{(1-x^2)}}\, dx = \frac{\pi}{n} \sum_{k=1}^{n} f\left(\cos \frac{2k-1}{2n}\pi\right) + R, \tag{5.32}$$

where

$$R = \frac{2\pi}{2^{2n}(2n)!} f^{(2n)}(\xi), \qquad -1 < \xi < 1.$$

We conclude this section by applying the Gauss–Legendre formula (5.29) to evaluate $\int_0^{\frac{1}{2}\pi} \sin x\, dx$. First we must make a substitution which transforms the range of integration to $[-1, 1]$. The substitution, in the general case of $\int_a^b f(x)\, dx$, is $x = \frac{1}{2}[(b-a)u + (b+a)]$, hence

$$\int_0^{\frac{1}{2}\pi} \sin x\, dx = \frac{\pi}{4} \int_{-1}^{1} \sin\left(\frac{\pi}{4}u + \frac{\pi}{4}\right) du,$$

$$\doteqdot \frac{\pi}{4}\left(0{\cdot}555\,555\,6 \sin (0{\cdot}225\,403\,3)\frac{\pi}{4} + 0{\cdot}888\,888\,9 \sin \frac{\pi}{4} + 0{\cdot}555\,555\,6 \sin (1{\cdot}774\,596\,7)\frac{\pi}{4}\right),$$

using (5.29) with $n = 3$ (the Gauss–Legendre three-point formula). Thus

$$\int_0^{\frac{1}{2}\pi} \sin x\, dx \doteqdot 1{\cdot}000\,008\,2.$$

Also, from (5.29) the error is certainly less in modulus than

$$[2^7(3!)^4/7(6!)^3] \sin 1 < 5 \times 10^{-5}.$$

In practice one compares results for successive n and carries on computations until these differ in modulus by less than some pre-assigned amount. One must be careful, however, since the Gauss–Legendre three-point and four-point formulae give respectively the consistent values 1·585 026 and 1·585 060 for

$$\int_{-1}^{1} \frac{dx}{x^4 + x^2 + 0{\cdot}9}$$

compared with the correct value of 1·582 233. Thus, for some problems it would be wiser to split the range, use the same order formula on each section and to compare with the result obtained for the full range.

5.6 The method of Chebyshev expansion

This method is based on results which are given in chapter 4, Exercises 6, 7, 8, and 12. Suppose $f(x)$ is a continuous, differentiable, bounded function in $[-1, 1]$ which can be expressed as

$$f(x) = \tfrac{1}{2}a_0 + a_1T_1(x) + a_2T_2(x) + \ldots + a_NT_N(x), \tag{5.33}$$

where $T_r(x)$ is the Chebyshev polynomial of degree r. The coefficients $a_0, a_1, \ldots,$ satisfy the equations

$$a_r = \frac{2}{N+1}\sum_{k=0}^{N} f(x_k)T_r(x_k), \qquad x_k = \cos\left(\frac{2k+1}{N+1}\frac{\pi}{2}\right). \tag{5.34}$$

Further

$$\int_{-1}^{1} f(x)\,dx = 2(C_1 + C_3 + C_5 + \ldots), \tag{5.35}$$

where $\quad C_{N+1} = a_N/2(N+1), \qquad C_N = a_{N-1}/2N,$

and $\quad C_i = (a_{i-1} - a_{i+1})/2i, \qquad i = 1, 2, \ldots, N-1.$

In practical applications we start off from $N = 4$, say, and we consider the size of $C_{N+1}, \alpha C_{N-1}, \alpha^2 C_{N-3}$ for some α in $(0, 1)$, say $\alpha = \frac{1}{8}$. If these are small enough then we terminate the procedure by evaluating (5.35) otherwise we continue calculations with N replaced by $2N$. To give some idea of the size of coefficients in (5.35) we find for

$$\int_{-1}^{1} \frac{dx}{x^4 + x^2 + 0{\cdot}9}$$

and $N = 4$ that $C_5 = 0{\cdot}006\,095$. For $N = 8$, $C_9 = -0{\cdot}000\,073$ and for $N = 16$,

$C_{17} = 0{\cdot}000\,000_2$. Further, $\frac{1}{8}C_{15} = 0{\cdot}000\,000_0$, $\frac{1}{64}C_{13} = -0{\cdot}000\,000_1$, and thus $N = 16$ is a suitable place to terminate calculations for six decimal place accuracy. We find

$$\int_{-1}^{1} \frac{\mathrm{d}x}{x^4 + x^2 + 0{\cdot}9} \mp 1{\cdot}582\,233.$$

Note that the integral $\int_a^b f(x)\,\mathrm{d}x$ can be transformed by the substitution $x = \frac{1}{2}[(b-a)u + (b+a)]$ into $\int_{-1}^{1} g(u)\,\mathrm{d}u$, and the above method is now applicable. Alternatively we can change the range of integration to (0, 1) and carry out a similar operation using shifted Chebyshev polynomials. For more details about the method of Chebyshev expansion the reader should consult Clenshaw and Curtis [**3**].

5.7 Singular integrals

Although not explicitly stated earlier, it has been assumed (except in the case of Gauss–Chebyshev integration) that the integrand is finite, differentiable, and continuous over the range of integration, i.e., the integrand can be represented by a Taylor series over the region of integration. If, however, the integrand (or its derivatives) are infinite for one or more values of x in the range then the previous methods no longer apply until the singularity has been removed. This can be done either by making a substitution or by subtracting out the singularity. For example, the singularity at $x = 0$ in

$$\int_0^1 f(x)x^{-\frac{1}{2}}\,\mathrm{d}x,$$

can be removed by the substitution $x = t^2$, provided $f(x)$ is suitably well-behaved. Likewise the singularities at $x = \pm 1$ in

$$\int_{-1}^{1} \frac{f(x)\,\mathrm{d}x}{\surd(1-x^2)}$$

can be removed by the substitution $x = \sin\theta$, and so on. (This integral can also be evaluated by formula (5.32).) To illustrate the method of subtracting out a singularity consider

$$I = \int_0^2 \frac{\mathrm{e}^{-x}}{1-x}\,\mathrm{d}x,$$

where the integrand has a singularity at $x = 1$, and we have to understand the integral in the sense of a Cauchy principal value, that is as

$$\lim_{\xi\to 0}\left(\int_0^{1-\xi} \frac{\mathrm{e}^{-x}}{1-x}\,\mathrm{d}x + \int_{1+\xi}^{2} \frac{\mathrm{e}^{-x}}{1-x}\,\mathrm{d}x\right).$$

Now we can write I in the form

$$I = \int_0^2 \frac{e^{-x} - e^{-1}}{1-x}\,dx + e^{-1}\int_0^2 \frac{dx}{1-x}$$

where the second integral has also to be understood as a Cauchy principal value.

The second integral can be expressed as

$$\lim_{\xi\to 0}\left(\int_0^{1-\xi} \frac{dx}{1-x} + \int_{1+\xi}^{2} \frac{dx}{1-x}\right) = \lim_{\xi\to 0}\left(\int_1^{\xi} \frac{-du}{u} + \int_{\xi}^{1} \frac{du}{-u}\right) = 0,$$

making the substitutions $1 - x = u$ and $1 - x = -u$ respectively. Thus

$$I = \int_0^2 \frac{e^{-x} - e^{-1}}{1-x}\,dx,$$

and our integral is no longer singular and the methods of previous sections can be employed to evaluate the integral.

Finally we mention a method due to Fox [4], of adapting Romberg integration to evaluate singular integrals such as

$$\int_0^1 x^{\frac{1}{2}}\,dx, \qquad \int_0^1 x \ln x\,dx, \qquad \int_0^1 x^{\frac{1}{2}} \ln x\,dx,$$

and, by extension,

$$\int_0^1 x^{\frac{1}{2}} g(x)\,dx, \qquad \int_0^1 x \ln x\, g(x)\,dx, \qquad \int_0^1 x^{\frac{1}{2}} \ln x\, g(x)\,dx.$$

Clearly in all these examples the singularity is at $x = 0$. We perform our integration by considering separately the ranges (x_0, x_1) and (x_1, x_N). We have, by Taylor's theorem,

$$f_0 = f_1 - h_N f'_1 + \frac{h_N^2}{2!} f''_1 - \ldots, h_N = (x_N - x_0)/N = 1/N \tag{5.36}$$

and also

$$\int_{x_0}^{x_1} f(x)\,dx = h_N f_1 - \frac{h_N^2}{2!} f'_1 + \frac{h_N^3}{3!} f''_1 - \ldots, \tag{5.37}$$

and, elimination of f'_1 between (5.36) and (5.37) produces

$$\int_{x_0}^{x_1} f(x)\,dx = \tfrac{1}{2} h_N (f_0 + f_1) - \frac{h_N^3}{2.3!} f''_1 + \frac{2h_N^4}{2.4!} f'''_1 - \frac{3h_N^5}{2.5!} f_1^{(iv)} + \ldots. \tag{5.38}$$

Further, for the range (x_1, x_N) we obtain (see (6.13))

$$\int_{x_1}^{x_N} f(x)\,\mathrm{d}x = h_N\{\tfrac{1}{2}f_1 + f_2 + \ldots + f_{N-1} + \tfrac{1}{2}f_N\}$$
$$- \tfrac{1}{12}h_N^2(f'_N - f'_1) + \tfrac{1}{720}h_N^4(f'''_N - f'''_1) - \ldots . \quad (5.39)$$

Addition of (5.38) and (5.39) gives

$$\int_{x_0}^{x_N} f(x)\,\mathrm{d}x = I_T(N) + \tfrac{1}{12}h_N^2 f'_1 - \tfrac{1}{12}h_N^3 f''_1 + \tfrac{29}{720}h_N^4 f'''_1 - \ldots$$
$$- \tfrac{1}{12}h_N^2 f'_N + \tfrac{1}{720}h_N^4 f'''_N - \ldots, \quad (5.40)$$

where $I_T(N)$ is the trapezium rule approximation with $h_N = (x_N - x_0)/N = \frac{1}{N}$. Now for $f(x) = x^{\frac{1}{2}}$, $f'(x) = \frac{1}{2}x^{-\frac{1}{2}}$, $f''(x) = -\frac{1}{4}x^{-\frac{3}{2}}$, etc., and in particular, $h_N^2 f'_1 = \frac{1}{2}h_N^{\frac{3}{2}}$, $h_N^3 f''_1 = -\frac{1}{4}h_N^{\frac{3}{2}}, \ldots, h_N^2 f'_N = \frac{1}{2}h_N^2, \ldots$. Thus, for $f(x) = x^{\frac{1}{2}}$, (5.40) can be written as

$$\int_0^1 x^{\frac{1}{2}}\,\mathrm{d}x = I_T(N) + Ah_N^{\frac{3}{2}} + Bh_N^2 + Ch_N^4 + \ldots . \quad (5.41)$$

A similar expression to (5.41) holds for the mid-point rule approximation $I_M(N)$. The expansion (5.41) is similar to (5.14) with the exception of the term $Ah_N^{\frac{3}{2}}$. The coefficients A, B, C can be eliminated in exactly the same manner as the coefficients in (5.14), giving the better approximations

$$I_T(1,2) = \{2\sqrt{2}I_T(2) - I_T(1)\}/(2\sqrt{2} - 1) \quad (5.42)$$

$$I_T(1,2,4) = \tfrac{1}{21}\{(32 + 8\sqrt{2})I_T(4) - (12 + 10\sqrt{2})I_T(2) + (1 + 2\sqrt{2})I_T(1)\}, \quad (5.43)$$

etc. Thus for $\int_0^1 x^{\frac{1}{2}}\,\mathrm{d}x$, Fox obtained

$$I_T(4) = 0{\cdot}643\,283, \qquad I_T(8) = 0{\cdot}658\,130, \qquad I_T(16) = 0{\cdot}663\,581_1,$$
$$I_T(4,8) = 0{\cdot}666\,250, \qquad I_T(8,16) = 0{\cdot}666\,562,$$

and finally,

$$I_T(4,8,16) = 0{\cdot}666\,667.$$

The reader should consult Fox's paper, or Fox and Mayers **[10]**, for a complete set of results although a number are given in the exercises.

5.8 Integration over a multidimensional hypercube

The most general problem is integration over the hypercube $-1 \leqslant x_1 \leqslant 1$, $-1 \leqslant x_2 \leqslant 1, \ldots, -1 \leqslant x_n \leqslant 1$, i.e., the evaluation of

$$\int_{-1}^{1} \ldots \int_{-1}^{1} f(x_1, x_2, \ldots, x_n)\, \mathrm{d}x_1\, \mathrm{d}x_2 \ldots \mathrm{d}x_n. \tag{5.44}$$

For simplicity we shall consider the problem of integrating

$$\int_{-1}^{1} \int_{-1}^{1} f(x, y)\, \mathrm{d}x\, \mathrm{d}y. \tag{5.45}$$

An obvious extension of the method will give results for (5.44). We first consider integration over a square of side $2h$ and centre the origin (Fig. 5.2).

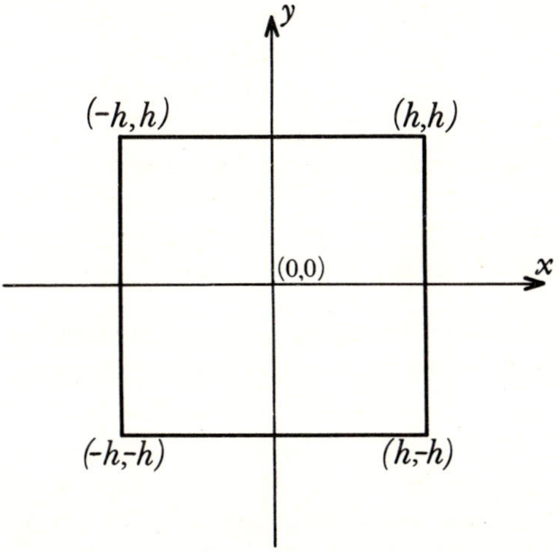

Fig. 5.2 Square of side $2h$ centre the origin

Taylor's expansion gives

$$f_{jk} = f(jh, kh) = f_{00} + h\left(j\frac{\partial}{\partial x} + k\frac{\partial}{\partial y}\right) f_{00} + \frac{h^2}{2!}\left(j\frac{\partial}{\partial x} + k\frac{\partial}{\partial y}\right)^2 f_{00} + \ldots \tag{5.46}$$

for $j = \pm 1$, $k = \pm 1$. Further

$$\int_{-h}^{h}\int_{-h}^{h} f(x,y)\, \mathrm{d}x\, \mathrm{d}y = \int_{-h}^{h}\int_{-h}^{h}\left\{f_{00} + \left(x\frac{\partial}{\partial x} + y\frac{\partial}{\partial y}\right) f_{00} + \right.$$

$$\left. + \frac{1}{2!}\left(x\frac{\partial}{\partial x} + y\frac{\partial}{\partial y}\right)^2 f_{00} + \ldots\right\} \mathrm{d}x\, \mathrm{d}y$$

$$= 4h^2\left\{f_{00} + \frac{h^2}{3!}\left(\frac{\partial^2}{\partial x^2} + \frac{\partial^2}{\partial y^2}\right) f_{00} + O(h^4)\right\}. \tag{5.47}$$

However, from (5.46),

$$h^2\left(\frac{\partial^2}{\partial x^2}+\frac{\partial^2}{\partial y^2}\right)f_{00}=\tfrac{1}{2}(f_{11}+f_{1,-1}+f_{-1,1}+f_{-1,-1}-4f_{00})+O(h^4)$$

and substitution in (5.47) gives the quadrature formula

$$\int_{-h}^{h}\int_{-h}^{h} f(x,y)\,\mathrm{d}x\,\mathrm{d}y=\tfrac{1}{3}h^2\{f_{11}+f_{1,-1}+f_{-1,1}+f_{-1,-1}+8f_{00}\}+O(h^6). \quad (5.48)$$

Evaluation of (5.45) is now possible by dividing up the square $-1 \leqslant x \leqslant 1$, $-1 \leqslant y \leqslant 1$ into a grid and applying formula (5.48) to each square of the grid, in the same way that the elementary formulae in section 5.1 are used to integrate over a range. Romberg integration, or the Lyness–McHugh variation, can then be used to improve results. An alternative approach is that of Gaussian quadrature. One method of approach is to treat each variable separately and to apply a one-dimensional formula. For instance, with $n = 2$, we have the formula

$$\int_{-1}^{1}\int_{-1}^{1} f(x,y)\,\mathrm{d}x\,\mathrm{d}y \doteqdot f\left(-\frac{1}{\sqrt{3}},-\frac{1}{\sqrt{3}}\right)+f\left(-\frac{1}{\sqrt{3}},\frac{1}{\sqrt{3}}\right)+f\left(\frac{1}{\sqrt{3}},-\frac{1}{\sqrt{3}}\right)+f\left(\frac{1}{\sqrt{3}},\frac{1}{\sqrt{3}}\right).$$

Further Gaussian methods can be found in Phillips [8].

5.9 Summary and Notes

The choice of method for any particular problem is usually a matter of experience. This is often dictated by circumstances. For instance, if only a desk machine is available one would probably recommend finite difference formulae, and in particular central difference formulae, or the elementary formulae in conjunction with Romberg integration, the Lyness and McHugh variant, or series expansion if appropriate. For automatic computation, Romberg integration, Gauss quadrature, and Chebyshev expansion prove to be very powerful methods for the evaluation of most integrals although series expansions have their uses. With regard to Chebyshev expansions the reader is referred to recent papers by Cooper [5] and O'Hara and Smith [6] concerning the calculation of the coefficients a_r and the size of error estimates respectively.

Other quadrature formulae, similar to Gaussian formulae and associated with the names of Lobatto and Radau are sometimes useful and an account of these can be found in Ralston [7].

If the integral is singular, one must either remove the singularity by substitution or 'subtracting out' or else, if suitable, use the extended Romberg integration (exercises 21–23). However, it is possible to adopt Gauss quadrature formulae to integrate functions with certain types of singularities (exercises 16–18).

Further methods of integration – by series summation, by asymptotic expansion, by integrating differential equations, and by Monte Carlo methods – are given in chapters 6, 7, 11, and 14 respectively and these chapters should be consulted in order to obtain a more complete picture of the methods available for numerical integration. Finally, it is advisable to note that even if it is possible to evaluate an integral explicitly it is often more laborious to calculate the resulting function than it is to use a numerical method. This assertion is borne out quite well by

$$\int_0^x \frac{t^2\,dt}{a^4+t^4} = -\frac{1}{4a\sqrt{2}}\ln\left\{\frac{x^2+ax\sqrt{2}+a^2}{x^2-ax\sqrt{2}+a^2}\right\} + \frac{1}{2a\sqrt{2}}\tan^{-1}\left\{\frac{ax\sqrt{2}}{a^2-x^2}\right\} \qquad (a>0)$$

for which, for given x, evaluation of the right-hand side involves as much computation as the evaluation of the left-hand side by a suitable numerical method.

REFERENCES

[1] *Interpolation and Allied Tables*, H.M.S.O., London (1956).

[2] Lyness, J. N. and McHugh, B. J. J., 'Integration over Multidimensional Hypercubes, I. A progressive procedure', *Computer J.*, **6** (1963) 264–70.

[3] Clenshaw, C. W. and Curtis, A. R., 'A method for Numerical Integration on an Automatic Computer', *Num. Math.*, **2** (1960) 197–205.

[4] Fox, L., 'Romberg Integration for a Class of Singular Integrals', *Computer J.*, **10** (1967) 87–93.

[5] Cooper, G. J., 'The Evaluation of the Coefficients in a Chebyshev Expansion', *Computer J.*, **10** (1967) 94–100.

[6] O'Hara, H. and Smith, F. J., 'Error estimation in the Clenshaw–Curtis quadrature formula', *Computer J.*, **11** (1968) 213–18.

[7] Ralston, A., *A First Course in Numerical Analysis*, McGraw-Hill Book Co., New York (1965).

[8] Phillips, G. M., 'Numerical Integration over an *N*-dimensional Rectangular Region', *Computer J.*, **10** (1967) 297–99.

[9] Abramowitz, M. and Stegun, I. A., *Handbook of Mathematical Functions*, Dover Publications Inc., New York (1965).

[10] Fox, L. and Mayers, D. F., *Computing Methods for Scientists and Engineers*, Oxford University Press, London (1968).

Exercises

1. The velocity, v, of a rocket launched from the ground is measured (in ft/s) every 5 s and the data is recorded for $t = 0(5)50$ in the table below. Estimate the height of the rocket after 50 s assuming it has been travelling vertically.

t	0	5	10	15	20	25	30
v	0	60·6	180·1	341·6	528·4	723·6	910·6

t	35	40	45	50
v	1072·5	1192·6	1254·1	1240·2

2. Prove that

$$\int \frac{dx}{a^3 - x^3} = \frac{1}{6a^2} \ln\left(\frac{a^2 + ax + x^2}{(a-x)^2}\right) + \frac{1}{a^2\sqrt{3}} \tan^{-1} \frac{2x + a}{a\sqrt{3}} + C.$$

Hence evaluate
$$\int_2^4 \frac{dx}{1 - x^3}.$$

Evaluate the same integral by a numerical method.

3. Evaluate

$$(a) \int_0^1 \frac{dx}{\sqrt{(1 - \frac{1}{4}x^2)}}, \qquad (b) \int_0^{\frac{1}{2}\pi} \frac{dx}{\sqrt{(1 - \frac{1}{4}\sin^2 x)}}.$$

(Hint: In (*b*) assume Wallis' formula for $\int_0^{\frac{1}{2}\pi} \sin^n x \, dx$.)

4. Show that

$$\int_0^1 x^n (\ln x)^m \, dx = \frac{-m}{n+1} \int_0^1 x^n (\ln x)^{m-1} \, dx.$$

Use this result to evaluate

$$\int_0^1 x^n (\ln x)^n \, dx,$$

and hence deduce the numerical value (to 4 decimal places) of

$$\int_0^1 x^x \, dx.$$

5. Obtain an integration formula from Everett's interpolation formula.

6. Find the equation of the three-point Lagrangian polynomial which passes through the points $(0, f_0)$, $(1, f_1)$, $(2, f_2)$. Show that the value of $f(x)$ at $x = p\,(<1)$ using the first three terms of the Newton interpolation formula,

$$f_p = f_0 + p\Delta_0 + \frac{p(p-1)}{2}\Delta_0^2 + \ldots,$$

is the same as that given by the Lagrangian polynomial. Show also that if $\int_0^2 f(x)\,dx$ is evaluated approximately using the above polynomial then the result is the same as that obtained using Simpson's rule.

7. If $f(x)$ is continuous and differentiable and

$$I = \int_{x_{-2}}^{x_2} f(x)\,dx$$

show that

$$I = \tfrac{4}{3}h(2f_{-1} - f_0 + 2f_1) + h(\tfrac{14}{45}\delta^4 + \ldots)f_0.$$

Use this result to evaluate $\int_0^4 f(x)\,dx$ from the following table specifying a quartic

x	0	1	2	3	4
$f(x)$	0	1	0	−2	0

8. Derive Laplace's integration formula

$$\int_{x_0}^{x_1} f(x)\,dx = f_0 + \tfrac{1}{2}\Delta_0 - \tfrac{1}{12}\Delta_0^2 + \tfrac{1}{24}\Delta_0^3 - \tfrac{19}{720}\Delta_0^4 + \dots .$$

9. Tabulate $y = x^2 + (\sin x/x)$ to five decimal places for $x = 0{\cdot}1(0{\cdot}1)1{\cdot}0$. Hence evaluate

$$\int_{0\cdot3}^{0\cdot8} \frac{x^3 + \sin x}{x}\,dx,$$

using a difference formula.

10. Find the volume of revolution obtained when the curve $y = xe^{-x}\cos x$ is rotated about the x-axis between $x = 0$ and $x = \pi$.

11. From the Taylor series expansion for $f(x)$ in the range (x_0, x_1) show that, assuming $f(x)$ has continuous derivatives,

$$\int_{x_0}^{x_1} f(x)\,dx = hf_0 + \tfrac{1}{2}h^2 f'_0 + \tfrac{1}{6}h^3 f''(\xi_1), \qquad x_0 < \xi_1 < x_1.$$

Show also that

$$\tfrac{1}{2}h(f_0 + f_1) = hf_0 + \tfrac{1}{2}h^2 f'_0 + \tfrac{1}{4}h^3 f''(\xi_2), \qquad x_0 < \xi_2 < x_1.$$

Hence deduce that

$$\int_{x_0}^{x_1} f(x)\,dx = \tfrac{1}{2}h(f_0 + f_1) - \tfrac{1}{12}h^3 f''(\xi), \qquad x_0 < \xi < x_1,$$

and the error in the trapezium rule approximation is less in modulus than $\frac{1}{12}h^3 \max\limits_{x_0<\xi<x_1} |f''(\xi)|$.

12. Show, by the method of the previous example, that for the mid-point rule

$$\int_{x_0}^{x_1} f(x)\,dx = hf_{\frac{1}{2}} + \tfrac{1}{24}h^3 f''(\xi), \qquad x_0 < \xi < x_1$$

and for Simpson's rule

$$\int_{x_0}^{x_2} f(x)\,dx = \tfrac{1}{3}h(f_0 + 4f_1 + f_2) - \tfrac{1}{90}h^5 f^{iv}(\xi), \qquad x_0 < \xi < x_2.$$

13. Find the order of the errors in using the Newton–Cotes formulae

$$\int_{x_0}^{x_4} f(x)\,\mathrm{d}x \mp \tfrac{1}{3}h(f_0 + 4f_1 + 2f_2 + 4f_3 + f_4),$$

and $$\int_{x_0}^{x_5} f(x)\,\mathrm{d}x \mp \tfrac{5}{24}h(11f_1 + f_2 + f_3 + 11f_4).$$

14. Show that if $f(x)$ is continuous and differentiable,

$$I = \int_{x_0}^{x_N} f(x)\,\mathrm{d}x,$$

and $I_T(N), I_M(N), I_S(N)$ denote the trapezium, mid-point, and Simpson rule approximations respectively then

$$I_T(N) = I + \alpha_2 h_N^2 + \alpha_4 h_N^4 + \alpha_6 h_N^6 + \ldots,$$
$$I_M(N) = I + \beta_2 h_N^2 + \beta_4 h_N^4 + \beta_6 h_N^6 + \ldots,$$
$$I_S(N) = I + \gamma_4 h_N^4 + \gamma_6 h_N^6 + \ldots.$$

(Hint: The first result follows from (6.13).)

15. Evaluate $$\int_0^1 (0{\cdot}92 \cosh x - \cos x)\,\mathrm{d}x$$

(a) analytically,
(b) by a finite difference method with $h = \frac{1}{8}$,
(c) by Romberg integration.

16. Deduce the nth order Gaussian quadrature formula of the form

$$\int_0^1 \frac{f(x)}{\sqrt{x}}\,\mathrm{d}x = \sum_{k=1}^{n} \alpha_k f(x_k) + R,$$

where x_k are the roots of $P_{2n}(\sqrt{x}) = 0$ and α_k is twice the value of the corresponding weight in the Gauss–Legendre formula of order $2n$. What is the value of the remainder term R?

17. Find a Gaussian quadrature formula for

$$\int_0^1 \sqrt{x}\, f(x)\,\mathrm{d}x.$$

18. Find a Gauss type formula for integrating a function of the form

$$\int_0^1 \left(\frac{x}{1-x}\right)^{\frac{1}{2}} f(x)\,\mathrm{d}x.$$

19. Can Romberg integration be used in conjunction with Gauss quadrature formulae? If it can be used evaluate by this technique

$$\int_0^{\frac{1}{2}\pi} x \cos x \, dx.$$

20. Evaluate

$$y = \frac{1}{\pi}\int_0^{\pi} \cos (x \sin \theta) \, d\theta$$

to four decimal places by quadrature for $x = 0(\frac{1}{4}\pi)5\pi$. From your results estimate the smallest positive value of x for which $y = 0$.

21. Let

$$I_1 = \int_0^1 x \ln x \, dx, \qquad I_2 = \int_0^1 x^{\frac{1}{2}} \ln x \, dx.$$

Show that

$$I_1 = I_T(N) + Ah_N{}^2 \ln h_N + Bh_N{}^2 + Ch_N{}^4 + \ldots,$$

$$I_2 = I_T(N) + Ah_N{}^{\frac{3}{2}} \ln h_N + Bh_N{}^{\frac{3}{2}} + Ch_N{}^4 + \ldots,$$

(with the same form of expression for the mid-point rule) while

$$I_1 = I_S(N) + Ah_N{}^2 + Bh_N{}^4 + Ch_N{}^6 + \ldots,$$

$$I_2 = I_S(N) + Ah_N{}^{\frac{3}{2}} \ln h_N + Bh_N{}^{\frac{3}{2}} + Ch_N{}^4 + Dh_N{}^6 + \ldots .$$

22. Extend Fox's method of integrating singular integrals to deal with

$$I = \int_0^1 x^{\frac{1}{2}} (1 - x)^{\frac{1}{2}} \, dx,$$

whose first derivative is singular at $x = 0$ and $x = 1$.

23. Evaluate, by using Fox's method of extended Romberg integration

$$\int_0^1 x^{-\frac{1}{2}} \tan^{-1} x \, dx, \qquad \int_0^1 \sin x \ln x \, dx.$$

(Hint: To obtain the appropriate expansion in terms of h_N it is sufficient to assume that the functions $\tan^{-1} x$, $\sin x$ can be expanded as convergent power series and to consider the integrals

$$\int_0^1 x^{r-\frac{1}{2}} \, dx \qquad \text{and} \qquad \int_0^1 x^r \ln x \, dx \qquad (r = 1, 3, 5, \ldots).)$$

24. Evaluate

$$\int_0^1 \frac{\ln x}{1 + e^x} \, dx$$

by subtracting out the singularity.

25. By means of the substitution $y^2 = 1 - x$ remove the singularity in

$$\int_0^1 \frac{dx}{\sqrt{(1-x^2)}}$$

and hence evaluate the integral. Evaluate the integral also by subtracting out the singularity at $x = 1$.

26. Evaluate by numerical methods

$$\int_0^{\frac{1}{2}\pi} \ln(\sin x)\, dx, \qquad \int_0^{\frac{1}{2}\pi} x \ln(\sin x)\, dx.$$

What are the exact values of these integrals?

(Hint: For the second part use $\int_0^a f(x)\, dx = \int_0^a f(a-x)\, dx$.)

27. Evaluate $f(1/n) = \int_0^1 \frac{\sin nt}{t}\, dt$ for $n = 6\pi, 10\pi, 14\pi, 18\pi$. By using a method of extrapolation evaluate $f(0)$ and hence obtain an approximation for π. (Compare Salzer's method for series summation—exercise 16, chapter 6.)

28. Suggest a method for tabulating the integral

$$\int_0^\infty \frac{e^{-\lambda x^2}}{1+x^2}\, dx, \qquad (\lambda > 0)$$

as a function of λ.

29. Find a formula for

$$\int_{-h}^{h}\int_{-h}^{h} f(x,y)\, dx\, dy$$

of the form

$$h^2[Af(0,0) + Bf(0,h) + Cf(0,-h) + Df(h,0) + Ef(-h,0)] + O(h^6).$$

30. Evaluate

$$\int_0^1\int_0^1 e^{xy}\, dx\, dy$$

numerically. What is the exact value of this integral?

31. (Filon's formula for oscillating functions – *Proc. Roy. Soc. Edin.* (A) **49** (1928) 38.)

$$\int_a^b f(x) \sin mx\, dx = h(\alpha_1 S_1 + \alpha_2 S_2 + \alpha_3 S_3) + R_n$$

where

$$h = 1/2n \qquad (n \text{ a suitable integer})$$

$$\alpha_1 = \frac{1}{mh} + \frac{\sin 2mh}{2m^2h^2} - \frac{2\sin^2 mh}{m^3h^3},$$

$$\alpha_2 = \frac{1+\cos^2 mh}{m^2h^2} - \frac{\sin 2mh}{m^3h^3},$$

$$\alpha_3 = \frac{4\sin mh}{m^3h^3} - \frac{4\cos mh}{m^2h^2},$$

$$S_1 = f(a)\cos ma - f(b)\cos mb,$$

$$S_2 = -f(a)\sin ma - f(b)\sin mb + 2\sum_{k=0}^{n} f(a+2kh)\sin(am+2kmh)$$

$$S_3 = \sum_{k=1}^{n} f(a+(2k-1)h)\sin(am+(2k-1)mh)$$

and $$R_n = (b-a)\frac{h^3}{12}\left(1 - \frac{1}{16\cos\frac{1}{4}mh}\right)\sin\tfrac{1}{2}mh\, f^{(4)}(\xi), \qquad a<\xi<b.$$

Use this formula to evaluate

$$\int_0^1 x^2 e^x \sin 2x\,dx.$$

32. Evaluate

$$\int_0^\infty \frac{e^{-x^2}}{x+1}\,dx.$$

33. (S.I.A.M. Review, problem 68–8*) Evaluate

$$\int_1^\infty x^{m-x}\,dx, \qquad (m \text{ a given integer}),$$

$$\int_1^\infty \frac{x^{-x}}{1+x^2}\,dx.$$

$$\int_0^\infty x^{-x} \begin{matrix}\cos\\ \sin\end{matrix}\left\{\frac{\pi x}{2}\right\} dx, \qquad \int_0^\infty \begin{matrix}\sin\\ \cos\end{matrix}\left\{x\ln x\right\} dx.$$

6 Series summation

6.1 Summation by rearrangement

We shall assume that all the series we examine are strictly convergent. For an alternating series

$$v_1 - v_2 + v_3 - \ldots + (-1)^{k-1} v_k + \ldots, \tag{6.1}$$

this means that we require

$$0 < v_{k+1} < v_k \qquad \text{for all } k > k_0 \ \ (k_0 \text{ given}),$$

and $\quad v_k \to 0$ as $k \to \infty$.

Under these conditions we know (see e.g., Hardy [1]) that the sequence of partial sums $\{s_k\}$ where

$$s_k = v_1 - v_2 + v_3 - \ldots + (-1)^{k-1} v_k, \tag{6.2}$$

tends to a finite sum s, as $k \to \infty$, which is the sum of the infinite series (6.1). Clearly, taking $k_0 = 1$ from now on,

$$s < v_1,$$

and

$$|s - s_k| < v_{k+1}.$$

If we apply these results to the evaluation of

$$s = 1 - \tfrac{1}{2} + \tfrac{1}{3} - \tfrac{1}{4} + \ldots = \sum_{1}^{\infty} (-1)^{n-1}/n, \tag{6.3}$$

then, clearly, $s < 1$ and

$$|s - s_{9999}| < 1/10\,000 = 0{\cdot}000\,1.$$

This means that, by direct summation of the first 9999 terms of (6.3), we can expect at most four decimal place accuracy – and this is *with* exact arithmetic! Thus direct summation may not be a very practical way to carry out the summation of a series.

A classical result, which rearranges the terms of the series (6.1) but which leaves the sum unaltered, enables us to speed up convergence. The result states: If the series (6.1) converges to s, and $v_k \to 0$ as $k \to \infty$, then the series

$$\tfrac{1}{2}v_1 + \tfrac{1}{2}(v_1 - v_2) - \tfrac{1}{2}(v_2 - v_3) + \ldots = \tfrac{1}{2}v_1 + \tfrac{1}{2}\sum_{1}^{\infty}(-1)^{k-1}(v_k - v_{k+1}) \qquad (6.4)$$

converges to s. This can be applied to evaluate the sum of the series (6.3) in the following manner. We have

$$s = \frac{1}{2} + \frac{1}{2}\left\{\left(1 - \frac{1}{2}\right) - \left(\frac{1}{2} - \frac{1}{3}\right) + \ldots\right\},$$

$$= \frac{1}{2} + \frac{1}{2}\sum_{k=1}^{\infty}\frac{(-1)^{k-1}}{k(k+1)}.$$

$\sum_{1}^{\infty}(-1)^{k-1}/k(k+1)$ is also an alternating series and consequently (6.4) can be applied again. This gives

$$s = \frac{1}{2} + \frac{1}{8} + \frac{1}{2}\sum_{k=1}^{\infty}\frac{(-1)^{k-1}}{k(k+1)(k+2)},$$

$$= \frac{1}{2} + \frac{1}{8} + \frac{1}{24} + \frac{3}{4}\sum_{k=1}^{\infty}\frac{(-1)^{k-1}}{k(k+1)(k+2)(k+3)}, \qquad (6.5)$$

by a further application of (6.4). This process can be repeated if desired. However, retention of only five terms in the last sum in (6.5) ensures that

$$0{\cdot}693\,06 < s < 0{\cdot}693\,30.$$

The exact value of s is, of course, $\ln 2 \doteqdot 0{\cdot}693\,147$.

For a series of positive terms

$$v_1 + v_2 + \ldots + v_k + \ldots \qquad (6.6)$$

we require

$$0 < v_{k+1} < v_k \qquad \text{for all } k > k_0 \ (k_0 \text{ given})$$

and

$$v_k = O(1/k^{1+\delta}), \qquad \delta > 0$$

to ensure convergence. One method of summation of such series is by means of identities and known results. As an example, consider the evaluation of

$$\sum_{1}^{\infty} 1/(k^2 + 1).$$

Here we can make use of the identity

$$\frac{1}{k^2+1}=\frac{1}{k^2}-\frac{1}{k^2(k^2+1)},$$

which can be continually re-applied to give

$$\begin{aligned}\frac{1}{k^2+1}&=\frac{1}{k^2}-\frac{1}{k^2}\left[\frac{1}{k^2}-\frac{1}{k^2(k^2+1)}\right]\\&=\frac{1}{k^2}-\frac{1}{k^4}+\frac{1}{k^4(k^2+1)},\\&=\frac{1}{k^2}-\frac{1}{k^4}+\frac{1}{k^6}-\frac{1}{k^6(k^2+1)},\end{aligned}$$

and so on, if required. For our purposes it is sufficient to use the last expression and we obtain

$$\sum_1^\infty\frac{1}{k^2+1}=\sum_1^\infty\frac{1}{k^2}-\sum_1^\infty\frac{1}{k^4}+\sum_1^\infty\frac{1}{k^6}-\sum_1^\infty\frac{1}{k^6(k^2+1)}.$$

Now (see exercise 14)

$$\sum_1^\infty\frac{1}{k^2}=\frac{\pi^2}{6},\qquad\sum_1^\infty\frac{1}{k^4}=\frac{\pi^4}{90},\qquad\sum_1^\infty\frac{1}{k^6}=\frac{\pi^6}{945},\tag{6.7}$$

($\pi=3{\cdot}141\,59\ldots$) so that the evaluation of $\sum_1^\infty 1/(k^2+1)$ can be performed by evaluation of the constants in (6.7) and the faster convergent series $\sum_1^\infty 1/k^6(k^2+1)$.

This latter series we sum directly, the number of terms we take being dictated by estimates of the error in omitting the remaining terms. The following theorem (Hardy [1]) is most helpful in estimating the error.

THEOREM 6-1 *If $f(x)$ is a positive decreasing function of x, and $\int_K^\infty f(x)\,dx$ exists for some integer K, then $\sum_{K+1}^\infty f(k)$ converges and*

$$\int_{K+1}^\infty f(x)\,dx<\sum_{K+1}^\infty f(k)<\int_K^\infty f(x)\,dx.\tag{6.8}$$

It follows from Theorem 6-1 that if we sum only the first 10 terms of $\sum_1^\infty 1/k^6(k^2+1)$ the error in omitting the remaining terms is less than

$$\int_{10}^\infty\frac{dx}{x^6(x^2+1)}<\int_{10}^\infty dx/x^8=\frac{1}{7}\times10^{-7}.$$

Another idea which is sometimes useful for speeding up convergence, of both alternating series and series with positive terms, is that of expansion in a power series. To demonstrate this method consider the summation of

$$s = 1 - \tfrac{1}{3} + \tfrac{1}{5} - \tfrac{1}{7} + \ldots + (-1)^k/(2k+1) + \ldots .$$

We can write

$$\begin{aligned} s &= 1 - \sum_{k=1}^{\infty} \left(\frac{1}{4k-1} - \frac{1}{4k+1} \right), \\ &= 1 - 2 \sum_{k=1}^{\infty} \frac{1}{16k^2 - 1}, \\ &= 1 - 2 \sum_{k=1}^{\infty} \frac{1}{16k^2} \left(1 - \frac{1}{16k^2} \right)^{-1}, \\ &= 1 - \frac{1}{8} \sum_{k=1}^{\infty} \left(\frac{1}{k^2} + \frac{1}{16k^4} + \frac{1}{16^2 k^6} + \ldots \right), \\ &= 1 - \frac{\pi^2}{48} - \frac{\pi^4}{11\,520} - \frac{\pi^6}{1\,935\,360} - \ldots, \qquad \text{by (6.7).} \end{aligned}$$

This series converges very quickly and the error in terminating at any term is easily calculated. Thus, if we terminate at $\pi^6/1\,935\,360$, the error is

$$-\frac{1}{8} \left(\frac{1}{16^3} \sum_1^{\infty} \frac{1}{k^8} + \frac{1}{16^4} \sum_1^{\infty} \frac{1}{k^{10}} + \ldots \right),$$

i.e.,

$$-\frac{1}{8} \cdot 2 \left(\frac{1}{16^3} + \frac{1}{16^4} + \ldots \right) < E < 0$$

$$-\frac{1}{4} \cdot \frac{1}{16^3} \left(1 + \frac{1}{16} + \frac{1}{16^2} + \ldots \right) < E < 0$$

i.e.,

$$0 > E > -\frac{1}{4} \cdot \frac{1}{16^3} \cdot \frac{1}{1 - 1/16} = -\frac{1}{15\,360}.$$

It is also possible to rearrange a series of positive terms into an alternating series, which can then be summed by (6.4). This method, due to Van Wijngaarden, is outlined in exercise 1. It must be pointed out, however, that the methods given in this section have only a limited range of application. In the next section we consider methods which can be applied generally.

6.2 General methods for series summation

First, we consider the summation of a series of alternating terms

$$v_1 - v_2 + v_3 - \ldots + (-1)^{n-1} v_n + \ldots .$$

In order to sum this series it is convenient to consider the evaluation of the sum

$$s^* = x v_1 - x^2 v_2 + x^3 v_3 - \ldots + (-1)^{n-1} x^n v_n + \ldots .$$

We make the transformation $x = y/(1-y)$, noting that $x = 1$ when $y = 1/2$. We have

$$s^* = \frac{y}{1-y}v_1 - \frac{y^2}{(1-y)^2}v_2 + \frac{y^3}{(1-y)^3}v_3 - \ldots + (-1)^{n-1}\frac{y^n}{(1-y)^n}v_n + \ldots .$$

For $|y| < 1$, it is permissible to expand each of the terms $(1-y)^{-1}, (1-y)^{-2}, \ldots$ by the binomial theorem. We have

$$s^* = y(1 + y + y^2 + \ldots)v_1 - y^2(1 + 2y + 3y^2 + \ldots)v_2 + y^3(1 + 3y + 6y^2 + \ldots)v_3 - \ldots$$

Now let us examine the coefficient of y^k ($k = 1, 2, \ldots$). It equals

$$v_1 - \binom{k-1}{1}v_2 + \binom{k-1}{2}v_3 - \ldots + (-1)^{k-1}v_k$$

$$= (-\Delta)^{k-1}v_1, \text{ where } \Delta \text{ is the forward difference operator.}$$

Thus, assuming it is allowed to rearrange the original series by grouping coefficients of y,

$$s^* = yv_1 - y^2\Delta v_1 + y^3\Delta^2 v_1 - \ldots + y^n(-\Delta)^{n-1}v_1 + \ldots .$$

In particular, if we take $y = 1/2$, we have the result that the series (6.1) and

$$\frac{1}{2}v_1 - \frac{1}{4}\Delta v_1 + \frac{1}{8}\Delta^2 v_1 - \ldots + \frac{1}{2^n}(-\Delta)^{n-1}v_1 + \ldots = \frac{1}{2}\sum_1^\infty \left(-\frac{1}{2}\Delta\right)^{n-1}v_1 \tag{6.9}$$

which is called Euler's summation formula, tend to the same sum.

At this point it is worth while to stress that a rearrangement of an alternating series, such as (6.3), which is only conditionally convergent (that is $\sum_k (-1)^{k-1}v_k$ converges but $\sum_k v_k$ does not ($v_k > 0$)), may lead to a series which does not even converge. Care must therefore be taken to ensure that in any method of rearrangement the rearranged series does converge and, moreover, converges to the sum of the given series. In the case of the Euler summation formula, it can be shown that the partial sums of (6.9) do in fact tend to the sum of the given series. This result will not be shown here, as the principal aim in this chapter is to demonstrate the application of the methods. The interested reader is referred to Hardy [2] for a proof.

We now illustrate the application of formula (6.9) to the summation of the series (6.3). We sum the first six terms directly. This gives

$$1 - \tfrac{1}{2} + \tfrac{1}{3} - \tfrac{1}{4} + \tfrac{1}{5} - \tfrac{1}{6} = 0{\cdot}616\,66,$$

to five decimal places. Next we construct the difference table, Table 6.1. From the table we find that the sum of the series (6.3) is

$$s \doteqdot 0{\cdot}616\,66 + 0{\cdot}076\,48 = 0{\cdot}693\,14.$$

Table 6.1 Illustration of the Euler transformation

k	$v_k = \dfrac{1}{k+6}$	Δ	Δ^2	Δ^3	Δ^4	Δ^5	$\left(-\frac{1}{2}\Delta\right)^{k-1}\left(\frac{1}{2}v_1\right)$
1	0·142 86						0·071 43
		−1786					
2	0·125 00		397				0·004 47
		−1389		−119			
3	0·111 11		278		43		0·000 50
		−1111		−76		−18	
4	0·100 00		202		25		0·000 07
		−909		−51		−8	
5	0·090 91		151		17		0·000 01
		−758		−34		Total	0·076 48
6	0·083 33		117				
		−641					
7	0·076 92						

A useful check is provided by summing the first seven terms directly (=0·759 52) and then applying the Euler summation formula. For this we require the extra boxed terms in Table 6.1. Formula (6.9) gives us as the estimate for $\frac{1}{8} - \frac{1}{9} + \frac{1}{10} - \ldots$ the value 0·066 38. Hence

$$s \doteqdot 0{\cdot}759\,52 - 0{\cdot}066\,38 = 0{\cdot}693\,14$$

which checks with the result obtained previously. Moreover, the error in this result is less than the maximum rounding error incurred in the summation. (Why?)

Now consider the summation of a series of positive terms. Here we can employ a result obtained in chapter 5, namely (5.11), which can be rewritten in the form

$$h\sum_{r=0}^{N} f_r = \int_{x_0}^{x_N} f(x)\,dx + \frac{1}{2}h(f_N + f_0) + h\left(\frac{1}{12}\mu\delta - \frac{11}{720}\mu\delta^3\right)(f_N - f_0) - \sum_{1}^{N} R_i. \quad (6.10)$$

Again, from chapter 3, it is known that

$$\mu\delta_i = hf'_i + \tfrac{1}{6}h^3 f'''_i + \tfrac{1}{120}h^5 f^{(V)}(\xi'_i), \qquad x_{i-1} < \xi'_i < x_{i+1}$$

$$\mu\delta_i^3 = h^3 f'''_i + \tfrac{1}{4}h^5 f^{(V)}(\xi''_i), \qquad x_{i-2} < \xi''_i < x_{i+2}.$$

Substitution in (6.10) gives

$$h\sum_{r=0}^{N} f_r = \int_{x_0}^{x_N} f(x)\,dx + \frac{1}{2}h(f_N + f_0) + \frac{1}{12}h^2(f'_N - f'_0) - \frac{1}{720}h^4(f'''_N - f'''_0) + R$$

where

$$R = -\sum_{1}^{N} R_i + \frac{1}{1440}h^6(f^{(V)}(\xi'_N) - f^{(V)}(\xi'_0)) - $$

$$- \frac{11}{2880}h^6(f^{(V)}(\xi''_N) - f^{(V)}(\xi''_0)). \quad (6.11)$$

In particular, if we take $x_0 = 1, h = 1, f_r = \nu_{r+1}$, we obtain the summation formula

$$\sum_1^{N+1} \nu_r = \int_1^{N+1} \nu(x)\,\mathrm{d}x + \frac{1}{2}(\nu_{N+1} + \nu_1) + \frac{1}{12}(\nu'_{N+1} - \nu'_1) - \frac{1}{720}(\nu'''_{N+1} - \nu'''_1) + R. \tag{6.12}$$

We illustrate the application of formula (6.12) by estimating $\sum_1^{\infty} 1/n^2$, whose exact value was given in (6.7). Direct summation of the first nine terms gives 1·539 768 to six decimal places. For the remainder of the series, which we can write as $\sum_1^{\infty} 1/(n+9)^2$, we have from (6.12),

$$\sum_1^{\infty} \frac{1}{(n+9)^2} \doteq \int_1^{\infty} \frac{\mathrm{d}x}{(x+9)^2} + \frac{1}{2}\left(0 + \frac{1}{10^2}\right) + \frac{1}{12}\left(0 - \frac{-2}{10^3}\right) - \frac{1}{720}\left(0 - \frac{-24}{10^5}\right),$$

$$= 0{\cdot}1 + 0{\cdot}005 + 0{\cdot}000\,166_7 - 0{\cdot}000\,000_3$$

$$= 0{\cdot}105\,166, \qquad \text{to six decimal places.}$$

Hence $$\sum_1^{\infty} \frac{1}{n^2} \doteq 1{\cdot}539\,768 + 0{\cdot}105\,166 = 1{\cdot}644\,934.$$

Since $\nu^{(5)}(x) = -720/(x+9)^7$, $\nu^{(6)}(x) = 5040/(x+9)^8$ are functions which decrease in modulus as x increases, it follows from (6.11) that the truncation error is certainly less in modulus than

$$\frac{1}{1440} \times \frac{720}{7^7} + \frac{11}{2880} \times \frac{720}{7^7} + \frac{191}{60\,480} \sum_1^{\infty} \frac{5040}{(n+6)^8}.$$

The last series converges very quickly and a rough estimate is obtainable from Theorem 6-1. We find that certainly

$$|R| < 10^{-4}.$$

In fact, this is an over-estimate for the truncation error as the result obtained agrees to six decimal places with the value of $\pi^2/6$ obtained by direct computation (see exercise 5).

We remark that formula (6.12) can be extended to give a summation formula, known as the Euler–Maclaurin summation formula, which can be written in the form

$$\sum_1^{N+1} \nu_r = \int_1^{N+1} \nu(x)\,\mathrm{d}x + \tfrac{1}{2}(\nu_{N+1} + \nu_1) + \sum_{k=1}^{m} B_{2k}\{\nu_{N+1}^{(2k-1)} - \nu_1^{(2k-1)}\}/(2k)! + R_m \tag{6.13}$$

where the numbers $B_2 = \frac{1}{6}$, $B_4 = \frac{-1}{30}$, $B_6 = \frac{1}{42}$, $B_8 = \frac{-1}{30}$, $B_{10} = \frac{5}{66}$, . . . are the Bernoulli numbers, which are the coefficients in the expansion

$$\frac{x}{\mathrm{e}^x - 1} = 1 - \frac{1}{2}x + \sum_{k=1}^{\infty} \frac{B_{2k} x^{2k}}{(2k)!}. \tag{6.14}$$

The form of the remainder term R_m is given in a large number of books (e.g., Hardy [2]) but, in many cases, it is unnecessary to know this. For instance, if it can be established that $v^{(2k+2)}(x)$ is of fixed sign for $k \geqslant K$ (K to be determined),

$$\int_1^{\infty} |v^{(2k+2)}(x)|\, dx \text{ is finite and } f^{(2k+1)}(x) \to 0 \text{ as } x \to \infty,$$

then the right-hand side of (6.13) is an asymptotic expansion and the error does not exceed the last term retained in the series (see section 7.4). The reader should use this result to obtain another (and more realistic) estimate for the truncation error in the summation of $\sum_1^{\infty} 1/(n+9)^2$, which was obtained previously.

We conclude this section by giving another formula which behaves asymptotically. The formula, for the summation of an alternating series, is

$$\sum_k (-1)^{k-1} v_k = \frac{1}{2}\left(v_1 - \frac{1}{2}v'_1 + \frac{1}{24}v'''_1 - \frac{1}{240}v_1^{(5)} + \frac{17}{40\,320}v_1^{(7)} - \ldots\right). \quad (6.15)$$

We shall apply this formula to evaluating the sum (6.3). If, following Hildebrand [3], we sum the first four terms of (6.3) separately, and use (6.15) to estimate the sum of the remaining terms we find

$$S = \frac{7}{12} + \frac{1}{10} + \frac{1}{100} - \frac{1}{500} + \frac{1}{62\,500} - \frac{17}{6\,250\,000} + \ldots . \quad (6.16)$$

Retention of five terms of the transformed series yields an approximation to ln 2 with an error of less than 6×10^{-7}. However, if more terms are taken it is found that after the ninth term they increase unboundedly. Note that Euler's formula (6.9) converges, in this instance, no matter how many terms are taken.

6.3 Summary and notes

This chapter has mainly been concerned with the summation of series as a problem in its own right but, in general, this problem arises as a by-product of integration. Consequently, this chapter should be read in conjunction with chapter 5 (and also chapter 7).

There is no easy method, in general, for deciding how to select methods of rearrangement for any particular problem. It might be fair to say that experience is the best guide in this matter. The general methods given in section 6.2 are the most useful in practice and the most widely used. It is also worth pointing out that, while it is possible in every method (as has been seen in the text) to obtain error estimates, it is also instructive to check results by using a different method of summation.

A number of additional methods, such as Salzer's and Van Wijngaarden's, are given in the exercises.

REFERENCES

[1] Hardy, G. H., *A Course of Pure Mathematics*, 10th edn., Cambridge University Press, London (1952).

[2] Hardy, G. H., *Divergent Series*, Oxford University Press, London (1949).

[3] Hildebrand, F. B., *Introduction to Numerical Analysis*, McGraw-Hill Book Co. Ltd., New York (1956).

Exercises

1. (Van Wijngaarden's transformation) Let

$$S = v_1 + v_2 + v_3 + \dots \qquad (v_i > 0).$$

Define

$$u_r = v_r + 2v_{2r} + 4v_{4r} + 8v_{8r} + \dots .$$

Show that

$$v_r = u_r - 2u_{2r},$$

and hence

$$S = u_1 - u_2 + u_3 - u_4 + \dots .$$

Use this transformation to sum the following series (correct to four decimal places)

$$\sum_1^\infty \frac{1}{n^2}, \qquad \sum_1^\infty \frac{1}{n^3}, \qquad \sum_1^\infty \frac{1}{n^{\frac{3}{2}}}.$$

2. From (6.15) derive the formula

$$\sum_1^\infty (-1)^{n-1} u_n = \tfrac{1}{2}(u_1 - \tfrac{1}{2}\mu\delta u_1 + \tfrac{1}{8}\mu\delta^3 u_1 - \dots + (-1)^{r-1} 2^{-2r-1} \mu\delta^{2r+1} u_1 + \dots).$$

3. Sum the first four terms of (6.3) separately then apply the result of exercise 2 to obtain an approximation for ln 2.

4. Find a finite difference formula for the sum of the series

$$\sum_1^\infty (-1)^{k-1} x^{k-1} u_k, \qquad 0 < x < 1.$$

Hence evaluate approximately

$$\sum_1^\infty (-0{\cdot}9)^{n-1} n^{-\frac{3}{2}}.$$

5. Find the value of $\sum_1^\infty 1/n^2$ by summing the first 15 terms directly and then applying (6.12). Compare the error estimate with that obtained in the text.

6. Prove Gregory's formula

$$\int_{x_0}^{x_n} y\,dx = h\Big[(y_0 + y_1 + \ldots + y_n) - \tfrac{1}{2}(y_0 + y_n) - \tfrac{1}{12}(\Delta y_{n-1} - \Delta y_0) - \tfrac{1}{24}(\Delta^2 y_{n-2} + \Delta^2 y_0)$$
$$- \tfrac{19}{720}(\Delta^3 y_{n-3} - \Delta^3 y_0) - \tfrac{3}{160}(\Delta^4 y_{n-4} + \Delta^4 y_0) - \ldots\Big].$$

(Hint: Start from (6.10).)

7. Find the value of

$$\int_{-1}^{1} e^{-x^2}\,dx,$$

using Gregory's formula.

8. Sum the following series by two different (numerical) methods

$$1 - \frac{1}{2^2} + \frac{1}{3^2} - \frac{1}{4^2} + \ldots,$$

$$\frac{1}{\ln 2} - \frac{1}{\ln 3} + \frac{1}{\ln 4} - \frac{1}{\ln 5} + \ldots,$$

$$1 - \frac{1}{\sqrt{2}} + \frac{1}{\sqrt{3}} - \frac{1}{\sqrt{4}} + \ldots.$$

Give your answers correct to four decimal places.

9. Sum the following series by two different (numerical) methods

$$\sum_{1}^{\infty} \frac{1}{n^{\frac{3}{2}}}, \qquad \sum_{0}^{\infty} \frac{1}{n^4 + \frac{1}{2}}, \qquad \sum_{0}^{\infty} \frac{\ln[(2n+1)\pi/2]}{(2n+1)^2\pi^2 - 4}.$$

Give your answers correct to four decimal places.

10. Let $f(z)$ be an analytic function of z and C a closed contour including the points $m, m+1, \ldots, n$ in which $f(z)$ is regular except for simple poles at a finite number of points $a_1, \ldots, a_k$ with residues $b_1, \ldots, b_k$. Show that

$$\int_C \pi \cot \pi z\, f(z)\,dz = 2\pi i \Big\{ f(m) + f(m+1) + \ldots + f(n)$$
$$+ b_1 \pi \cot \pi a_1 + \ldots + b_k \pi \cot \pi a_k \Big\}.$$

Hence deduce that

$$\sum_{0}^{\infty} \frac{1}{n^4 + a^4} = \frac{1}{2a^4}\left\{1 + \frac{b(\sin 2b + \sinh 2b)}{\cosh 2b - \cos 2b}\right\}, \qquad b = \frac{a\pi}{\sqrt{2}}$$

and verify your numerical results for $\sum_{0}^{\infty} \frac{1}{n^4 + \frac{1}{2}}$ (exercise 9).

11. By integrating $\int_C \frac{\tanh z \log z}{z^2 + 1}\,dz$ round the contour C consisting of the square with vertices $n\pi(\pm 1 \pm i)$ and a cut along the negative real axis with a circular path enclosing the logarithmic singularity show that

$$\int_0^{\infty} \frac{\tanh x}{x^2 + 1}\,dx = 8 \sum_{0}^{\infty} \frac{\ln[(2n+1)\pi/2]}{(2n+1)^2\pi^2 - 4}.$$

Hence verify the result obtained in exercise 9 by using a numerical method to integrate

$$\int_0^\infty \frac{\tanh x}{x^2+1}\,dx.$$

12. Show that

$$\pi z \cot \pi z = 1 + z \sum_{n=1}^{\infty} \left(\frac{1}{z-n} + \frac{1}{z+n}\right).$$

(Hint: Use the result of exercise 10.)

13. Deduce, from (6.14), that

$$z \cot z = 1 - B_2 \frac{(2z)^2}{2!} + B_4 \frac{(2z)^4}{4!} - B_6 \frac{(2z)^6}{6!} + \ldots .$$

14. Use the results of exercises 12 and 13 to prove the assertions (6.7) and to deduce the general result

$$\sum_1^\infty \frac{1}{n^{2k}} = \frac{(-1)^{k+1}(2\pi)^{2k}}{2(2k)!} B_{2k}.$$

15. Show that

$$\sum_{k=1}^{n} k^2 = \tfrac{1}{6} n(n+1)(2n+1).$$

(Hint: Use (6.12) with $\nu(x) = x^2$.) Hence deduce that

$$\sum_{k=1}^{\infty} \frac{1}{1^2 + 2^2 + \ldots + k^2} = \sum_{k=1}^{\infty} \left[\frac{-6}{k(k+1)} + \frac{6}{k(k+\frac{1}{2})}\right],$$

and find the sum of this series.

16. (Salzer's method) Let S_n be the nth partial sum of a slowly convergent series and let $f(1/n) = S_n$. Take $n = 5, 10, 20, 25$ and evaluate $f(1/n)$. The sum $S = f(0)$ can now be found by extrapolation using Lagrange's formula. Use this method to evaluate $\sum_1^\infty (1/n^2)$.

7 Function approximation

7.1 Representation of functions by truncated series. Taylor series and economized polynomials

It is natural for a mathematician to express a real differentiable function $f(x)$, possessing no singularities in some neighbourhood of $x = \alpha$, in the form of a power series and a remainder term $R_n(x)$, viz.,

$$f(x) = a_0 + a_1(x-\alpha) + a_2(x-\alpha)^2 + \ldots + a_{n-1}(x-\alpha)^{n-1} + R_n(x),$$

since this form of statement is central in the theory of classical analysis. In fact we can, from Taylor's theorem, give values to the coefficients a_i and the remainder $R_n(x)$. We have

$$f(x) = f(\alpha) + f'(\alpha)(x-\alpha) + \frac{f''(\alpha)}{2!}(x-\alpha)^2 + \ldots + \frac{f^{(n-1)}(\alpha)}{(n-1)!}(x-\alpha)^{n-1} + R_n(x) \qquad (7.1)$$

where $f^{(k)}(\alpha)$, $(k = 1, 2, \ldots, n-1)$ denotes the kth derivatives of $f(x)$ evaluated at $x = \alpha$, and

$$R_n(x) = \frac{f^{(n)}(\alpha + \theta h)h^n}{n!}, \qquad h = x - \alpha, \qquad 0 < \theta < 1. \qquad (7.2)$$

If $R_n(x)$ tends to zero quickly as n increases, and the coefficients $f^{(k)}(\alpha)$ are readily calculable, the formula (7.1) provides an efficient method for calculating $f(x)$, with only a small truncation error $R_n(x)$. For example, for the function $f(x) = \mathrm{e}^x$ in the range $-1 \leqslant x \leqslant 1$, we have from (7.1) with $\alpha = 0$,

$$\mathrm{e}^x = 1 + x + \frac{x^2}{2!} + \ldots + \frac{x^{10}}{10!} + R_{11}(x). \qquad (7.3)$$

Since $f^{(n)}(x) = \mathrm{e}^x$, all integer $n > 0$,

$$\begin{aligned} |R_{11}(x)| &\leqslant \mathrm{e}^\theta/11!, \qquad 0 < \theta < 1, \\ &< \mathrm{e}/11! < 7 \times 10^{-8}, \end{aligned} \qquad (7.4)$$

and consequently 7×10^{-8} will be the maximum error incurred (in modulus) in omitting $R_{11}(x)$ in (7.3). If we neglect $R_{11}(x)$ we are left with a polynomial of degree 10 and, since e^x is required often in computational problems, it would be helpful to find some method for reducing the degree of our polynomial without introducing any large additional truncation error (cf. section 1.4). This can be done by resorting to the minimax principle (see chapter 4). We replace x^{2r} and x^{2r+1}, $r = 0, 1, 2, \ldots$, in (7.3) by their corresponding Chebyshev series defined by (4.5) and we obtain the Fourier–Chebyshev series

$$e^x = \tfrac{1}{2}a_0 T_0 + a_1 T_1 + \ldots + a_{10}T_{10} + R_{11}(x). \tag{7.5}$$

By virtue of Theorem 4-4 the coefficient a_{10} will be $1/2^9 \times 10!$ and, since $|T_{10}(x)| < 1$, the increase in the truncation error by omitting the term $a_{10}T_{10}$ in (7.5) is less than $0{\cdot}6 \times 10^{-9}$. We can similarly treat the term $a_9 T_9$, etc., and obtain an 'economized' Chebyshev series. Since the rth Chebyshev polynomial T_r is a polynomial of degree r this series can be reconverted to a polynomial of lower degree than (7.3), i.e., we have an 'economized' polynomial.

We can also approach the result (7.5) from a different angle by noting that the coefficients a_r in (7.5) can be determined from the formula

$$a_r = \frac{2}{\pi}\int_{-1}^{1} \frac{f(x)T_r(x)}{\sqrt{(1-x^2)}}\,dx, \tag{7.6}$$

$$= \frac{2}{\pi}\int_{-1}^{1} \frac{e^x T_r(x)}{\sqrt{(1-x^2)}}\,dx, \qquad \text{in our case,}$$

$$= \frac{2}{\pi}\int_{0}^{\pi} e^{\cos\theta} \cos r\theta \, d\theta = 2I_r(1),$$

where $I_r(x)$ is the modified Bessel function of order r.

Note that if our range of x was (0, 1) then we would use the shifted Chebyshev polynomials $T_r^*(x)$ to obtain our economized series/polynomial.

7.2 The minimax approximation

The representation of functions by polynomials, such as curtailed Taylor series or economized polynomials, is in many ways highly desirable, particularly in view of the following theorem of Weierstrass, a proof of which may be found in Titchmarsh [7].

THEOREM 7-1 (*Weierstrass's approximation theorem*) *If $f(x)$ is a continuous function in (a, b), and ϵ is a given positive number, there is a polynomial $p(x)$ such that*

$$|f(x) - p(x)| < \epsilon, \qquad a \leqslant x \leqslant b.$$

Weierstrass's theorem is a pure existence theorem and gives no indication of the order of the polynomial satisfying the approximation. In practice it is usually desirable to choose the degree of the approximating polynomial and to have some a priori estimate of the magnitude of the maximum error in the interval (a, b). The aim of an approximation to a function is to make the maximum error as small as possible – the approximation with the minimum 'maximum error' being termed the minimax approximation.

Many authors (e.g. Ralston [8]) use the term minimax to refer only to rational approximations where the numerator and denominator are polynomials of the same degree. There are advantages in rational approximations, one being that functions with poles can be satisfactorily approximated.

It is not possible here to enlarge on the topic of minimax approximation and we shall concentrate in the rest of the chapter on specific methods suitable for various types of problems. We remark, however, that it follows from Theorem 4-4 that $2^{-n+1}T_n(x)$ is an approximation in the minimax sense to x^n and this provides the basis for desiring approximations in terms of Chebyshev polynomials. Details of the construction of minimax approximations can be found in Ralston's book [8].

7.3 The Lanczos τ-method for summation of power series

In section 7.1 the function under discussion possessed a rapidly convergent Taylor series. Many functions, e.g., $\ln(1 + x)$, $1/(1 + x)$, have slowly convergent Taylor series in $(0, 1)$ (or $(-1, 1)$) and, even if we convert our series into the economized shifted Chebyshev series, slow convergence still persists. This difficulty can be overcome if the function $f(x)$ which we are evaluating in the range $(-1, 1)$ or $(0, 1)$ can be expressed in the form $P(x)/Q(x)$, where $P(x)$ and $Q(x)$ are polynomials in x of degree p and q respectively. If we assume $f(x)$ can be put in the form

$$f(x) = \tfrac{1}{2}a_0T_0 + a_1T_1 + \ldots + a_nT_n, \tag{7.7}$$

then

$$P(x) = Q(x)\,(\tfrac{1}{2}a_0T_0 + a_1T_1 + \ldots + a_nT_n) \tag{7.8}$$

and, since both $P(x)$ and $Q(x)$ can be expressed as finite Chebyshev series, (7.8) can be written as

$$\sum_{r=0}^{p} b_rT_r(x) = \sum_{r=0}^{n+q} c_rT_r(x). \tag{7.9}$$

By equating the coefficients of $T_r(x)$ on both sides we obtain a system of algebraic equations for the determination of the a_r. The number of equations is $n + q + 1$, assuming $n + q > p$, while the number of unknown coefficients is only $n + 1$ so that we cannot, in general, satisfy this system by an approximation such as (7.7). Lanczos [1] gets round this problem by adding terms of the form τ_sT_s $(s > n)$ to the left-hand side of (7.8) so that we now have a modified form of (7.9) and sufficient equations to enable both the a_r and τ_s to be determined uniquely. The error in the computed expression $\frac{1}{2}a_0 T_0 + \sum_1^n a_r T_r$ for $P(x)/Q(x)$ is then

$$(\Sigma \tau_s T_s)/Q(x) < (\Sigma |\tau_s|)/\min Q(x).$$

To illustrate the method, consider the expansion of $1/(1+x)$ in $(0, 1)$, and suppose

$$1/(1+x) = \tfrac{1}{2} a_0 T_0^* + a_1 T_1^* + \ldots + a_n T_n^*.$$

Then we have

$$T_0^* = (1+x)(\tfrac{1}{2}a_0 T_0^* + a_1 T_1^* + \ldots + a_n T_n^*), \tag{7.10}$$

and, substituting in (7.10),

$$xT_0^* = \tfrac{1}{2}(T_0^* + T_1^*),$$
$$xT_k^* = \tfrac{1}{4}(T_{k-1}^* + 2T_k^* + T_{k+1}^*), \qquad k \geqslant 1,$$

we obtain the following equations for a_r by comparison of the coefficients of T_r^* on either side of the equation

$$\left.\begin{aligned} &\tfrac{3}{4}a_0 + \tfrac{1}{4}a_1 = 1, \\ &\tfrac{3}{2}a_r + \tfrac{1}{4}(a_{r-1} + a_{r+1}) = 0, \qquad r = 1, \ldots, n-1, \\ &\tfrac{3}{2}a_n + \tfrac{1}{4}a_{n-1} = 0, \\ &\tfrac{1}{4}a_n = 0. \end{aligned}\right] \tag{7.11}$$

These equations are indeterminate as they stand but, by the introduction of the term τT_{n+1}^* on the left of (7.10) the last equation of (7.11) becomes

$$\tfrac{1}{4}a_n = \tau,$$

and we now have a determinate system of equations. If we take $n = 5$ we find

$$a_5 = 4\tau, \quad a_4 = -24\tau, \quad a_3 = 140\tau, \quad a_2 = -816\tau, \quad a_1 = 4756\tau,$$
$$a_0 = -27\,720\tau,$$

and hence from the first equation of (7.11), $\tau = -1/19\,601$. Our Chebyshev series approximation for $1/(1+x)$ is thus

$$\frac{1}{1+x} \doteq \frac{1}{19\,601}(13\,860 T_0^* - 4756 T_1^* + 816 T_2^* - 140 T_3^* + 24 T_4^* - 4 T_5^*) \tag{7.12}$$

with an error of at most $\frac{1}{19\,601}\,T_6^*(x)/(1+x) < 1/19\,601$. Equation (7.12) can also be expressed in the polynomial form

$$\frac{1}{1+x} \doteqdot \frac{1}{19\,601}(19\,600 - 19\,528x + 18\,688x^2 - 15\,104x^3 + 8192x^4 - 2048x^5) \quad (7.13)$$

with the same error. It is worth noting that (7.13) converges at $x = 1$ while, for this value of x, the Taylor series expansion

$$1/(1+x) = 1 - x + x^2 - \ldots$$

does not converge in the ordinary sense (it does however converge to the correct value $\frac{1}{2}$ in the Cesàro sense).

We assumed in (7.7) a representation for $f(x)$ as a finite linear combination of Chebyshev polynomials. We could equally as well have considered an infinite Chebyshev series. Thus, for example, consider the expansion of the function $1/(1+x)$ as an infinite series of Chebyshev polynomials. The coefficients a_r do not now satisfy all the conditions (7.11) but satisfy instead

$$\left.\begin{aligned} &\tfrac{3}{4}a_0 + \tfrac{1}{4}a_1 = 1, \\ &\tfrac{3}{2}a_r + \tfrac{1}{4}(a_{r-1} + a_{r+1}) = 0, \qquad r \geqslant 1. \end{aligned}\right] \quad (7.14)$$

The difference equation (see chapter 11),

$$\tfrac{3}{2}a_r + \tfrac{1}{4}(a_{r-1} + a_{r+1}) = 0$$

has the solution

$$a_r = A\lambda_1^r + B\lambda_2^r, \quad (7.15)$$

where λ_1, λ_2 are the roots of the quadratic equation

$$\lambda^2 + 6\lambda + 1 = 0,$$

viz., $\lambda_1 = -3 + 2\sqrt{2}$, $\lambda_2 = -3 - 2\sqrt{2}$. In order for the Chebyshev series to be convergent the coefficients a_r defined by (7.15) must tend to zero as r increases. Since $|\lambda_2|^r \to \infty$ as $r \to \infty$, this will not be the case unless the coefficient B is zero. If this condition is satisfied, and also the first equation of (7.14), we find $A = \sqrt{2}$ and our series for $1/(1+x)$ follows, namely

$$\frac{1}{1+x} = \sqrt{2}\,(\tfrac{1}{2}T_0^* + \lambda_1 T_1^* + \lambda_1^2 T_2^* + \lambda_1^3 T_3^* + \ldots). \quad (7.16)$$

Moreover, since $\lambda_1 \doteqdot -0{\cdot}17$, and $|T_n^*| < 1$, we need only take ten terms of the series (7.16) in order to obtain a value of $1/(1+x)$ correct to seven decimal places. Further, since

$$\ln(1+x) = \int_0^x \frac{1}{1+x}\,\mathrm{d}x,$$

we have, by integration of (7.16) and application of the results of chapter 4 (exercise 10),

$$\ln(1+x) = \ln\frac{3+2\sqrt{2}}{4}T_0^* - 2(\lambda_1 T_1^* + \frac{1}{2}\lambda_1^2 T_2^* + \frac{1}{3}\lambda_1^3 T_3^* + \ldots). \tag{7.17}$$

7.4 Asymptotic expansions

In sections, 7.1, 7.3 we were occupied with the expansion of the function $f(x)$ for small values of x. It often happens that we are required to evaluate $f(x)$ for large values of x. This can sometimes be effected by making the substitution $x = 1/u$ and obtaining a Taylor series expansion in u. Thus, to give a simple example, we can evaluate $1/(1+x)$ for large x by evaluating the series in u given by

$$\frac{1}{1+x} = \frac{u}{1+u} = u - u^2 + u^3 - \ldots.$$

In a large number of problems, especially in the evaluation of functions defined by convergent infinite integrals, the function evaluation for large x can be carried out by means of an asymptotic expansion. This is defined for real x by

$$S(x) = A_0 + \frac{A_1}{x} + \frac{A_2}{x^2} + \ldots + \frac{A_n}{x^n} + \ldots \tag{7.18}$$

where, if $S_n(x)$ is the sum of the terms up to A_n/x^n and $R_n(x) = f(x) - S_n(x)$, we require

$$\lim_{|x|\to\infty} x^n R_n(x) = 0 \qquad (n \text{ fixed}). \tag{7.19}$$

If (7.19) is satisfied, we write

$$f(x) \sim S(x). \tag{7.20}$$

To illustrate how we obtain our asymptotic expansion we consider as an example the function $f(x)$ defined by

$$f(x) = \int_0^\infty \frac{e^{-xu}}{1+u}\,du, \qquad x \gg 0.$$

Integration by parts gives

$$f(x) = \frac{1}{x} - \frac{1}{x}\int_0^\infty \frac{e^{-xu}}{(1+u)^2}\,du,$$

$$= \frac{1}{x} - \frac{1}{x^2} + \frac{2!}{x^3} - \ldots + (-1)^{n-1}\frac{(n-1)!}{x^n} + (-1)^n \frac{n!}{x^n}\int_0^\infty \frac{e^{-xu}}{(1+u)^{n+1}}\,du,$$

by repetition of the integration procedure. Denoting the last integral by $R_n(x)$,

$$|R_n(x)| = \frac{n!}{x^n}\left|\int_0^\infty \frac{e^{-xu}}{(1+u)^{n+1}}\,du\right| < \frac{n!}{x^n}\left|\int_0^\infty e^{-xu}\,du\right| = \frac{n!}{x^{n+1}}.$$

Hence

$$\lim_{x\to\infty} |x^n R_n(x)| < \lim_{x\to\infty} |n!/x| = 0, \qquad \text{since } n \text{ is fixed.}$$

The Poincaré condition, (7.19), is satisfied and our asymptotic expansion is

$$f(x) \sim \frac{1}{x} - \frac{1}{x^2} + \frac{2!}{x^3} - \ldots + (-1)^{n-1}\frac{(n-1)!}{x^n} + \ldots.$$

As far as the numerical computation of $f(x)$ is concerned we note that, as n increases, $R_n(x)$ alternates in sign and consequently the error incurred in neglecting this term is always less than $|S_n(x) - S_{n+1}(x)|$. Since the terms in the expansion first decrease and then increase again when $m = n - 1 > x \geqslant n - 2$, the error in our approximation to $f(x)$ is going to be least if we terminate our summation with $S_{m-2}(x)$. Further, $\frac{1}{2}\{S_{m-2}(x) + S_{m-1}(x)\}$ is a still better approximation to $f(x)$ than $S_{m-2}(x)$. (Why?) Thus, in the above example,

$$\begin{aligned} f(10) &\sim \tfrac{1}{2}\{S_{10}(10) + S_{11}(10)\} \\ &\sim \tfrac{1}{2}(0{\cdot}091\,54_6 + 0{\cdot}091\,58_2) = 0{\cdot}091\,56_4. \end{aligned}$$

There is thus an error of at most two units in the fifth decimal place. (The digits at a lower level indicate the next significant digit in the calculation of that term.)

7.5 Iterative methods

Many functions are often best calculated by the repeated use of a single formula. Thus, for example, suppose we want to calculate $1/y$, $(y > 0)$, using a four decimal digit floating point computer similar to that described in chapter 1 but lacking the built-in facility of division (otherwise the exercise is trivial). Then, since

$$\begin{aligned} y &= a\,.\,10^b, \qquad b \text{ an integer}, \qquad 0{\cdot}1000 \leqslant a < 1, \\ 1/y &= (1/a)10^{-b}, \end{aligned}$$

and the problem of evaluating $1/y$ is now reduced to evaluating $1/a$. This can be done by means of Newton's method (section 2.4) applied to the function $f(x) = a - 1/x$. This leads to the iterative formula

$$x_{n+1} = x_n(2 - ax_n), \tag{7.21}$$

and, by taking $x_0 = a$, say, as a first approximation we will eventually arrive at a good approximation to $1/a$ and hence, by suitable scaling, to a value of y^{-1}.

The main snag about this method is the poor (in the sense that a large number of iterations may be required) initial approximation $x_0 = a$. To obtain a better initial approximation it will help us if we examine a diagram of the function $1/a$ in $(0{\cdot}1, 1)$ (Fig. 7.1). One approach that springs to mind is to replace $1/a$ by a polynomial function of low order which passes through selected points of the curve $f(a) = 1/a$. Thus,

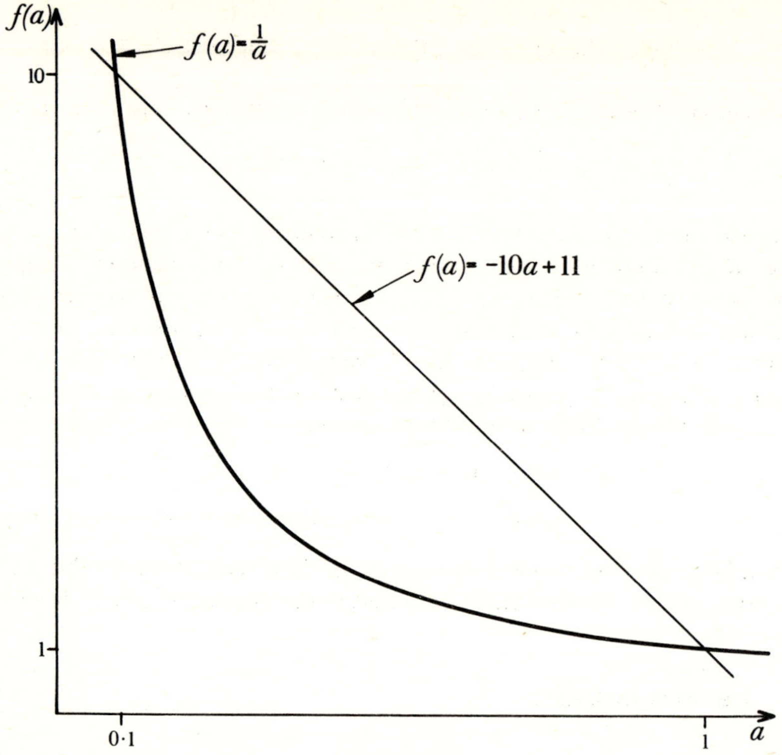

Fig. 7.1 Graph of $1/a$ in $(0{\cdot}1, 1)$

for example, we could take $x_0 = -10a + 11$, which gives the value $1/a$ exactly at the end points of the range of a. This method unfortunately fails for a in the range

$$(11 - \sqrt{41})/20 \leqslant a \leqslant (11 + \sqrt{41})/20,$$

since in this region $2 - ax_0 \leqslant 0$ and our iterative sequence (7.21) only converges (exercise 12) if both $2 - ax_0 > 0$ and $x_0 > 0$. A little thought shows that by dividing up the range $(0{\cdot}1, 1)$ suitably we can obtain, in the manner already suggested a linear approximation for each division which proves a good starting value. For instance a fairly good starting approximation is given by the formula

$$x_0 = \begin{cases} 12 - 20a, & 0{\cdot}1 \leqslant a < 0{\cdot}5 \\ 3 - 2a, & 0{\cdot}5 \leqslant a < 1. \end{cases} \tag{7.22}$$

By use of more sophisticated techniques we can obtain even better starting approximations (exercise 12) but the speed of convergence, as compared with (7.22), is not substantially improved.

7.6 Functions defined by integral formula and differential equations

In a large number of practical problems it is beneficial to use a combination of the techniques described in this chapter in conjunction with methods described in other chapters, particularly chapters 5, 6, and 11. A good example of this is provided by the function

$$f(x) = \int_0^\infty \frac{e^{-u^2}}{u+x}\, du, \tag{7.23}$$

tabulated by Goodwin and Staton [2]. For small values of x, $f(x)$ is readily computed from

$$f(x) + e^{-x^2} \ln x = e^{-x^2}\left[\sqrt{\pi} \sum_0^\infty \frac{x^{2n+1}}{n!(2n+1)} - \sum_1^\infty \frac{x^{2n}}{n!2n} - \frac{1}{2}\gamma\right] \tag{7.24}$$

where γ is Euler's constant 0·577 . . . . For large values of x we can employ the asymptotic expansion

$$f(x) \sim \frac{1}{2}\sqrt{\pi}\left(\frac{1}{x} + \frac{1}{2x^3} + \frac{1.3}{4x^5} + \frac{1.3.5}{8x^7} + \ldots\right) - \frac{1}{2}\left(\frac{1}{x^2} + \frac{1}{x^4} + \frac{2!}{x^6} + \frac{3!}{x^8} + \ldots\right), \tag{7.25}$$

while for intermediate values we can integrate the differential equation (see chapter 11)

$$f' + 2xf = \sqrt{\pi} - (1/x). \tag{7.26}$$

Note that $f(x)$ has a logarithmic singularity at $x = 0$ which is removed in (7.24) by tabulating $f(x) + e^{-x^2} \ln x$.

Finally, we give an example which illustrates how Laplace transforms can be employed to express a function in a form more suitable for computational purposes. The function in question, $I(t)$, defined by

$$I(t) = \int_0^\infty \frac{x^{\frac{1}{2}} \sin tx}{1+x}\, dx, \tag{7.27}$$

is to be evaluated for $t > 0$. With the usual bar notation for Laplace transforms,

$$\begin{aligned}\bar{I}(s) &= \int_0^\infty e^{-st} I(t)\, dt,\\ &= \int_0^\infty \left(\int_0^\infty \frac{x^{\frac{1}{2}} \sin tx}{1+x}\, dx\right) e^{-st}\, dt,\\ &= \int_0^\infty \frac{x^{\frac{3}{2}}}{(1+x)(s^2+x^2)}\, dx,\end{aligned}$$

inversion of the order of integration being permissible if we assume $Rs > 0$. This integral can be evaluated by using the known result (Whittaker and Watson [3], p. 118)

$$\int_0^\infty x^{a-1} Q(x)\,dx = \pi \operatorname{cosec}(a\pi) \sum \text{(residues of } Q(x)),$$

or, alternatively, making the substitution $x = u^2$ and integrating round a standard semicircular contour. Both methods give

$$\bar{I}(s) = \frac{\pi}{s^2+1} + \frac{\pi}{\sqrt{2}} \frac{s^{\frac{1}{2}}(s-1)}{s^2+1}. \tag{7.28}$$

In (7.28), let $\bar{f}(s) = s^{\frac{1}{2}}(s-1)/(s^2+1)$. We can rewrite $\bar{f}(s)$ in the form

$$\bar{f}(s) = s^{-\frac{1}{2}} - s^{-\frac{1}{2}}\left\{\frac{s}{s^2+1} + \frac{1}{s^2+1}\right\}.$$

The second term on the right-hand side is the product of the transforms of $1/\sqrt{(\pi t)}$ and $(\cos t + \sin t)$. This implies, by the convolution theorem for transforms,* that $f(t)$ (the inverse transform of $\bar{f}(s)$) satisfies

$$f(t) = \frac{1}{\sqrt{(\pi t)}} - \int_0^t \{\cos(t-u) + \sin(t-u)\}/\sqrt{(\pi u)}\,du.$$

This simplifies to

$$f(t) = \frac{1}{\sqrt{(\pi t)}} - \frac{2}{\sqrt{\pi}}\{(\cos t + \sin t)C(\sqrt{t}) + (\sin t - \cos t)S(\sqrt{t})\}, \tag{7.29}$$

where $C(t)$, $S(t)$ are the Fresnel integrals

$$C(t) = \int_0^t \cos(u^2)\,du, \qquad S(t) = \int_0^t \sin(u^2)\,du.$$

Since, from (7.28), $I(t) = \pi \sin t + (\pi/\sqrt{2}) f(t)$, we obtain

$$I(t) = \pi \sin t + \sqrt{(\pi/2t)} - \sqrt{(2\pi)}\{(\cos t + \sin t)C(\sqrt{t}) + (\sin t - \cos t)S(\sqrt{t})\}. \tag{7.30}$$

*This states: If $\bar{f}_1(s)$ is the Laplace transform of $f_1(t)$ and $\bar{f}_2(s)$ is the Laplace transform of $f_2(t)$ then $\bar{f}_1(s)\bar{f}_2(s)$ is the Laplace transform of $\int_0^t f_1(t-u) f_2(u)\,du$.

Equation (7.30) is a sum of functions in a form very amenable to computation. We note $\sin t$, $\cos t$, $t^{-\frac{1}{2}}$ are elementary standard functions and, for small t, $C(\sqrt{t})$ and $S(\sqrt{t})$ can be evaluated by a series expansion. For large t,

$$\begin{matrix} C \\ S \end{matrix}(\sqrt{t}) = \int_0^\infty \begin{matrix} \cos \\ \sin \end{matrix} (u^2)\, du - \int_{\sqrt{t}}^\infty \begin{matrix} \cos \\ \sin \end{matrix} (u^2)\, du = \frac{1}{2}\sqrt{\left(\frac{\pi}{2}\right)} - \int_{\sqrt{t}}^\infty \begin{matrix} \cos \\ \sin \end{matrix} (u^2)\, du,$$

and we can find asymptotic expansions for

$$\int_{\sqrt{t}}^\infty \sin(u^2)\, du, \qquad \int_{\sqrt{t}}^\infty \cos(u^2)\, du.$$

In particular, we now evaluate $I(t)$ from the formula

$$I(t) = \sqrt{(\pi/2t)} + \sqrt{(2\pi)}\left\{(\cos t + \sin t) \int_{\sqrt{t}}^\infty \cos(u^2)\, du + (\sin t - \cos t) \int_{\sqrt{t}}^\infty \sin(u^2)\, du\right\}.$$

7.7 Summary and notes

We have concentrated on the basic standard methods of function approximation in this chapter although there has been a certain amount of digression in section 7.5. It is sometimes desirable to express functions in other ways such as a ratio of polynomial functions, a continued fraction, a Fourier series, a sum of exponentials, or a spline function and the reader is referred to Lanczos [1], Walsh [4], Fröberg [5], and Barrowdale and Young [6] for more details of these methods and for further references.

Again, no mention has been made here about the approximation of functions which are given as discrete data. Some ideas on this problem can be gleaned by studying the methods of interpolation (chapter 3), and of least squares (chapter 13).

REFERENCES

[1] Lanczos, C., *Applied Analysis*, Pitman, London (1957).

[2] Goodwin, E. T. and Staton, J., 'Table of $\int_0^\infty \frac{e^{-u^2}}{u+x}\, du$', *Quart. J. Mech.*, **1** (1948) 319–26.

[3] Whittaker, E. T. and Watson, G. N., *A Course of Modern Analysis*, 4th edn., Cambridge University Press, London (1962).

[4] Walsh, J. (ed), *Numerical Analysis: An Introduction*, Academic Press, London (1966).
[5] Fröberg, C. E., *Introduction to Numerical Analysis*, Addison–Wesley, Reading, Mass. (1965).
[6] Barrowdale, I. and Young, A., 'A Note on Numerical Procedures for Approximation by Spline Functions', *Computer J.*, **9** (1966) 318–20.
[7] Titchmarsh, E. C., *The Theory of Functions*, 2nd edn., Oxford University Press, London (1939).
[8] Ralston, A., *A First Course in Numerical Analysis*, McGraw-Hill Book Co. Ltd., New York (1965).

Exercises

1. Obtain a series expansion for $(1-x)^{-\frac{1}{2}}$, valid for $-1 < x < 1$, and estimate the sum of the series when $x = 1/50$ (to seven decimal places). Hence deduce the value of $\sqrt{2}$ correct to six decimal places.

2. Find a series expansion for $\log\{(1+x)/(1-x)\}$. Hence deduce a formula for $\log(n/m)$, where n and m are integers such that $n > m$. Given $\ln 2 = 0{\cdot}693\,147$, use the above result to evaluate ln 3.

3. Show that

$$\operatorname{erf} x = \frac{2}{\sqrt{\pi}}\int_0^x e^{-u^2}\,du,$$

$$= \frac{2}{\sqrt{\pi}}\left(x - \frac{1}{3}x^3 + \frac{1}{2!5}x^5 - \ldots + (-1)^n \frac{x^{2n+1}}{n!(2n+1)} + \ldots\right).$$

4. Show that

$$\int_x^\infty e^{-t^2}\,dt = \frac{\sqrt{\pi}}{2}(1 - \operatorname{erf} x).$$

5. Prove that

$$\int_0^1 x^k \log x\,dx = -\frac{1}{(k+1)^2}, \qquad (k = 0, 1, 2, \ldots).$$

Hence, by expanding $\exp(-tx)$ as a power series, evaluate

$$\int_0^1 e^{-tx}\log x\,dx, \qquad t = 0(0{\cdot}1)1.$$

6. Given a table of values of e^x for $x = 0(0{\cdot}1)5$, to four decimal places, use the result $e^x = e^{x_0+h} = e^{x_0}e^h$ to obtain a formula for evaluating e^x at non-tabular points. Hence estimate the value of $\exp 3{\cdot}25$. Give error bounds for your answer.

7. Find series expansions for

$$C(t) = \int_0^t \cos(u^2)\,du, \qquad S(t) = \int_0^t \sin(u^2)\,du.$$

8. Evaluate the function $I(t)$ defined by (7.27) for $t = 1(0{\cdot}1)2$.

9. Find, by the Lanczos τ-method, an expansion of the form

$$f(x) = \tfrac{1}{2}a_0T_0 + a_1T_1 + \ldots + a_nT_n \qquad (n \leqslant 10)$$

for the function $f(x) = 1/(1 + x^2)$.

10. Find an infinite Chebyshev expansion for the function $f(x)$ of the previous exercise and hence deduce an expansion for $\tan^{-1}x$.

11. Show that

$$\cos \pi x = \tfrac{1}{2}a_0 + \sum_{r=1} a_{2r}T_{2r}(x), \qquad (-1 \leqslant x \leqslant 1),$$

where $$a_{2r} = (-1)^r 2J_{2r}(\pi),$$

and $J_{2r}(x)$ represents the Bessel function of the first kind of order $2r$ (see exercise 21).

12. Show that the result of exercise 22, chapter 13, is a suitable first approximation to $1/a$ when $\frac{1}{2} \leqslant a \leqslant 1$ (in the sense that the iteration formula (7.21) converges). Show how this can be used to evaluate $1/x$ on a t-digit binary floating point computer. Show that an initial approximation, x_0, must satisfy $x_0 > 0$, $2 - ax_0 > 0$ for convergence.

13. By considering Newton's formula (2.13) applied to the function $f(x) = a - (1/x^2)$, $a > 0$, show that

$$x_{n+1} = \tfrac{1}{2}x_n(3 - ax_n^2)$$

is an iterative formula which converges to $1/\sqrt{a}$ provided that $0 < x_0 < \sqrt{(3/a)}$. How could you use this formula to evaluate $\sqrt{a}$?

14. Prove that

$$\int_x^\infty e^{-\frac{1}{2}t^2}\,dt \sim e^{-\frac{1}{2}x^2}\left(\frac{1}{x} - \frac{1}{x^3} + \frac{3}{x^5} - \ldots + (-1)^n \frac{(2n)!}{2^n n!}\frac{1}{x^{2n+1}} + \ldots\right).$$

Find also an asymptotic expansion for $\int_x^\infty e^{-t^2}\,dt$ and evaluate this latter integral when $x = 5$.

15. Derive asymptotic expansions for the Fresnel integrals

$$\int_t^\infty \cos(u^2)\,du, \qquad \int_t^\infty \sin(u^2)\,du.$$

16. Obtain a series expansion for

$$\int_0^x \frac{\sin t}{t}\,dt,$$

and an asymptotic expansion for

$$\int_x^\infty \frac{\sin t}{t}\,dt.$$

Show that

$$\int_x^\infty \frac{\sin t}{t}\,dt = \frac{\pi}{2} - \int_0^x \frac{\sin t}{t}\,dt.$$

17. The exponential integral ei(z) is defined by

$$\text{ei}(z) = \int_z^\infty (e^{-u}/u)\,du.$$

Prove that

$$\text{ei}(z) = -\gamma - \ln z + z - \frac{z^2}{2.2!} + \frac{z^3}{3.3!} - \ldots + (-1)^{n-1}\frac{z^n}{n.n!} + \ldots,$$

where γ is Euler's constant (=0·577 215 665 . . .).

18. Show that the asymptotic expansion of ei(z) is

$$\text{ei}(z) \sim e^{-z}\left(\frac{1}{z} - \frac{1}{z^2} + \frac{2!}{z^3} - \frac{3!}{z^4} + \ldots + (-1)^n \frac{n!}{z^{n+1}} + \ldots\right).$$

Compare the values of ei(5) obtained using the asymptotic expansion and the series expansion in exercise 17.

19. Define $\quad z! = \Gamma(z+1) = \int_0^\infty e^{-t} t^z\,dt \quad$ (the gamma or factorial function)

and $\quad \psi(z) = \frac{d}{dz}(\log z!) \quad$ (the psi or digamma finction).

Prove that

$$\psi'(z) = \int_0^\infty \tfrac{1}{2}u\,e^{-uz}(\coth \tfrac{1}{2}u - 1)\,du$$

and $$\psi'(z) \sim \frac{1}{z} - \frac{1}{2z^2} + \sum_{m=1} \frac{(-1)^{m-1}B_{2m}}{z^{2m+1}}$$

where B_{2m} denotes the mth Bernoulli number (see chapter 6). Hence deduce that

$$\psi(z) \sim \log z + \frac{1}{2z} + \sum_{m=1} \frac{(-1)^m B_{2m}}{2m z^{2m}}$$

and Stirling's asymptotic formula

$$z! \sim \sqrt{(2\pi)}\, z^{z+\frac{1}{2}} e^{-z} \left(1 + \frac{1}{12z} + \frac{1}{288z^2} - \frac{139}{51\,840z^3} + \ldots\right).$$

20. Given that ln 10 = 2·302 585 092 994 use the asymptotic expansion for $\psi(z)$ (previous exercise) to find $\psi(10)$ correct to 12 decimal places. Hence deduce that

$$\gamma \doteqdot 0{\cdot}577\,215\,664\,90.$$

(Hint: Re-define $\psi(z)$ by means of Gauss's definition of $z!$,

$$z! = \lim_{n\to\infty} \frac{n!n^z}{(z+1)\ldots(z+n)}.\Big)$$

21. The Bessel function of the first kind of order p is denoted by $J_p(x)$ and satisfies the differential equation

$$x^2y'' + xy' + (x^2 - p^2)y = 0, \qquad (x > 0).$$

Show that

$$J_p(x) = \sum_{n=0}^{\infty} \frac{(-1)^n}{\Gamma(n+1)\Gamma(n+p+1)} \left(\frac{x}{2}\right)^{2n+p}.$$

Tabulate $J_p(x)$ for $p = 0(0{\cdot}1)1$ and $x = 0(0{\cdot}5)10$.

22. Find the first four values of x for which the Bessel function $J_0(x)$ is equal to zero.

23. Evalulate by some method of numerical integration

$$I_n = \int_{n\pi/t}^{(n+1)\pi/t} \frac{x^{\frac{1}{2}} \sin tx}{1+x}\,dx \qquad (n = 0, 1, \ldots, 9;\ t = 1(0{\cdot}1)2).$$

Show that successive terms I_n and I_{n+1} are of different sign and

$$I(t) = \sum_{n=0}^{\infty} I_n$$

where $I(t)$ is defined by (7.27). By using a method for the summation of alternating series use the above results to estimate $I(t)$, $t = 1(0{\cdot}1)2$, and compare your results with those previously obtained in exercise 8. Use this method to evaluate

$$\int_0^{\infty} \frac{x^{\beta}}{1+x^{2\beta}} \sin tx\,dx, \qquad \beta = 0{\cdot}2(0{\cdot}2)1,\ t = 1(0{\cdot}1)2.$$

24. Evaluate

$$\int_0^x e^{-t^2}\,dt \qquad \text{for } x = 3(0{\cdot}1)5,$$

by using some numerical method (see chapter 11) to solve the differential equation

$$y' = e^{-x^2}, \qquad y(5) = \tfrac{1}{2}\sqrt{\pi} - \int_5^\infty e^{-t^2}\,dt,$$

(see exercise 14).

8 Direct methods for the solution of simultaneous linear equations

8.1 Notation and summary of basic matrix properties

We shall write $A = (a_{ij})$ to denote a matrix whose (i, j) element is a_{ij}. The ith row of A will be denoted by the vector $\mathbf{r}_i(A)$ and the jth column by $\mathbf{c}_j(A)$. With exact arithmetic matrix addition is commutative and associative. That is, if A, B, and C have the same order,

$$\begin{aligned} A + B &= B + A \\ A + (B + C) &= (A + B) + C. \end{aligned} \tag{8.1}$$

The product of a matrix A and a scalar k is denoted by

$$kA = (ka_{ij}). \tag{8.2}$$

If the order of A is $(m \times n)$ and the order of B is $(n \times p)$, the product matrix $X = AB$ of order $(m \times p)$ can be formed where

$$x_{ij} = \mathbf{r}_i(A)\mathbf{c}_j(B) = \sum_{k=1}^{n} a_{ik} b_{kj}. \tag{8.3}$$

Assuming that the products exist, matrix multiplication satisfies the associative and distributive laws

$$\begin{aligned} A(BC) &= (AB)C \\ (A + B)C &= AC + BC \\ A(B + C) &= AB + AC. \end{aligned} \tag{8.4}$$

Since arithmetic operations on a computer are inexact (chapter 1) these laws only hold approximately in practice.

The transpose of the matrix A, denoted by A', satisfies

$$A' = (a_{ji}) \tag{8.5}$$

or

$$\mathbf{r}_i(A') = \mathbf{c}_i(A)'. \tag{8.6}$$

If $A' = A$ the matrix A is termed symmetric. We prove the following useful theorem on the transpose of a product.

THEOREM 8-1 $\quad (AB)' = B'A' \qquad (8.7)$

Proof The (i, j) element of $(AB)'$
$$\begin{aligned} &= (j, i) \text{ element of } AB, \\ &= \mathbf{r}_j(A)\mathbf{c}_i(B), \\ &= \mathbf{c}_i(B)'\mathbf{r}_j(A)', \qquad \text{(Why?)} \\ &= \mathbf{r}_i(B')\mathbf{c}_j(A') \\ &= (i, j) \text{ element of } B'A'. \end{aligned}$$

Also $(AB)'$ and $B'A'$ are of the same order. Thus

$$(AB)' = B'A'.$$

We shall be concerned mainly with $(n \times n)$ or square matrices. Diagonal matrices, apart from the unit matrix I and the zero matrix O, will be denoted by D or diag $(d_1, \ldots, d_n)$ where d_i denotes the diagonal element in the ith row and column. Upper triangular matrices will be denoted by U and lower triangular matrices by L. If all the diagonal elements of an upper (or lower) triangular matrix are zero the matrix will be called strictly upper (or lower) triangular.

A square matrix A is called singular if

$$\det A = 0, \tag{8.8}$$

and non-singular if $\det A \neq 0$. Other properties of determinants which we shall use are

$$\det A' = \det A \tag{8.9}$$

$$\det (AB) = \det A \det B. \tag{8.10}$$

We note that $\det A = 0$ if, and only if, there exist numbers $k_1, \ldots, k_n$ (not all zero) such that

$$k_1\mathbf{r}_1(A) + \ldots + k_n\mathbf{r}_n(A) = \boldsymbol{\sigma} \qquad \text{(the zero vector).} \tag{8.11}$$

If $\det A \neq 0$ then the linear equations $A\mathbf{x} = \mathbf{b}$ have a unique solution

$$\mathbf{x} = A^{-1}\mathbf{b} \tag{8.12}$$

although we must emphasize that, to compute $\mathbf{x}$, it is not necessary to compute A^{-1}. Some useful results on inverses are

$$(A^{-1})' = (A')^{-1} \tag{8.13}$$

and

$$(AB)^{-1} = B^{-1}A^{-1}. \tag{8.14}$$

We also remark that a matrix A is positive definite if $\mathbf{x}'A\mathbf{x} > 0$ for all $\mathbf{x} \neq \boldsymbol{\sigma}$ or, alternatively, if the leading minors of A are strictly positive.

Proofs of the above results can be found in references [1] and [2].

Finally, for large matrices, it is often convenient to partition the matrix into smaller matrices, called submatrices, as in the following example. The partitioning is carried out by drawing horizontal and vertical lines between rows and columns. The two given partitioned matrices are multiplied together by treating each submatrix as an element.

$$\left[\begin{array}{cc|cc|c} 2 & 3 & 1 & 0 & 0 \\ 3 & -4 & 0 & 1 & 0 \\ \hline 4 & 2 & 3 & 1 & 5 \\ 1 & -3 & 2 & -1 & -2 \end{array}\right] \left[\begin{array}{c|cc} 2 & 1 & 0 \\ -1 & 0 & 1 \\ \hline 0 & 5 & 0 \\ -3 & 0 & 2 \\ \hline 3 & -2 & -1 \end{array}\right] = \left[\begin{array}{c|cc} 1 & 7 & 3 \\ 7 & 3 & -2 \\ \hline 18 & 9 & -1 \\ 2 & 15 & -3 \end{array}\right]$$

The multiplication in partitioned form can be summarized in the equation

$$\begin{bmatrix} A_{11} & A_{12} & A_{13} \\ A_{21} & A_{22} & A_{23} \end{bmatrix} \begin{bmatrix} B_{11} & B_{12} \\ B_{21} & B_{22} \\ B_{31} & B_{32} \end{bmatrix} = \begin{bmatrix} A_{11}B_{11} + A_{12}B_{21} + A_{13}B_{31}, & A_{11}B_{12} + A_{12}B_{22} + A_{13}B_{32} \\ A_{21}B_{11} + A_{22}B_{21} + A_{23}B_{31}, & A_{21}B_{12} + A_{22}B_{22} + A_{23}B_{32} \end{bmatrix}$$

Note that, for the multiplication to be possible, the columns of the first matrix and the rows of the second matrix must be partitioned in the same way.

8.2 The solution of simultaneous linear equations by Gaussian elimination

Let the equations to be solved be

$$A\mathbf{x} = \mathbf{b},$$

where A is a $(n \times n)$ matrix. If A is non-singular we have a unique solution

$$\mathbf{x} = A^{-1}\mathbf{b}, \tag{8.15}$$

which could be determined by calculating A^{-1} directly, but this is usually a laborious method of solution. The two methods considered in this chapter, *Gaussian elimination* and *triangular resolution*, are direct ones, i.e., they do not involve successive approximations (cf. chapter 10). They are the methods normally used in computer programmes and are also good for desk machines. Both methods involve reduction of the equations to triangular form, i.e., a form in which the matrix of coefficients A is triangular. For example, if A is an upper triangular matrix, the equations $A\mathbf{x} = \mathbf{b}$ are

$$\left.\begin{aligned} a_{11}x_1 + a_{12}x_2 + a_{13}x_3 + \ldots + a_{1,n-1}x_{n-1} + a_{1n}x_n &= b_1' \\ a_{22}x_2 + a_{23}x_3 + \ldots + a_{2,n-1}x_{n-1} + a_{2n}x_n &= b_2 \\ a_{33}x_3 + \ldots + a_{3,n-1}x_{n-1} + a_{3n}x_n &= b_3 \\ \cdot \quad \cdot \quad \cdot \quad \cdot \quad \cdot \quad \cdot \quad \cdot \quad \cdot \quad \cdot \quad & \cdot \\ a_{n-1,\,n-1}x_{n-1} + a_{n-1,\,n}x_n &= b_{n-1} \\ a_{nn}x_n &= b_n \end{aligned}\right] . \tag{8.16}$$

These are solved in the obvious way, computing first x_n, then x_{n-1}, then x_{n-2}, etc. This process is known as a *back-substitution.*

We consider first Gaussian elimination without selection of pivots and we illustrate the process by solving four equations in four unknowns, viz.:

$$\left.\begin{aligned} (\alpha)\quad a_{11}x_1 + a_{12}x_2 + a_{13}x_3 + a_{14}x_4 &= b_1 \\ (\beta)\quad a_{21}x_1 + a_{22}x_2 + a_{23}x_3 + a_{24}x_4 &= b_2 \\ (\gamma)\quad a_{31}x_1 + a_{32}x_2 + a_{33}x_3 + a_{34}x_4 &= b_3 \\ (\delta)\quad a_{41}x_1 + a_{42}x_2 + a_{43}x_3 + a_{44}x_4 &= b_4 \end{aligned}\right] . \tag{8.17}$$

Stage 1. We add suitable multiples of equation (8.17α) to equations (β), (γ), and (δ) to reduce the coefficients of x_1 in them to zero. In the general case, if m_{k1} is the multiple of the first equation to be added to the kth equation, we have $m_{k1} = -a_{k1}/a_{11}$. Equation (α), which remains unaltered, is called the *pivotal* equation. The coefficient of the unknown being eliminated in this equation is called the pivot which, at this stage, is a_{11}. Let the modified equations (β), (γ), and (δ) at this stage be

$$\left.\begin{aligned} (\beta')\quad a_{22}^{(1)}x_2 + a_{23}^{(1)}x_3 + a_{24}^{(1)}x_4 &= b_2^{(1)} \\ (\gamma')\quad a_{32}^{(1)}x_2 + a_{33}^{(1)}x_3 + a_{34}^{(1)}x_4 &= b_3^{(1)} \\ (\delta')\quad a_{42}^{(1)}x_2 + a_{43}^{(1)}x_3 + a_{44}^{(1)}x_4 &= b_4^{(1)} \end{aligned}\right] .$$

Stage 2. This is similar to stage 1 except that now (β′) is the pivotal equation and x_2 is the variable being eliminated. In the general case we multiply the equation (β′) by $m_{k2} = -a_{k2}^{(1)}/a_{22}^{(1)}$ and add to the kth equation to eliminate x_2 in that equation. After this stage (γ′) and (δ′) are reduced to

$$\begin{aligned} (\gamma'')\quad a_{33}^{(2)}x_3 + a_{34}^{(2)}x_4 &= b_3^{(2)} \\ (\delta'')\quad a_{43}^{(2)}x_3 + a_{44}^{(2)}x_4 &= b_4^{(2)} \end{aligned}$$

Stage 3. This reduces (δ'') to (δ''') by eliminating x_3. In the general case we have an obvious extension and increase in the number of stages, reducing eventually to an equation $a_{nn}^{(n-1)}x_n = b_n^{(n-1)}$

$$(\delta''')\quad a_{44}^{(3)}x_4 = b_4^{(3)}.$$

Stage 4. Equations (α), (β'), (γ''), (δ''') are triangular in form and can be solved by a back-substitution.

Now suppose we apply the above method to solve the equations

$$\left.\begin{aligned} x_1 + x_2 + x_3 + x_4 &= 4 \\ x_1 + x_2 + 3x_3 + 2x_4 &= 7 \\ x_1 + 2x_2 + 4x_3 + 3x_4 &= 10 \\ 2x_1 + 3x_2 + 4x_3 + 5x_4 &= 14 \end{aligned}\right] \tag{8.18}$$

For this case the method we described breaks down at stage 2 because the pivot is zero and the multipliers are therefore infinite. It is not necessary, however, to use the first available equation as the pivotal one and, in practice, it is found advisable to select pivots in one of two ways both of which overcome the difficulty of infinite multipliers.

(a) Partial pivoting

In this method we eliminate the unknowns in order, starting with x_1, but at each stage we select as the pivotal equation that equation having the largest coefficient (in modulus) of the unknown being eliminated. Thus, in the solution of (8.18) by partial pivoting, we would select the last equation as the pivotal equation at the first stage since this equation has the largest coefficient associated with x_1 (namely, $a_{41} = 2$).

(b) Complete pivoting

In this method we select as the pivotal equation the equation containing the largest coefficient (in modulus) and we proceed to eliminate the variable associated with this pivot. Thus, in the solution of (8.18) by complete pivoting, we would select the last equation as the pivotal equation since the largest coefficient overall is $a_{44} = 5$ and we would proceed to eliminate x_4 from the remaining equations. Similarly at the next stage, and so on.

These procedures not only avoid zero pivots, but also ensure that all multipliers are numerically less than one, thus avoiding excessive accumulation of rounding errors (see section 1.5). This is irrelevant in the case of the solution of (8.18) where the answers are exact ($x_1 = x_2 = x_3 = x_4 = 1$) but is relevant in the solution of the equations (8.19) where the errors are disastrous if the basic method is used although none of the pivots actually vanish. The equations we require to solve are

$$\left.\begin{aligned} 197x_1 + 305x_2 - 206x_3 - 840x_4 &= 136 \\ 46{\cdot}8x_1 + 71{\cdot}3x_2 - 47{\cdot}4x_3 + 52{\cdot}0x_4 &= 11{\cdot}7 \\ 88{\cdot}6x_1 + 76{\cdot}4x_2 - 10{\cdot}8x_3 + 802x_4 &= 25{\cdot}1 \\ 1{\cdot}45x_1 + 5{\cdot}90x_2 + 6{\cdot}13x_3 + 36{\cdot}5x_4 &= 6{\cdot}60 \end{aligned}\right] . \tag{8.19}$$

We solve these equations by complete pivoting, after first carrying out a process of normalization by which we mean, in this context, reducing the coefficients in each row and column of the matrix to a number which is less (in modulus) than one. If we do not normalize, pivoting may not limit errors effectively (see section 8.5).

The calculations have been performed in Table 8.1. The column of row sums is for checking purposes – we are merely using the fact that the same operations carried out on the row sums as on the rows should (apart from a small rounding error perhaps) give the row sums of the next stage. Another point worth noting is that each pivot, apart from the last, is used more than once as a divisor and, since division usually takes much longer than multiplication, it is thus worthwhile to evaluate the reciprocals of the pivots initially and then to use these as multipliers.

The reduction of the equations $A\mathbf{x} = \mathbf{b}$ to upper triangular form $U\mathbf{x} = \mathbf{y}$ by Gaussian elimination (without pivoting) can be expressed conveniently in matrix

Table 8.1 Solution of equations by complete pivoting (pivots are underlined)

Equation	*Multiplier*	x_1	x_2	x_3	$x'_4(=10x)$	*Right-hand side*	*Row sum*
(α)	−0·2223	0·197	0·305	−0·206	−0·084	0·136	0·348
(β)	−0·5282	0·468	0·713	−0·474	0·052	0·117	0·876
(γ)		<u>0·886</u>	0·764	−0·108	0·802	0·251	2·595
(δ)	−0·1637	0·145	0·590	0·613	0·365	0·660	2·373
(α′)	0·2886		0·1352	−0·1820	−0·2623	0·0802	−0·2289
(β′)	0·6612		0·3095	−0·4170	−0·3716	−0·0156	−0·4947
(δ′)			0·4649	<u>0·6307</u>	0·2337	0·6189	1·9482
(α″)	−0·4367		0·2694		−0·1949	0·2588	0·3333
(β″)			<u>0·6169</u>		−0·2171	0·3936	0·7934
(α‴)					<u>−0·1001</u>	0·0869	−0·0132

Solution: $x_1 = 0{\cdot}9111$, $x_2 = 0{\cdot}3326$, $x_3 = 1{\cdot}0578$, $x_4 = \frac{1}{10}x'_4 = -0{\cdot}0868$.

form. Let $A = A_0$, $\mathbf{b} = \mathbf{b}_0$ and let the equations at the end of stage k be $A_k\mathbf{x} = \mathbf{b}_k$ where $A_k = (a_{ij}^{(k)})$ and $\mathbf{b}_k = (b_i^{(k)})$. A_k is derived from A_{k-1} by the operations

$$\mathbf{r}_i(A_k) = \mathbf{r}_i(A_{k-1}) \qquad (i \leqslant k)$$

$$\mathbf{r}_i(A_k) = \mathbf{r}_i(A_{k-1}) + m_{ik}\mathbf{r}_k(A_{k-1}) \qquad (i > k)$$

where
$$m_{ik} = -a_{ik}^{(k-1)}/a_{kk}^{(k-1)}$$

is the multiple of the pivotal equation to be added to the ith equation. These equations can be written in the form

$$A_k = M_k A_{k-1},$$

where

$$M_k = \begin{bmatrix} 1 & & & & & \\ & \cdot & & & & \\ & & \cdot & & & \\ & & & 1 & & \\ & & & m_{k+1,k} & \cdot & \\ & & & \vdots & & \cdot \\ & & & m_{nk} & & & 1 \end{bmatrix}$$

Combining the $(n-1)$ stages we have

$$U = A_{n-1} = M_{n-1}M_{n-2} \ldots M_2 M_1 A_0$$

or
$$A_0 = A = LU \qquad (8.20)$$

where
$$L = (M_{n-1} \ldots M_1)^{-1} = M_1^{-1} M_2^{-1} \ldots M_{n-1}^{-1}.$$

Since M_k^{-1} is of the same form as M_k with the non-diagonal elements reversed in sign it follows that L can be written in the form

$$L = \begin{bmatrix} 1 & & & & \\ -m_{21} & 1 & & & \\ -m_{31} & -m_{32} & 1 & & \\ \vdots & \vdots & \vdots & \ddots & \\ -m_{n1} & -m_{n2} & \cdot & & 1 \end{bmatrix}$$

In the same way we can show that

$$\mathbf{b} = L\mathbf{y}. \qquad (8.21)$$

8.3 Triangular resolution

In the last section we saw that the solution of $A\mathbf{x} = \mathbf{b}$ by Gauss elimination was equivalent to the steps

$$A = LU \tag{8.22}$$

$$L\mathbf{y} = \mathbf{b} \tag{8.23}$$

$$U\mathbf{x} = \mathbf{y}\,. \tag{8.24}$$

In the method of triangular resolution the given equations are solved by first finding L and U from (8.22) then solving (8.23) for $\mathbf{y}$ and finally (8.24) for $\mathbf{x}$. The 'factorization' (8.22) is not unique but can be made so by choosing the diagonal elements of L to be unity. We can then write

$$L = L_0 + I, \qquad U = U_0 + D$$

where L_0 and U_0 are respectively strictly lower and upper triangular matrices, I is the unit matrix and $D = \text{diag}\,(u_{ii})$ is a non-singular diagonal matrix. It follows that

$$A = LU = L_0U + U = L_0U_0 + L_0D + U$$

i.e.,

$$U + L_0D = A - L_0U_0\,. \tag{8.25}$$

Equating the (i, j) elements in (8.25) gives

$$\left.\begin{aligned} u_{ij} &= a_{ij} - \mathbf{r}_i(L_0)\mathbf{c}_j(U_0) \qquad (i \leqslant j) \\ \text{and} \qquad u_{jj}l_{ij} &= a_{ij} - \mathbf{r}_i(L_0)\mathbf{c}_j(U_0) \qquad (i > j) \end{aligned}\right]. \tag{8.26}$$

Equation (8.26) enables us to obtain L and U systematically, calculating a row of U and then a row of L, noting that the first row of U is the same as that of A. The method clearly fails if one of the elements u_{jj} is zero but this difficulty can be overcome by using permutation matrices (exercise 12 and Wilkinson [3]). Computation can be eased by noting that if $\mathbf{b}$ and $\mathbf{y}$ are regarded as the $(n + 1)$th columns of A and U respectively then the calculation of $\mathbf{y}$ can be included in that of U. Again, for checking purposes – particularly with desk calculators – it is advisable to record the column vectors $\mathbf{s}$ and $\boldsymbol{\sigma}$ consisting of the row sums of A and U respectively. Our calculations will be correct if, after calculating the ith row of L we have

$$\mathbf{r}_i(L)\boldsymbol{\sigma} = s_i$$

satisfied. We now find the decomposition of

$$A = \begin{bmatrix} 10 & 9 & 7 & 5 \\ 9 & 10 & 8 & 6 \\ 7 & 8 & 10 & 7 \\ 5 & 6 & 7 & 5 \end{bmatrix}$$

in the form LU and our results are given in Table 8.2. For convenience of tabulation we have incorporated L_0 and U together in one matrix

Table 8.2 *LU* decomposition

A	$\mathbf{s}$	$L_0 + U$	$\boldsymbol{\sigma}$	$\mathbf{r}_i(L)\boldsymbol{\sigma}$
$\begin{bmatrix} 10 & 9 & 7 & 5 \\ 9 & 10 & 8 & 6 \\ 7 & 8 & 10 & 7 \\ 5 & 6 & 7 & 5 \end{bmatrix}$	$\begin{bmatrix} 31 \\ 33 \\ 32 \\ 23 \end{bmatrix}$	$\begin{bmatrix} 10 & 9 & 7 & 5 \\ 0{\cdot}9 & 1{\cdot}9 & 1{\cdot}7 & 1{\cdot}5 \\ 0{\cdot}7 & 0{\cdot}8947 & 3{\cdot}579 & 2{\cdot}158 \\ 0{\cdot}5 & 0{\cdot}7895 & 0{\cdot}6030 & 0{\cdot}0150 \end{bmatrix}$	$\begin{bmatrix} 31 \\ 5{\cdot}1 \\ 5{\cdot}737 \\ 0{\cdot}015 \end{bmatrix}$	$\begin{bmatrix} 31 \\ 33 \\ 32 \\ 23 \end{bmatrix}$

The method can be considerably simplified when the matrix A is symmetric and positive definite (exercise 15).

8.4 Matrix norms

We shall be requiring in later work on error analysis (section 8.5) some measure of the size of a (real) matrix. This measure we call a *norm* and we denote the norm of A by $\|A\|$. Norms satisfy the following relations

$$\|A\| > 0, \qquad A \neq O \quad (\|O\| = 0) \tag{8.27}$$

$$\|kA\| = |k|\,\|A\|, \qquad k \text{ a scalar} \tag{8.28}$$

$$\|A + B\| \leqslant \|A\| + \|B\| \tag{8.29}$$

$$\|AB\| \leqslant \|A\|\,\|B\|. \tag{8.30}$$

The types of norm that we shall use in this book are the column norm $\|A\|_1$, the row norm $\|A\|_\infty$, the spectral norm $\|A\|_2$, and the Euclidean or Schur or Frobenius norm $\|A\|_E$. These are defined as follows

$$\|A\|_1 = \max_j \sum_i |a_{ij}|; \qquad \|A\|_\infty = \max_i \sum_j |a_{ij}|;$$

$$\|A\|_2 = (\text{maximum eigenvalue of } A'A)^{\frac{1}{2}} \qquad \text{(see chapter 9)}$$

$$\|A\|_E = \sqrt{\Big(\sum_{i,j} |a_{ij}|^2\Big)}.$$

It is easily verified that these norms satisfy (8.27)–(8.30) and we shall demonstrate the validity of (8.30) for column norms as an example. We let

$$C = AB$$

Then

$$c_{ij} = \mathbf{r}_i(A)\mathbf{c}_j(B) = \sum_s a_{is} b_{sj}$$

and $$\|C\|_1 = \max_j \sum_i |c_{ij}| = \max_j \sum_i |\sum_s a_{is} b_{sj}|$$

$$\leqslant \max_j \sum_i \sum_s |a_{is}|\,|b_{sj}| = \max_j \sum_s \sum_i |a_{is}|\,|b_{sj}|$$

$$\leqslant \max_j \sum_s \|A\|_1\,|b_{sj}| = \|A\|_1 \max_j \sum_s |b_{sj}| = \|A\|_1 \|B\|_1.$$

This completes the proof.

From the basic properties we can deduce results which are of importance in an error analysis. For instance, from (8.27) we have

$$\|-A\| = \|A\|,$$

and hence from (8.28) it follows that

$$\|A + B\| \geqslant \|A\| - \|B\|. \tag{8.31}$$

Now let I be the unit matrix and S a square matrix such that $\|S\| < 1$. We can easily show (cf. exercise 2) that $I + S$ is non-singular. Further, since

$$I = (I + S)^{-1}(I + S) = (I + S)^{-1} + (I + S)^{-1} S$$

it follows from (8.31) that, provided our norm is not Euclidean (in which case $\|I\|_E = n^{\frac{1}{2}}$),

$$\begin{aligned} 1 = \|I\| &\geqslant \|(I + S)^{-1}\| - \|(I + S)^{-1} S\| \\ &\geqslant \|(I + S)^{-1}\| - \|(I + S)^{-1}\|\,\|S\| \end{aligned}$$

from which we have immediately that

$$\|(I + S)^{-1}\| \leqslant 1/(1 - \|S\|). \tag{8.32}$$

We can also prove in a similar fashion (this is left as an exercise to the reader) that

$$\|I - (I + S)^{-1}\| \leqslant \|S\|/(1 - \|S\|).$$

Finally we note that if the matrix A is a $(n \times 1)$ column vector $\mathbf{x}$ then the definitions we have given for norms reduce to

$$\begin{aligned} \|\mathbf{x}\|_1 &= |x_1| + |x_2| + \ldots + |x_n|, \\ \|\mathbf{x}\|_2 &= \|\mathbf{x}\|_E = \sqrt{(x_1^2 + x_2^2 + \ldots + x_n^2)}, \\ \|\mathbf{x}\|_\infty &= \max_i |x_i|. \end{aligned}$$

8.5 Errors in the solution of equations

Errors can arise from two sources, (*a*) inexact data, (*b*) rounding errors during computation. We shall consider both these problems separately.

(a) Inexact data

We suppose that computation is exact but that the items of data A and $\mathbf{b}$ contain errors δA and $\delta\mathbf{b}$ giving rise to an error $\delta\mathbf{x}$ in $\mathbf{x}$. The actual equation we solve is the perturbed equation

$$(A + \delta A)(x + \delta\mathbf{x}) = \mathbf{b} + \delta\mathbf{b}. \tag{8.33}$$

Since $A\mathbf{x} = \mathbf{b}$ we have, ignoring second order terms,

$$\delta\mathbf{x} \doteq A^{-1}(\delta\mathbf{b} - \delta A\mathbf{x}). \tag{8.34}$$

From this equation we see that, if A^{-1} contains large elements, small errors δA and $\delta\mathbf{b}$ could result in large errors $\delta\mathbf{x}$ in $\mathbf{x}$. Take for instance the equation

$$A\mathbf{x} = \mathbf{b}$$

where

$$A = \begin{bmatrix} 1 & \frac{1}{2} & \frac{1}{3} \\ \frac{1}{2} & \frac{1}{3} & \frac{1}{4} \\ \frac{1}{3} & \frac{1}{4} & \frac{1}{5} \end{bmatrix}, \qquad \mathbf{b} = \begin{bmatrix} 0{\cdot}3 \\ 1 \\ 1 \end{bmatrix};$$

the exact solution is $\mathbf{x} = (-3{\cdot}3, 1{\cdot}2, 9)'$. Now suppose A* is still measured exactly but an error of 1 per cent is made in recording b_3 which now has the value $1{\cdot}01$. The exact solution of the 'perturbed' equation

$$A\mathbf{x} = \mathbf{b} + \delta\mathbf{b}$$

is $\mathbf{x} = (-3, -0{\cdot}6, 10{\cdot}8)'$ and, clearly, we have substantial errors in our solution. Equations such as that just described are examples of ill-conditioned equations and we now investigate ill-conditioning in greater detail. Our findings may be summarized by

THEOREM 8-2 *A bound for the relative error* rel $\mathbf{x} = \|\delta\mathbf{x}\|/\|\mathbf{x}\|$ *in* (8.33) *is given by*

$$\text{rel}\,\mathbf{x} \leqslant \frac{\kappa(\text{rel}\,A + \text{rel}\,\mathbf{b})}{1 - \kappa\,\text{rel}\,A} \tag{8.35}$$

where rel $A = \|\delta A\|/\|A\|$, rel $\mathbf{b} = \|\delta\mathbf{b}\|/\|\mathbf{b}\|$ *and* $\kappa = \|A\|\,\|A^{-1}\|$. *The number* κ *is called the condition number of the matrix A and is always greater than or equal to* 1.

*The matrix A is the 3 x 3 Hilbert matrix. The $n \times n$ Hilbert matrix has elements $a_{ij} = 1/(i+j-1)$.

Proof We assume that the norms being used are row norms or column norms. We have from (8.33)

$$\delta\mathbf{x} = (A + \delta A)^{-1}(\delta\mathbf{b} - \delta A\mathbf{x})$$
$$= (I + S)^{-1}(A^{-1}\delta\mathbf{b} - S\mathbf{x}), \qquad \text{where } S = A^{-1}\delta A.$$

Therefore from (8.30),

$$\|\delta\mathbf{x}\| \leqslant \|(I + S)^{-1}\|.\|A^{-1}\delta\mathbf{b} - S\mathbf{x}\|.$$

Now

$$\|A^{-1}\delta\mathbf{b} - S\mathbf{x}\| \leqslant \|A^{-1}\|\cdot\|\delta\mathbf{b} - \delta A\mathbf{x}\|$$
$$\leqslant \|A^{-1}\|(\|\delta\mathbf{b}\| + \|\delta A\|\cdot\|\mathbf{x}\|)$$

and therefore from (8.32)

$$\|\delta\mathbf{x}\| \leqslant \frac{\|A^{-1}\|(\|\delta\mathbf{b}\| + \|\delta A\|.\|\mathbf{x}\|)}{1 - \|S\|}$$
$$\leqslant \frac{\kappa\{(\|\delta\mathbf{b}\|/\|A\|) + \|\mathbf{x}\|\text{rel } A\}}{1 - \kappa \text{ rel } A}$$

and thus

$$\text{rel } \mathbf{x} \leqslant \frac{\kappa(\text{rel } A + \{\|\delta\mathbf{b}\|/\|A\|.\|\mathbf{x}\|\})}{1 - \kappa \text{ rel } A}.$$

Finally, we again use the fact that $A\mathbf{x} = \mathbf{b}$ so that

$$\|\mathbf{b}\| \leqslant \|A\|.\|\mathbf{x}\|,$$

and thus (8.35) holds.

We see from the theorem that ill-conditioning is dependent on the size of the condition number κ and rel A, rel $\mathbf{b}$. For the 3 x 3 Hilbert matrix, the exact value of A^{-1} is given by

$$A^{-1} = \begin{bmatrix} 9 & -36 & 30 \\ -36 & 192 & -180 \\ 30 & -180 & 180 \end{bmatrix}$$

and so, for row or column norms, $\kappa = \|A\|.\|A^{-1}\| = 408 \times \frac{11}{6} = 748$. Further rel $\mathbf{b}$ is fairly large in the problem we discussed and this gives a large bound for rel $\mathbf{x}$. This was confirmed by the exact results obtained.

(b) Errors during computation

For this investigation we shall consider the method of Gaussian elimination, with some restriction on the pivots, and the floating-point mode of computation. It is convenient to consider first the case of back substitution where we are solving the set of equations

$$U\mathbf{x} = \mathbf{b},$$

where U is an upper triangular matrix and $n = 5$, say. The computed values of x_i are obtained from the equations

$$x_5 = fl(b_5/a_{55})$$

$$x_i = fl\left\{\frac{-u_{i,\,i+1}x_{i+1} - \ldots - u_{i5}x_5 + b_i}{u_{ii}}\right\}, \qquad i = 4, 3, 2, 1.$$

It follows that (see section 1.10)

$$x_5 = b_5(1 + \epsilon_1)/u_{55}$$

$$x_4 = (-u_{45}(1 + \epsilon_2) + b_4)(1 + \epsilon_3)(1 + \epsilon_4)/u_{44}$$

$$x_3 = \{[-u_{34}x_4(1 + \epsilon_5) - u_{35}x_5(1 + \epsilon_6)](1 + \epsilon_7) + b_3\}(1 + \epsilon_8)(1 + \epsilon_9)/a_{33},$$

etc., and the computed solution satisfies exactly the perturbed equations

$$(U + \delta U)\mathbf{x} = \mathbf{b},$$

where, if second order terms are neglected, the matrix δU is such that

$$|\delta U| \leqslant \epsilon \begin{bmatrix} 2|u_{11}| & 4|u_{12}| & 4|u_{13}| & 3|u_{14}| & 2|u_{15}| \\ 0 & 2|u_{22}| & 3|u_{23}| & 3|u_{24}| & 2|u_{25}| \\ 0 & 0 & 2|u_{33}| & 2|u_{34}| & 2|u_{35}| \\ 0 & 0 & 0 & 2|u_{44}| & |u_{45}| \\ 0 & 0 & 0 & 0 & |u_{55}| \end{bmatrix}$$

ϵ being the maximum relative error in magnitude for an arithmetic operation (for a four decimal digit floating point computer $\epsilon = 0{\cdot}55 \times 10^{-3}$) and by $|\delta U|$ we mean the matrix whose elements are the elements of δU in magnitude. Thus

$$\|\delta U\|_\infty \leqslant 15\epsilon \times \max|u_{ij}|. \tag{8.36}$$

For a normalized matrix U, with its largest element less than one, (8.36) is a reassuring result for it shows that the error in a back-substitution is relatively small (in the sense of a backward error analysis).

Now consider the reduction of a set of equations to triangular form. We shall assume that at each stage the largest element in each row and column is numerically less than one and is initially of the order of 1, that is, the equations have been normalized. Further, we adopt some form of pivoting such as partial or complete pivoting so that the multipliers m_{ij} are numerically less than unity. Then we can show that the exact solution of the triangularized equations we have obtained satisfies the perturbed equation

$$(A + \delta A)\mathbf{x} = \mathbf{b} + \delta\mathbf{b},$$

where δA and $\delta\mathbf{b}$ are relatively small. More precisely, with the same notation as section 8.2, we shall show that the computed values of L and U (denoted by $\overline{L}$ and $\overline{U}$) satisfy

$$\overline{L}\overline{U} = A + \delta A,$$

and we shall obtain bounds for the elements of δA. The bounds for the elements of $\delta \mathbf{b}$ can be determined in a similar manner and will also be quoted.

We first trace the history of the elements $\overline{a}_{ij}^{(k)}$ of the matrix $\overline{A}_k$ and we have to distinguish between two cases according as $i \leqslant j$ and $i > j$.

Case 1 $\quad i \leqslant j$

The elements are modified in each transformation until A_{i-1} is reached and thereafter remain constant. We have

$$\overline{a}_{ij}^{(r)} = \overline{a}_{ij}^{(r-1)} - \overline{m}_{ir}\overline{a}_{rj}^{(r-1)} + \epsilon_{ij}^{(r)}, \qquad r = 1, \ldots, i-1 \tag{8.37}$$

where $\epsilon_{ij}^{(k)}$ is the difference between the exact value of $a_{ij}^{(k)}$ calculated from the computed $\overline{a}_{ij}^{(k-1)}$, $\overline{m}_{ik}$, $a_{kj}^{(k-1)}$, and the computed value. Adding equations (8.37) we have

$$\overline{a}_{ij}^{(i-1)} = a_{ij}^{(0)} - \sum_{r=0}^{i-2} \overline{m}_{i,r+1}\overline{a}_{r+1,j}^{(r)} + \delta a_{ij}$$

where

$$\delta a_{ij} = \sum_{r=1}^{i-1} \epsilon_{ij}^{(r)}. \tag{8.38}$$

Case 2 $\quad i > j$

Elements are modified in exactly the same way as Case 1 until A_{j-1} is reached. $\overline{a}_{ij}^{(j-1)}$ is then used to compute $\overline{m}_{ij}$ and the elements $a_{ij}^{(j)}, \ldots, a_{ij}^{(n-1)}$ are all taken to be zero. We obtain

$$\overline{a}_{ij}^{(r)} = \overline{a}_{ij}^{(r-1)} - \overline{m}_{ir}\overline{a}_{rj}^{(r-1)} + \epsilon_{ij}^{(r)}, \qquad r = 1, \ldots, j-1 \tag{8.39}$$

and we can introduce an additional equation

$$0 = \overline{a}_{ij}^{(j-1)} - \overline{m}_{ij}\overline{a}_{ij}^{(j-1)} + \epsilon_{ij}^{(j)}, \tag{8.40}$$

by virtue of the fact that

$$\overline{m}_{ij} = (\overline{a}_{ij}^{(j-1)}/\overline{a}_{jj}^{(j-1)}) + \xi_{ij},$$

where ξ_{ij} is the rounding error in the division and

$$\epsilon_{ij}^{(j)} = \xi_{ij}\overline{a}_{jj}^{(j-1)}.$$

Addition of equations (8.39) and (8.40) gives

$$0 = a_{ij}^{(0)} - \sum_{r=0}^{j-1} \overline{m}_{i,r+1} a_{r+1,j}^{(r)} + \delta a_{ij},$$

where
$$\delta a_{ij} = \sum_{r=1}^{j} \epsilon_{ij}^{(r)} . \tag{8.41}$$

We next determine $\epsilon_{ij}^{(k)}$. We have

$$\begin{aligned}\overline{a}_{ij}^{(k)} &= fl\{\overline{a}_{ij}^{(k-1)} - \overline{m}_{ik}\overline{a}_{kj}^{(k-1)}\} \\ &= \{\overline{a}_{ij}^{(k-1)} - \overline{m}_{ik}\overline{a}_{kj}(1+\epsilon_1)\}(1+\epsilon_2).\end{aligned} \tag{8.42}$$

Thus
$$\begin{aligned}\epsilon_{ij}^{(k)} &= \overline{a}_{ij}^{(k)} - \{\overline{a}_{ij}^{(k-1)} - \overline{m}_{ik}\overline{a}_{kj}^{(k-1)}\} \\ &= \overline{a}_{ij}^{(k)} - \left(\frac{\overline{a}_{ij}^{(k)}}{1+\epsilon_2} + \overline{m}_{ik}\overline{a}_{kj}^{(k-1)}\,\epsilon_1\right) \\ &= \frac{\overline{a}_{ij}^{(k)}\epsilon_2}{1+\epsilon_2} - \overline{m}_{ik}\overline{a}_{kj}^{(k-1)}\epsilon_1\end{aligned}$$

and consequently
$$|\epsilon_{ij}^{(k)}| < 2{\cdot}1\, a\epsilon \tag{8.43}$$

where $a(\leqslant 1)$ is the maximum element in A_i (all i), ϵ is the maximum possible error in modulus in the floating point operations. This is true for all $\epsilon_{ij}^{(k)}$ except $\epsilon_{ij}^{(j)}$ where $i > j$. In this instance we have

$$\overline{m}_{ij} = fl\left(\frac{\overline{a}_{ij}^{(j-1)}}{\overline{a}_{jj}^{(j-1)}}\right) = \left(\frac{\overline{a}_{ij}^{(j-1)}}{\overline{a}_{jj}^{(j-1)}}\right)(1+\epsilon_3)$$

so that
$$|\epsilon_{ij}^{(j)}| = |\xi_{ij}\overline{a}_{jj}^{(j-1)}| = |\epsilon_3 a_{ij}^{(j-1)}| < a\epsilon$$

and $|\epsilon_{ij}^{(j)}| < 2{\cdot}1\, a\epsilon$, is true also for $\epsilon_{ij}^{(j)}$. From (8.38) and (8.41) we can estimate the maximum value in modulus of the elements in δA. Thus, for example, when n = 5, the maximum possible value of the elements of δA are

$$2{\cdot}1\epsilon \times \begin{bmatrix} 0 & 0 & 0 & 0 & 0 \\ 1 & 1 & 1 & 1 & 1 \\ 1 & 2 & 2 & 2 & 2 \\ 1 & 2 & 3 & 3 & 3 \\ 1 & 2 & 3 & 4 & 4 \end{bmatrix}.$$

In the same manner, we can show that the element δb_i in the ith row of $\delta\mathbf{b}$ is at most $2{\cdot}1\epsilon(i-1)$ in magnitude. Combination of these results with that for the back substitution gives us the complete perturbed solution

$$(A+E)\mathbf{x} = \mathbf{b} + \epsilon$$

and shows that $\|E\|$ and $\|\epsilon\|$ are small. Fuller details of error analyses concerned with the solution of linear equations may be found in Wilkinson [4], [7], and Rall [5].

8.6 Inverses and determinants

For a laige number of mathematical problems inverses and determinants are required. These can be found using methods developed in earlier sections. For instance, to evaluate $\det A$, we note that if

$$\begin{aligned} A &= LU, \\ \det A &= (\det L)(\det U), \\ &= \det U = u_{11}u_{22}\ldots u_{nn}, \end{aligned}$$

since the determinant of a triangular matrix is just the product of its diagonal elements. Thus, for the matrix A in Table 8.2 (p. 141),

$$\det A = 10 \times 1{\cdot}9 \times 3{\cdot}579 \times 0{\cdot}0150 \mp 1{\cdot}020.$$

To evaluate U^{-1} we could solve the equation

$$U\mathbf{x}_i = \mathbf{e}_i \qquad (i = 1, 2, \ldots, n)$$

successively, where $\mathbf{e}_i$ is the 'unit' vector, that is, the vector with all its elements zero except the ith. It follows that

$$U\{\mathbf{x}_1, \mathbf{x}_2, \ldots, \mathbf{x}_n\} = \{\mathbf{e}_1, \mathbf{e}_2, \ldots, \mathbf{e}_n\} = I$$

and consequently from (8.11)

$$U^{-1} = \{\mathbf{x}_1, \mathbf{x}_2, \ldots, \mathbf{x}_n\}$$

L^{-1} can be evaluated in a similar way. In fact, since L' is an upper triangular matrix, exactly the same procedure can be used to evaluate $(L')^{-1}$, and L^{-1} can then be evaluated from

$$L^{-1} = ((L')^{-1})'.$$

Finally

$$A^{-1} = U^{-1}L^{-1}$$

and this completes the process of determining the inverse. Alternatively, if q_{ij} are the elements of U^{-1}, they can be determined systematically from the relations (obtained by comparing coefficients in UU^{-1} and I)

$$\left.\begin{aligned} q_{ij} &= 0, \qquad i>j \\ q_{ii} &= 1/u_{ii} \\ q_{ij} &= -q_{jj}\left\{\sum_{k=i}^{j-1} q_{ik}u_{kj}\right\}, \quad i<j \end{aligned}\right] . \tag{8.44}$$

Thus, for the matrix U determined in Table 8.2,

$$U = \begin{bmatrix} 10 & 9 & 7 & 5 \\ & 1{\cdot}9 & 1{\cdot}7 & 1{\cdot}5 \\ & & 3{\cdot}579 & 2{\cdot}158 \\ & & & 0{\cdot}015\,00 \end{bmatrix}$$

we have from (8.44)

$$U^{-1} = \begin{bmatrix} 0{\cdot}1 & -0{\cdot}4738 & 0{\cdot}029\,48 & 9{\cdot}807 \\ & 0{\cdot}5263 & -0{\cdot}2500 & -16{\cdot}67 \\ & & 0{\cdot}2794 & -40{\cdot}20 \\ & & & 66{\cdot}67 \end{bmatrix}.$$

We can evaluate L^{-1} similarly, and hence obtain A^{-1}. The exact value of det A for the matrix in Table 8.2 is 1. Further, if U^{-1} had been obtained exactly from the computed U, we would expect $UU^{-1} = I$. In fact, the computed value of UU^{-1} is

$$\begin{bmatrix} 1 & -0{\cdot}0013 & 0{\cdot}0006 & -0{\cdot}01 \\ & 0{\cdot}999\,97 & -0{\cdot}000\,02 & -0{\cdot}008 \\ & & 0{\cdot}999\,726 & -0{\cdot}001\,94 \\ & & & 1{\cdot}000\,05 \end{bmatrix}.$$

This is not really a surprising result since, if we examine the norms,

$$\|U\|_1 \|U^{-1}\|_1 = 11{\cdot}279 \times 133{\cdot}347 \doteqdot 1504.$$

8.7 Improvement of the computed results

In view of the error analyses already carried out it is clear that a greater number of correct significant digits would be obtained by using double length arithmetic. Since computing time is greatly increased by the use of double length arithmetic it is something to be avoided as much as possible. By a judicious combination of single and double length arithmetic we can improve results, with very little extra cost in time, as follows.

We suppose $\mathbf{x}_1$ is the single-length computed solution of $A\mathbf{x} = \mathbf{b}$, where A and $\mathbf{b}$ are exact. We next compute with double length arithmetic, the residual vector $\mathbf{r}$,

$$\mathbf{r} = A\mathbf{x}_1 - \mathbf{b}.$$

It follows that if $\delta\mathbf{x} = \mathbf{x}_1 - \mathbf{x}$ is the error in $\mathbf{x}$ then approximately

$$A\delta\mathbf{x} = \mathbf{r}. \tag{8.45}$$

This system of equations is now solved by single length arithmetic – it is best to normalize $\mathbf{r}$ and round its elements to give a single length vector – making use of the fact that, if we are using triangular resolution for example, L and U have already been determined in obtaining $\mathbf{x}_1$. Adding the computed value $\delta\mathbf{x}$ ($\delta\mathbf{x}$ has to be modified by the scale factor used to normalize $\mathbf{r}$) to $\mathbf{x}_1$ gives a new approximation $\mathbf{x}_2$ which can be improved, if need be, in exactly the same way.

8.8 Summary and notes

We have seen in section 8.2 that equations in triangular form are easy to solve. Thus it is most natural to seek methods which reduce a general set of equations to triangular form and this is the case with the two related methods, Gauss elimination and triangular resolution, discussed in the text. Both these methods require the same amount of work namely $\frac{1}{3}n(n^2 + 3n - 1)$ multiplications and n reciprocals and both methods are satisfactory with regard to error analysis (see Wilkinson [4] for error analysis of triangular resolution). For special types of equations such as those in which the matrices are symmetric or in band form shorter methods exist. Thus, for example, Choleski's square root method only involves $\frac{1}{6}n(n^2 + 6n - 1)$ multiplications and n square root reciprocals (exercise 15).

Several sets of equations having the same coefficients but different right-hand sides can be solved together using either triangular resolution or Gauss elimination. Since the matrices L and U or the multipliers and the coefficients in the triangular form are the same for each set these need only be calculated once. Each extra right-hand side involves an extra n^2 multiplications. Alternatively, we can find the inverse of A and then form $A^{-1}\mathbf{b}$ for each right-hand side (this also requires n^2 multiplications). However, since finding A^{-1} involves finding L and U plus their inverses, this method is longer.

In section 8.5 we have shown that the computed solution of $A\mathbf{x} = \mathbf{b}$ satisfies a perturbed equation

$$(A + \delta A)(\mathbf{x} + \delta\mathbf{x}) = \mathbf{b} + \delta\mathbf{b}$$

and that, provided the equations are well-conditioned (which can be determined by computing the condition number κ), the relative error in $\mathbf{x}$ will in general be small.

Even if the equations are ill-conditioned $\mathbf{x}$ can be determined more accurately if we are prepared to resort to multiple length arithmetic (section 8.7). In practice it is a little inconvenient to have to evaluate κ and a rule of thumb procedure which is usually satisfactory involves examination of the pivots. Thus, in Gauss elimination with partial or complete pivoting, if the ratio of the largest to the smallest pivot is 10^q, say ($q > 0$), then we can only expect at most $p - q$ correct significant decimal digits in our calculations, where p is the number of significant digits to which the data is correct. Another useful rule of thumb approximation concerns strongly diagonally dominant matrices. Here the condition number is very nearly $1/|\text{smallest pivot}|$, where initially the largest pivot is 1.

Finally, we mention that a good account of other direct methods for the solution of linear equations (Crout, Doolittle, etc.) can be found in Fox [**6**]. Indirect methods, which are extremely useful in solving many of the linear equation systems which approximate partial differential equations, will be discussed in chapter 10.

REFERENCES

[1] Nef, W., *Linear Algebra*, McGraw-Hill Publ. Co. Ltd., Maidenhead (1967).

[2] Birkhoff, G. and MacLane, S., *A Survey of Modern Algebra*, rev. edn., Macmillan Co., New York (1953).

[3] Wilkinson, J. H., *The Algebraic Eigenvalue Problem*, Oxford University Press, London (1965).

[4] Wilkinson, J. H., Rounding Errors in Algebraic Processes, *N.P.L. Notes on Applied Science*, No. 32, H.M.S.O., London (1963).

[5] Rall, L. B. (ed.), *Error in Digital Computation*, I and II, J. Wiley and Sons, New York (1965).

[6] Fox, L., *An Introduction to Numerical Linear Algebra*, Oxford University Press, London (1964).

[7] Wilkinson, J. H., 'Error Analysis of Direct Methods of Matrix Inversion', *A.C.M.*, **8** (1961) 281–330.

Exercises

1. Show that, for any matrix A, the product matrices AA' and $A'A$ are symmetric. If A is a symmetric matrix and X is a matrix such that $X'X = I$ (i.e., X is *orthogonal*) show that the matrix $X^{-1}AX$ is symmetric.

2. If S is a matrix for which $S' = -S$ (i.e., S is skew-symmetric) show that S^2 is symmetric. Show further that if $I + S$ is non-singular

$$(I - S)(I + S)^{-1}$$

is orthogonal. Deduce from chapter 9, exercise 16, that $I + S$ is always non-singular.

3. If

$$X = \begin{bmatrix} A & B \\ O & C \end{bmatrix}$$

where A and C are square matrices and O is a zero matrix of some order (not necessarily square) show that

$$\det X = \det A \,.\, \det C.$$

4. Let the matrix X be partitioned in the form

$$X = \begin{bmatrix} A & B \\ C & D \end{bmatrix}$$

where A and D are square matrices such that $\det A \neq 0$, $\det D \neq 0$. If the inverse of X is X^{-1} where

$$X^{-1} = \begin{bmatrix} \alpha & \beta \\ \gamma & \delta \end{bmatrix}$$

and $\alpha, \beta, \gamma, \delta$ are partitioned exactly like A, B, C, D respectively, express $\alpha, \beta, \gamma, \delta$ in terms of functions of A, B, C, and D. Hence find the inverses of

$$\begin{bmatrix} 2 & 1 & 0 & 0 \\ 1 & 2 & 1 & 0 \\ 0 & 0 & 2 & 1 \\ 1 & 0 & 1 & 2 \end{bmatrix} \quad \text{and} \quad \begin{bmatrix} 1 & 2 & 2 & 1 & 0 \\ 2 & 1 & 2 & 0 & 1 \\ 2 & 2 & 1 & 1 & 0 \\ 0 & 0 & 0 & 1 & 0 \\ 0 & 0 & 0 & 2 & 1 \end{bmatrix}.$$

5. If $A = (a_{ij})$ is a symmetric positive definite matrix of order 2 x 2 show that

$$|a_{12}| \leqslant \sqrt{(a_{11}\, a_{22})} \leqslant \tfrac{1}{2}(a_{11} + a_{22}).$$

Extend this result to show that the diagonal elements of a symmetric positive definite matrix are all positive and the numerically largest element lies on the main diagonal.

6. Solve the following simultaneous equations by Gauss elimination with complete pivoting.

$$\begin{aligned} 0{\cdot}5398x_1 + 0{\cdot}7161x_2 - 0{\cdot}5554x_3 - 0{\cdot}2982x_4 &= 0{\cdot}2058 \\ 0{\cdot}5257x_1 + 0{\cdot}6924x_2 + 0{\cdot}3565x_3 - 0{\cdot}6255x_4 &= -0{\cdot}0503 \\ 0{\cdot}6465x_1 - 0{\cdot}8187x_2 - 0{\cdot}1872x_3 + 0{\cdot}1291x_4 &= 0{\cdot}1070 \\ 0{\cdot}5814x_1 + 0{\cdot}9400x_2 - 0{\cdot}7779x_3 - 0{\cdot}4042x_4 &= 0{\cdot}1859. \end{aligned}$$

Comment on errors incurred during computation.

7. Solve the following simultaneous equations by triangular resolution using exact arithmetic.

$$\begin{aligned} x_1 + 5x_2 + 4x_3 + 3x_4 &= -5 \\ 2x_1 + 7x_2 + 6x_3 + 10x_4 &= 7 \\ 3x_1 + 3x_2 + 8x_3 + 25x_4 &= 65 \\ 2x_1 + x_2 - 2x_3 + 27x_4 &= 20. \end{aligned}$$

8. Solve the equations in exercise 6 by triangular resolution.

9. Normalize the equations in exercise 7 and solve them by Gauss elimination. Estimate errors and check against the known solution.

10. Solve the simultaneous equations

$$\begin{aligned} 3x + y &= 10\,000 \\ -0{\cdot}8000x + 0{\cdot}000\,124\,9y &= 0{\cdot}6 \end{aligned}$$

by complete pivoting (but without normalization) on a four decimal digit floating point computer. Explain why normalization is required for complete pivoting to work. Compare these equations with the equations in section 1.5.

11. If A is a matrix with $a_{ij} = 0$ for $|i-j| > m$ (i.e., A is a band matrix) show that for the triangular resolution $A = LU$,

$$l_{ij} = 0 \quad \text{and} \quad u_{ij} = 0 \quad \text{for} \quad |i-j| > m$$

and the property $a_{ij} = 0$ for $|i-j| > m$ holds also for the matrix UL.

12. Let P_{st} be the $n \times n$ square matrix such that

$$\begin{aligned} \mathbf{r}_i(P_{st}) &= \mathbf{r}_i(I), & i \neq s, t \\ \mathbf{r}_s(P_{st}) &= \mathbf{r}_t(I), & \mathbf{r}_t(P_{st}) = \mathbf{r}_s(I), \end{aligned}$$

i.e., P_{st} is the unit matrix with rows s and t interchanged or permuted (hence the name permutation matrix for P_{st}). Show that the matrix product $P_{st}A$ is the same as A except that rows s and t of A have been interchanged. How can this result be used to modify the decomposition into LU form of the matrix

$$\begin{bmatrix} 4 & 6 & 1 \\ 2 & 3 & 1 \\ 1 & 1 & 1 \end{bmatrix} ?$$

13. Use the fact that the factorization of a non-singular matrix into LU form is unique if the diagonal elements of L are unity to show that $L' = D^{-1}U$ if A is symmetric (where D is a diagonal matrix).

14. With the notation of exercise 13 and section 8.3, show that

$$DA^{-1} = L^{-1} - U_0 A^{-1}.$$

If $A^{-1} = (\alpha_{ij})$ and $L^{-1} = (\lambda_{ij})$ prove that

$$\alpha_{ij} = [\lambda_{ij} - \mathbf{r}_i(U_0)\mathbf{c}_j(A^{-1})]/u_{ii}$$

and show that L^{-1} need not be computed if A is symmetric.

15. (Choleski's square root method) Show that a symmetric non-singular matrix A can be expressed in the form $U'U$, and that the elements of U are given by

$$u_{ii} = \sqrt{\{a_{ii} - \mathbf{r}_i(U'_0)\mathbf{c}_i(U_0)\}}$$
$$u_{ij} = \{a_{ij} - \mathbf{r}_i(U'_0)\mathbf{c}_j(U_0)\}/u_{ii}.$$

Note that U is real only if A is positive definite. (It is sometimes preferable to consider the decomposition in the form LDL' and one then avoids complex arithmetic.)

16. For any column vector $\mathbf{x}$, show that

$$\|\mathbf{x}\|_\infty \leqslant \|\mathbf{x}\|_1, \qquad \|\mathbf{x}\|_\infty \leqslant \|\mathbf{x}\|_E$$

and for any square matrix A show that the condition numbers associated with the various types of norms satisfy

$$\kappa_1 \geqslant 1, \qquad \kappa_2 \geqslant 1, \qquad \kappa_\infty \geqslant 1, \qquad \kappa_E \geqslant \sqrt{n}.$$

Prove also that for any square matrix A

$$\|A\|_E^2 = \text{trace } (AA')$$
$$= \text{sum of diagonal elements of } AA'.$$

17. Find the inverses of the matrices.

$$\begin{bmatrix} 2 & 3 \\ 4 & 7 \end{bmatrix} \quad \begin{bmatrix} 1 & 0 & 0 \\ 0 & 0{\cdot}5 & 0 \\ 0 & 0 & 0{\cdot}1 \end{bmatrix} \quad \begin{bmatrix} 1 & 0 & 0 \\ \frac{1}{2} & 1 & 0 \\ \frac{1}{2} & \frac{1}{2} & 1 \end{bmatrix}$$

and their column condition numbers.

18. Find the condition number of the matrix of coefficients in exercise 6, using row or column norms, and use it to give error bounds for the solution assuming all the data is given correct to four significant digits.

19. If $F = (I + E)^{-1} - (I - E)$ show that

$$\|F\| \leqslant \|E\|^2/(1 - \|E\|).$$

20. Given that $(A + E)^{-1} = A^{-1} + F$, prove that

$$F + A^{-1}EF = -A^{-1}EA^{-1}$$

and deduce that

$$\|F\| \leqslant \|A^{-1}\|^2\|E\|/(1 - \|A^{-1}\|\|E\|).$$

21. Show that

$$\text{rel } A^{-1} \leqslant \kappa \text{ rel } A/(1 - \kappa \text{ rel } A),$$

where κ is the condition number of A.

22. If $E = I - AX$ show that

$$\frac{\|A^{-1} - X\|}{\|A^{-1}\|} \leqslant \|E\|.$$

How can this be used to estimate the accuracy of a computed inverse?

23. If X_n is a good approximation to A^{-1}, i.e., $X_n = A^{-1}(I + E_n)$. where $\|E_n\|$ is small show that

$$X_{n+1} = X_n(2I - AX_n)$$

is a better approximation. How small must $\|E_0\|$ be for the sequence $X_0, X_1, \ldots$ to converge to A^{-1}?

24. Given

$$A = \begin{bmatrix} a_1 & b \\ b & a_2 \end{bmatrix}$$

where $b = 24$, $a_1 = 26$, $a_2 = 26 + \epsilon$, and ϵ is a small error, find the approximate error in $\mathbf{x}'A^{-1}\mathbf{x}$ when $\mathbf{x} = (1, 1)'$.

25. Given that the matrix A and the vector $\mathbf{x}$ are stored exactly in a floating point computer estimate the error in evaluating $A\mathbf{x}$.

26. A is a symmetric positive definite matrix and the equations $A\mathbf{x} = \mathbf{b}$ are solved by the method of complete pivoting. With the notation of section 8.2 show that the pivot in A_k is less than the pivot in A_{k-1}.

27. If the elements of the matrix A satisfy $|a_{ij}| \leqslant 1$ and p_r denotes the absolute value of the rth pivot in Gaussian elimination show that

(a) $p_r \leqslant 2^{r-1}$ for partial pivoting;

(b) $p_{r+1}p_{r+2} \ldots p_n \leqslant p_{r+1}^{n-r} (n-r)^{\frac{1}{2}(n-r)}$,

for complete pivoting. Hence deduce that the largest pivot is less in modulus than

$$n^{\frac{1}{2}} (2^1 3^{\frac{1}{2}} \ldots n^{\frac{1}{n-1}})^{\frac{1}{2}}.$$

(Hint: Use Hadamard's theorem, chapter 9, exercise 8 – see Wilkinson [7].)

9 The algebraic eigenvalue problem

9.1 Introduction

Three aims have been foremost in the author's mind in the preparation of this chapter. First, to give the reader a good grounding in (to use Bodewig's terminology [5]) matrix calculus. Second, to present a numerically stable method, namely Householder's method, for the complete solution of the eigenvalue problem and, finally, to present a method for treating the specialized problem of finding just the largest and smallest eigenvalues of a matrix. With this background it is felt that the reader will be able to comprehend those methods which are not given in the text, such as the Jacobi and *L–R* methods (exercises 30 and 32), and those which may be found by consulting the references; for instance, Eberlein's method (Walsh [8]), the Givens and Lanczos methods (Fox [7] and Wilkinson [1]), and the Francis *Q–R* method (Wilkinson [1]). Further, it is hoped that the reader will be able to extend Householder's method to deal with the case of matrices with complex elements (exercise 19) and, where the matrix is Hermitian, reduce the complex problem to a real problem (exercise 17) – although this has the disadvantage of requiring more storage space and computation. Finally, we remark that it has been possible to present only the simplest error analyses relating to the eigenvalue problem. Thus it has not been possible to extend the analysis to treat the fact that the orthogonal matrix P of the similarity transformation is not obtained exactly (exercise 21) or to consider the effect of taking elements as zero which, due to rounding errors, are in fact non-zero. These and other similar problems are discussed in Wilkinson [1], [3], and Rall [6].

We consider in the next few sections the matrix calculus. For background reading the reader should also consult books which have a more theoretical approach such as Nef [9] or Birkhoff and MacLane [10].

9.2 Basic preliminaries

Let A be a $n \times n$ matrix. We shall investigate in this chapter the problem of finding a non-zero $n \times 1$ vector $\mathbf{x}$ and a scalar λ for which

$$A\mathbf{x} = \lambda\mathbf{x}$$

or

$$(A - \lambda I)\mathbf{x} = \mathbf{0} \tag{9.1}$$

It is known that a solution $\mathbf{x}(\neq \mathbf{0})$ of (9.1) exists provided

$$\det (A - \lambda I) = 0. \tag{9.2}$$

Equation (9.2) is a polynomial equation of degree n in λ called the characteristic equation of the matrix A. Its n zeros are called the *characteristic* roots or *latent* roots or *eigenvalues* of A. Corresponding to each distinct eigenvalue there exists a distinct vector $\mathbf{x}$ satisfying (9.1) called a characteristic vector or latent vector or *eigenvector* of A. For the sake of uniformity we shall only use the terms eigenvalue and eigenvector.

For the matrix A given by

$$A = \begin{bmatrix} 2 & 0 & 0 \\ 2 & 2 & 1 \\ 1 & 1 & 2 \end{bmatrix} \tag{9.3}$$

the characteristic equation is found to be

$$\lambda^3 - 6\lambda^2 + 11\lambda - 6 = 0$$

so that the eigenvalues are 1, 2, and 3. The corresponding eigenvectors are

$$\begin{bmatrix} 0 \\ 1 \\ -1 \end{bmatrix} \quad \begin{bmatrix} 1 \\ -1 \\ -2 \end{bmatrix} \quad \text{and} \quad \begin{bmatrix} 0 \\ 1 \\ 1 \end{bmatrix}.$$

The eigenvectors are not unique in the sense that, if $\mathbf{x}$ is a vector satisfying (9.1) then $\alpha\mathbf{x}$ also satisfies (9.1) where $\alpha \neq 0$. It is thus convenient to normalize $\mathbf{x}$ so that, for some vector norm, $\|\mathbf{x}\| = 1$. It must be emphasized that if the eigenvalues are not distinct, as in the matrix

$$\begin{bmatrix} 1 & 1 & 1 \\ 0 & 1 & 1 \\ 0 & 0 & 1 \end{bmatrix}$$

it is possible that the eigenvectors are also not distinct. However, it can be shown that a symmetric matrix of order n always possesses a complete system of n distinct eigenvectors whatever its eigenvalues. In what follows we shall assume the eigenvalues (and hence the eigenvectors) are all distinct, unless the contrary is stated.

The direct computation of the characteristic equation (9.2) is a numerical method to be avoided since the condition of the polynomial could be unsatisfactory, as the coefficients may have large errors (see section 2.2). We now proceed to give the basis of the indirect methods used.

9.3 Similar matrices

If $\mathbf{x}_r$ is the eigenvector corresponding to the eigenvalue λ_r of a $n \times n$ matrix A we have, from (9.1),

$$A\mathbf{x}_r = \lambda_r \mathbf{x}_r, \qquad r = 1, 2, \ldots, n.$$

Juxtaposition of all these column vectors gives

$$A(\mathbf{x}_1, \mathbf{x}_2, \ldots, \mathbf{x}_n) = (\lambda_1 \mathbf{x}_1, \lambda_2 \mathbf{x}_2, \ldots, \lambda_n \mathbf{x}_n),$$

or

$$AX = X\Lambda, \tag{9.4}$$

where X, the matrix of eigenvectors, is termed the *modal matrix* and Λ, the diagonal matrix whose elements are the eigenvalues, is termed the *spectral matrix*. To illustrate this point, the matrix A defined by (9.3) has spectral matrix, Λ,

$$\Lambda = \begin{bmatrix} 1 & 0 & 0 \\ 0 & 2 & 0 \\ 0 & 0 & 3 \end{bmatrix}$$

and modal matrix, X,

$$X = \begin{bmatrix} 0 & 1 & 0 \\ 1 & -1 & 1 \\ -1 & -2 & 1 \end{bmatrix}.$$

The eigenvectors $\mathbf{x}_r$ $(r = 1, 2, \ldots, n)$ can be shown to be linearly independent (Exercise 1) and it follows that $\det X \neq 0$ and X^{-1} exists. Premultiplication of (9.4) by X^{-1} gives

$$X^{-1}AX = \Lambda \tag{9.5}$$

and hence the result that $X^{-1}AX$ has the same eigenvalues as A. The matrices A and $X^{-1}AX$ are termed *similar* matrices. This property of similarity is not restricted to the case where X is the modal matrix and we have the more general result.

THEOREM 9-1 *If B is any $n \times n$ non-singular matrix, A and $B^{-1}AB$ have the same eigenvalues. Moreover, if $\mathbf{x}$ is an eigenvector of $B^{-1}AB$, then $B\mathbf{x}$ is an eigenvector of A.*

Proof If $\mathbf{x}$ is the eigenvector of $B^{-1}AB$ corresponding to an eigenvalue λ then

$$(B^{-1}AB)\mathbf{x} = \lambda \mathbf{x}.$$

Premultiplication by B gives

$$A(B\mathbf{x}) = \lambda(B\mathbf{x}),$$

which implies that $B\mathbf{x}$ is the eigenvector of A corresponding to the eigenvalue λ. This proves Theorem 9-1.

Theorem 9-1 is fundamental to many of the practical methods for determining eigenvalues and will be referred to later.

9.4 Orthogonal and symmetric matrices

If A is an unsymmetric matrix it is worthwhile to consider the eigenvalues and eigenvectors of the transpose matrix A'. Since

$$\det(A - \lambda I) = \det(A - \lambda I)' = \det(A' - \lambda I),$$

it follows that the eigenvalues of A and A' are the same, i.e. Λ, the spectral matrix of A, is also the spectral matrix of A'. In general the modal matrices of A and A', call them X and Y, are different and they satisfy the equations

$$\left.\begin{aligned} AX &= X\Lambda \\ A'Y &= Y\Lambda \end{aligned}\right]. \tag{9.6}$$

The second equation of (9.6) is equivalent to

$$Y'A = \Lambda Y', \tag{9.7}$$

and this suggests the designation of the columns of X as *eigencolumns* and the rows of Y' as *eigenrows* of A. If we premultiply the first equation of (9.6) by Y' and postmultiply (9.7) by X we have

$$Y'AX = (Y'X)\Lambda = \Lambda(Y'X). \tag{9.8}$$

The only matrices which commute with a diagonal matrix, whose elements are all different, are diagonal matrices. Hence $Y'X$ is a diagonal matrix and, moreover, it is non-singular since X and Y are non-singular. We can thus normalize individual vectors so that

$$Y'X = I, \tag{9.9}$$

and this implies the *orthogonality* relations of the eigenvectors, viz.:

$$\left.\begin{aligned} \mathbf{y}'_r\mathbf{x}_s &= 0, \qquad r \neq s \\ \mathbf{y}'_r\mathbf{x}_r &= 1. \end{aligned}\right] \tag{9.10}$$

Further, (9.8) implies

$$Y'AX = \Lambda,$$

and comparison with (9.5) gives

$$Y' = X^{-1}. \tag{9.11}$$

If we now assume that A is symmetric, i.e., $A' = A$, we have $Y = X$ and (9.9) now becomes

$$X'X = I. \tag{9.12}$$

A matrix X satisfying (9.12) is called an *orthogonal* matrix. From (9.12) it follows that the columns of an orthogonal matrix satisfy the relations

$$\mathbf{x}'_r\mathbf{x}_s = 0, \quad r \neq s; \qquad \mathbf{x}'_r\mathbf{x}_r = 1. \tag{9.13}$$

Another result which follows directly from (9.12) is

$$\det X = \pm 1.$$

If $\det X = 1$, X is termed a proper orthogonal matrix while if $\det X = -1$, X is termed an improper orthogonal matrix. The simplest example of a proper orthogonal matrix is

$$\begin{bmatrix} \cos\theta & \sin\theta \\ -\sin\theta & \cos\theta \end{bmatrix}.$$

In what follows we shall take the term 'orthogonal matrix' to mean 'proper orthogonal matrix'. With A still a symmetric matrix and P an orthogonal matrix it follows from Theorem 8-1 and (9.12) that the similar matrix $P^{-1}AP$ is also symmetric. This result is of considerable practical importance in a large number of methods for the determination of the eigenvalues of a symmetric matrix.

As symmetric matrices occur frequently in practice it is worthwhile to delve a little deeper into their properties. The following result about the eigenvalues and eigenvectors of a real symmetric matrix is particularly important.

THEOREM 9-2 *The eigenvalues of a real symmetric matrix A are real.*

Proof Let $\mathbf{x}$ denote the eigenvector corresponding to the eigenvalue λ, that is

$$A\mathbf{x} = \lambda\mathbf{x}. \tag{9.14}$$

Denote the complex conjugates of $\mathbf{x}$ and λ by $\tilde{\mathbf{x}}$ and $\tilde{\lambda}$. Then

$$A\tilde{\mathbf{x}} = \tilde{\lambda}\tilde{\mathbf{x}},$$

since A is real. Transposition gives

$$\tilde{\mathbf{x}}'A = \tilde{\lambda}\tilde{\mathbf{x}}',$$

since A is symmetric. Hence,

$$\tilde{\mathbf{x}}'A\mathbf{x} = \tilde{\lambda}\tilde{\mathbf{x}}'\mathbf{x}.$$

But, from (9.14),

$$\tilde{\mathbf{x}}'A\mathbf{x} = \lambda\tilde{\mathbf{x}}'\mathbf{x}.$$

Thus $$(\tilde{\lambda} - \lambda)\tilde{\mathbf{x}}'\mathbf{x} = 0.$$

But $$\tilde{\mathbf{x}}'\mathbf{x} = |x_1|^2 + |x_2|^2 + \ldots > 0 \qquad \text{(Why?)}$$

and so $\tilde{\lambda} = \lambda$ and the theorem is proved.

Theorem 9-2 is a particular example of a more general result for Hermitian matrices (exercise 3).

Now, consider (9.1) in the form

$$A\mathbf{x} = \lambda\mathbf{x}$$

where A is a real symmetric matrix. Premultiplication by $\mathbf{x}'$, and rearrangement, gives

$$\lambda = \frac{\mathbf{x}'A\mathbf{x}}{\mathbf{x}'\mathbf{x}}. \tag{9.15}$$

From (9.15) we see that, if $\mathbf{x}'A\mathbf{x} > 0$ for any non-zero eigenvector $\mathbf{x}$, that is the matrix A is positive definite, the eigenvalue λ will be positive.

If $\mathbf{y}$ is not an eigenvector then

$$\lambda_R = \frac{\mathbf{y}'A\mathbf{y}}{\mathbf{y}'\mathbf{y}} \tag{9.16}$$

is a number, called *Rayleigh's quotient*, which has important applications in applied mathematics. It is also extremely useful in finding a second-order approximation to an eigenvalue, λ_k, if a first-order approximation to the eigenvector $\mathbf{x}_k$ is known. For suppose

$$\mathbf{y} = \sum_{r=1}^{n} \alpha_r \mathbf{x}_r,$$

is the first-order approximation to $\mathbf{x}_k$, i.e., if ϵ is a small quantity then all the coefficients α_r are $O(\epsilon)$, except α_k which is of order unity. It follows that

$$\lambda_R - \lambda_k = \frac{\sum_{r=1}^{n} \alpha_r^2 \lambda_r}{\sum_{r=1}^{n} \alpha_r^2} - \lambda_k,$$

$$= \sum_{r=1}^{n} \alpha_r^2(\lambda_r - \lambda_k) \Big/ \sum_{r=1}^{n} \alpha_r^2 = O(\epsilon^2), \tag{9.17}$$

since the coefficient of α_k^2 is zero. Equation (9.17) thus implies that λ_R gives a second-order approximation to λ_k. Equation (9.17) also implies, if λ_k is taken to be the smallest eigenvalue, that $\lambda_R > \lambda_k$ since each term in the numerator is now positive and so is the denominator. This means that the Rayleigh quotient always overestimates the smallest eigenvalue and consequently the determination of eigenvalues can be considered a problem in minimization (section 13.8). Similarly it can be shown that λ_R underestimates the largest eigenvalue. If A is not symmetric then $\mathbf{y}'A\mathbf{x}/\mathbf{y}'\mathbf{x}$ plays an analogous role to the Rayleigh quotient (exercise 7).

9.5 Bounds and other results for the eigenvalues of a matrix

We can rewrite the characteristic equation (9.2) in the form

$$\lambda^n + s_1\lambda^{n-1} + s_2\lambda^{n-2} + \ldots + s_{n-1}\lambda + s_n = 0. \tag{9.18}$$

Comparison of the coefficients in the expansion of (9.2) with those of (9.18) gives, in particular,

$$s_1 = -(a_{11} + a_{22} + \ldots + a_{nn}) = -\text{trace } A$$
$$s_n = (-1)^n \det A.$$

Hence, from the algebraic properties of polynomial equations, we have the results

THEOREM 9-3 *If $\lambda_1, \lambda_2, \ldots, \lambda_n$ are the eigenvalues of the matrix A, we have*

$$\lambda_1 + \lambda_2 + \ldots + \lambda_n = a_{11} + a_{22} + \ldots + a_{nn} = \text{trace } A \tag{9.19}$$
$$\lambda_1\lambda_2 \ldots \lambda_n = \det A. \tag{9.20}$$

Equation (9.19) is the most useful of these two results since it provides a rapid and simple means of checking that the computed eigenvalues are correct (apart from a small computational error, of course). Thus for the specimen matrix (9.3) where our eigenvalues have been determined exactly, we have

$$\lambda_1 + \lambda_2 + \lambda_3 = 1 + 2 + 3 = 6, \qquad a_{11} + a_{22} + a_{33} = 2 + 2 + 2 = 6;$$

and

$$\lambda_1\lambda_2\lambda_3 = 1 \times 2 \times 3 = 6, \qquad \det A = 2(2 \times 2 - 1 \times 1) = 6.$$

Equation (9.20) is not very practical as approximately $(1/3)n^3$ multiplications are required for the determination of det A, for a $n \times n$ matrix A. We can derive from it, however, a bound for the smallest eigenvalue in modulus, call it λ_1. We have

$$\begin{aligned} |\det A| &= |\lambda_1\lambda_2 \ldots \lambda_n| \\ &\geqslant |\lambda_1|^n. \end{aligned}$$

But Hadamard's theorem (exercise 8) gives

$$|\det A| \leqslant \prod_{i=1}^{n} r_i,$$

where

$$r_i = \sqrt{\left(\sum_{j=1}^{n} |a_{ij}|^2\right)}.$$

Hence

$$|\lambda_1| \leqslant \left(\prod_{i=1}^{n} r_i\right)^{\frac{1}{n}}. \tag{9.21}$$

A more useful bound for eigenvalues (for practical purposes) is that obtained from the following theorem:

THEOREM 9-4 (Gerschgorin's Theorem). *The eigenvalues of the matrix A lie in the union of the circular discs*

$$|\lambda - a_{ii}| \leqslant \sum_{j \neq i} |a_{ij}|, \qquad i = 1, 2, \ldots, n. \tag{9.22}$$

Proof Let λ be any eigenvalue of A. Then there exists a non-zero vector $\mathbf{x}$ with the property (9.1), viz.,

$$(A - \lambda I)\mathbf{x} = \boldsymbol{\sigma}.$$

Suppose the ith element of $\mathbf{x}$ is the largest. Then we can normalize $\mathbf{x}$ so that $x_i = 1$, i.e.,

$$\mathbf{x}' = (x_1, \ldots, x_{i-1}, x_i = 1, x_{i+1}, \ldots, x_n)' \qquad \text{where } |x_j| \leqslant 1,\ j \neq i.$$

Since the ith component of the left-hand side of (9.1) must be zero we have

$$\sum_{j=1}^{n} a_{ij}x_j - \lambda x_i = 0,$$

and hence

$$\lambda - a_{ii} = \sum_{j \neq i}^{n} a_{ij}x_j,$$

i.e.,

$$|\lambda - a_{ii}| \leqslant \sum_{j \neq i} |a_{ij}||x_j| \leqslant \sum_{j \neq i} |a_{ij}|.$$

This proves Theorem 9-4. It follows immediately from the theorem that

$$\max_k |\lambda_k| \leqslant \max_i \left(\sum_j |a_{ij}| \right) = \|A\|_\infty . \tag{9.23}$$

Note that similar results to (9.22) and (9.23) obtain if we work with A' instead of A.

From (9.21) and (9.23) we have for the specimen matrix (9.3)

$$\min_k |\lambda_k| \leqslant (6\sqrt{6})^{\frac{1}{3}} = \sqrt{6} = 2{\cdot}4495 \qquad \text{(to four decimal places)},$$

and

$$\max_k |\lambda_k| \leqslant 5.$$

The exact values of these quantities were shown in section 9.2 to be 1 and 3, respectively.

Another important result of Gerschgorin helps to pinpoint the position of eigenvalues. The result states (see Wilkinson [1] for a proof):

THEOREM 9-5 *If s of the circular discs (9.22) form a connected domain which is isolated from the other discs, then there are precisely s eigenvalues of A within this connected domain.*

We now consider the application of Theorem 9-5 to the matrix A,

$$A = \begin{bmatrix} 5 & 1 & 1 \\ -1 & 1 & 0 \\ 0 & -0{\cdot}5 & 0 \end{bmatrix}.$$

The Gerschgorin discs are

$$|\lambda - 5| \leqslant 2, \qquad |\lambda - 1| \leqslant 1, \qquad |\lambda - 0| \leqslant 0{\cdot}5.$$

The discs are shown in Fig. 9.1. From the diagram it is clear that the disc $|\lambda - 5| \leqslant 2$ is isolated from the other discs and it follows from Theorem 9-5 that there is exactly one eigenvalue contained in this disc. Moreover, since complex eigenvalues occur in pairs, we can assert that this eigenvalue is real.

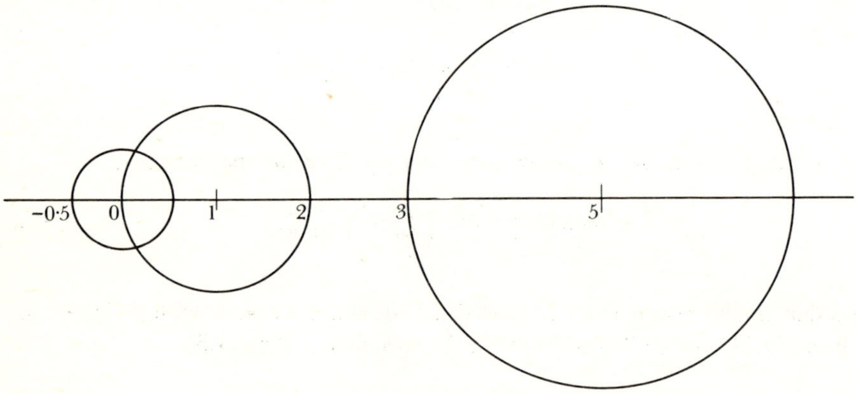

Fig. 9.1 An illustration of Gerschgorin discs.

All we know at present about the remaining two eigenvalues is that they lie in the union of the discs $|\lambda - 1| \leqslant 1$, $|\lambda| \leqslant 0{\cdot}5$. We can, however, obtain more information about them by means of a similarity transformation. Let D be the diagonal matrix diag(0·49, 1, 1). Then $D^{-1} = \text{diag}(2{\cdot}04\ldots, 1, 1)$, and, from Theorem 9-1,

$B = D^{-1}AD$ has the same eigenvalues as A. The effect of the transformation is to multiply the first row of A by 2·04 . . . and the first column by 0·49 so that

$$B = \begin{bmatrix} 5 & 2\cdot04\ldots & 2\cdot04\ldots \\ -0\cdot49 & 1 & 0 \\ 0 & -0\cdot5 & 0 \end{bmatrix}$$

and B has the Gerschgorin discs

$$|\lambda - 5| \leqslant 4\cdot08\ldots, \qquad |\lambda - 1| \leqslant 0\cdot49, \qquad |\lambda| \leqslant 0\cdot5.$$

The last disc is isolated and hence exactly one eigenvalue lies in the range $-0\cdot5 \leqslant \lambda \leqslant 0\cdot5$. Again, from the results already obtained, we see that the third eigenvalue must lie in the range $0\cdot51 \leqslant \lambda \leqslant 1\cdot49$.

We remark, finally, that if we had used A' instead of A we would have obtained $|\lambda - 5| \leqslant 1$, while use of B' shows the even stronger result $|\lambda - 5| \leqslant 0\cdot49$. The exact eigenvalues of A are 0·1085, 1·148, 4·786 to four significant figures.

9.6 Eigenvalues of symmetric tridiagonal matrices. Stürm sequences

The matrix $A = (a_{ij})$ is tridiagonal if $a_{ij} = 0$ for $|i-j| > 1$. The typical tridiagonal matrix thus has the form

$$A = \begin{bmatrix} a_1 & b_2 & 0 & 0 & \ldots & 0 & 0 \\ c_2 & a_2 & b_3 & 0 & \ldots & 0 & 0 \\ 0 & c_3 & a_3 & b_4 & \ldots & 0 & 0 \\ \cdot & \cdot & \cdot & \cdot & \ldots & \cdot & \cdot \\ 0 & 0 & 0 & 0 & \ldots & a_{n-1} & b_n \\ 0 & 0 & 0 & 0 & \ldots & c_n & a_n \end{bmatrix}$$

and it is symmetric if $b_i = c_i\,(i = 2, 3, \ldots, n)$. Now let A be a symmetric tridiagonal matrix and consider

$$\det(A - \lambda I) = \begin{vmatrix} a_1-\lambda & b_2 & 0 & \ldots & 0 & 0 \\ b_2 & a_2-\lambda & b_3 & \ldots & 0 & 0 \\ 0 & b_3 & a_3-\lambda & \ldots & 0 & 0 \\ \cdot & \cdot & \cdot & \ldots & \cdot & \cdot \\ \cdot & \cdot & \cdot & \ldots & \cdot & \cdot \\ \cdot & \cdot & \cdot & \ldots & \cdot & \cdot \\ 0 & 0 & 0 & \ldots & a_{n-1}-\lambda & b_n \\ 0 & 0 & 0 & \ldots & b_n & a_n-\lambda \end{vmatrix}. \qquad (9.24)$$

We can find the characteristic polynomial of the matrix A by means of a sequence of recurrence relations. If we let $p_r(\lambda)$ be the determinant of the matrix formed by the first r rows and r columns of (9.24) and let $p_0(\lambda) = 1$, we have

$$\left.\begin{aligned} &p_0(\lambda) = 1, \qquad p_1(\lambda) = a_1 - \lambda, \\ &p_{r+1}(\lambda) = (a_{r+1} - \lambda)p_r(\lambda) - b_{r+1}^2 p_{r-1}(\lambda), \qquad r = 1, \ldots, n-1 \end{aligned}\right] \tag{9.25}$$

(exercise 10). The final polynomial, $p_n(\lambda)$, is the characteristic polynomial of the matrix A. It is not necessary to evaluate explicitly the coefficients of the polynomial $p_n(\lambda)$ to determine the eigenvalues of A. We use instead a special property of the (Stürm) sequence (9.25) called a Stürm property, viz., for a particular value of λ, say, λ', the number of agreements of sign of the sequence

$$p_0(\lambda'),\ p_1(\lambda'), \ldots, p_n(\lambda'),$$

indicates the number of roots of $p_n(\lambda)$ (and consequently the number of eigenvalues of A) greater than λ'. The particular values of λ chosen are dictated by (9.23) and we use the method of bisection (section 2.3) to isolate the roots (see also section 9.10). The calculation can be completed by this method or, alternatively, we can use Newton's method (section 2.4) or interpolation (section 3.2) to obtain quadratic convergence.

Consider, for example, the evaluation of the largest eigenvalue of the tridiagonal matrix

$$A = \begin{bmatrix} 1 & 2 & 0 & 0 \\ 2 & 1 & 2 & 0 \\ 0 & 2 & 1 & 2 \\ 0 & 0 & 2 & 1 \end{bmatrix} \tag{9.26}$$

The Stürm sequence for the characteristic polynomial of A is

$$\begin{aligned} &p_0(\lambda) = 1, \qquad p_1(\lambda) = 1 - \lambda, \\ &p_{r+1}(\lambda) = (1 - \lambda)p_r(\lambda) - 4p_{r-1}(\lambda), \qquad r = 1, 2, 3. \end{aligned}$$

From (9.23), the eigenvalues of A lie in the range

$$-5 \leqslant \lambda \leqslant 5.$$

The operations to find the largest root might proceed as indicated in Table 9.1 (using a four decimal digit floating point computer similar to the one described in chapter 1).

Table 9.1 Application of the Stürm sequence property to evaluate an eigenvalue of a symmetric tridiagonal matrix

i	0	1	2	3	4	*Number of agreements of sign*
$p_i(5)$	1	−4	12	−32	80	0
$p_i(4)$	1	−3	5	−3	−11	1
$p_i(4{\cdot}5)$	1	−3·5	8·25	−14·88	19·08	0
$p_i(4{\cdot}22)$	1	−3·22	6·368	−7·62	−0·93	1

As guaranteed by Gerschgorin's theorem there is no eigenvalue λ such that $\lambda > 5$. At the next trial value, $\lambda' = 4$, there is one agreement of sign amongst the functions $p_i(\lambda')$ and hence the largest eigenvalue lies between 4 and 5. We have proceeded to use the method of bisection for one stage only to obtain $p_i(4{\cdot}5)$ and have then used quadratic interpolation to establish the approximate position of the eigenvalue to be 4·220. Re-evaluation of the Stürm sequence shows that the eigenvalue lies between 4·220 and 4·5. Further interpolation (using the already tabulated values) will bring us closer to the exact value which is $\lambda = 2 + \sqrt{5} \doteqdot 4{\cdot}236$.

Note that we have assumed throughout that our eigenvalues are distinct. If the tridiagonal matrix has an eigenvalue of multiplicity k then the tridiagonal matrix has at least $(k-1)$ of its off-diagonal elements zero. The matrix can then be partitioned into a direct sum of tridiagonal matrices of lower order to each of which the Stürm sequence property applies. While, of necessity, the possession of a multiple eigenvalue indicates the presence of zero off-diagonal elements, the converse is not true. For example, the tridiagonal matrix, A,

$$A = \begin{bmatrix} 2 & 1 & & \\ 1 & 2 & 0 & \\ & 0 & 3 & 4 \\ & & 4 & 5 \end{bmatrix}$$

does not have any eigenvalue repeated although it has zero off-diagonal elements.

9.7 Eigenvectors of a symmetric tridiagonal matrix

Suppose λ_1 is an exact eigenvalue. Then the corresponding eigenvector $\mathbf{x}_1$ satisfies (9.1),

$$(A - \lambda_1 I)\mathbf{x}_1 = \mathbf{0}.$$

Since $A - \lambda_1 I$ is singular one of its rows is linearly dependent on the other rows. Thus, to obtain the eigenvector $\mathbf{x}_1$ we omit one equation and solve the remaining $(n-1)$ equations as a set of equations in $(n-1)$ ratios (section 9.2).

In practice, due to rounding errors, λ_1 is not obtained exactly and we get an approximation $\lambda = \lambda_1 - \epsilon$ (ϵ small). This fact can be made use of in the following way, which is usually called inverse iteration. We select an arbitrary normalized vector $\mathbf{y}_0$ and we compute (by triangular resolution) the sequence of vectors $\mathbf{y}_r$ given by

$$(A - \lambda I)\mathbf{y}_{r+1} = \mathbf{y}_r. \tag{9.27}$$

If rounding errors are ignored we have, since an arbitrary vector $\mathbf{y}_0$ can be expressed as a linear combination of the eigenvectors $\mathbf{x}_i, \ldots, \mathbf{x}_n$, that is

$$\mathbf{y}_0 = \alpha_1 \mathbf{x}_1 + \alpha_2 \mathbf{x}_2 + \ldots + \alpha_n \mathbf{x}_n, \tag{9.28}$$

$$\mathbf{y}_r = \sum_{i=1}^{n} \alpha_i \left(\frac{1}{\lambda_i - \lambda}\right)^r \mathbf{x}_i, \tag{9.29}$$

(see section 9.12) and, in particular

$$\mathbf{y}_r = \frac{\alpha_1}{\epsilon^r} \mathbf{x}_1 + \sum_{2}^{n} \alpha_i \left(\frac{1}{\lambda_i - \lambda}\right)^r \mathbf{x}_i$$

$$= \frac{1}{\epsilon^r} \left[\alpha_1 \mathbf{x}_1 + \epsilon^r \sum_{2}^{n} \alpha_i \left(\frac{1}{\lambda_i - \lambda}\right)^r \mathbf{x}_i\right]. \tag{9.30}$$

If λ_1 is an isolated eigenvalue then $|\lambda_i - \lambda| \gg \epsilon$ for $i = 2, \ldots, n$ and it follows from (9.30) that $\mathbf{y}_r$ is eventually proportional to $\mathbf{x}_1$ (unless α_1 is 0 or very close to 0) and, after normalization, we obtain a very good approximation to the eigenvector after a few iterations. However, it often happens that $\mathbf{y}_0$ is deficient in the eigenvector $\mathbf{x}_1$ (i.e., $\alpha_1 = 0$). This can be overcome as follows. We take as our initial vector $\mathbf{y}'_0 = (1, 1, \ldots, 1)$ but, instead of (9.27), we compute the sequence defined by

$$U\mathbf{y}_1 = \mathbf{y}_0; \qquad (A - \lambda I)\mathbf{y}_{r+1} = \mathbf{y}_r, \qquad (r = 1, 2, \ldots), \tag{9.31}$$

where $A - \lambda I = LU$. This usually ensures that $\mathbf{y}_1$ has a component of the eigenvector $\mathbf{x}_1$ which is not small. In fact, only one more iteration is usually necessary to achieve an eigenvector whose accuracy is consistent with that of λ.

Consider, for example, the determination of the eigenvector of the matrix A (given by equation (9.26)) corresponding to the computed eigenvalue 4·236. For this matrix and eigenvalue

$$A - \lambda I = \begin{bmatrix} -3{\cdot}236 & 2 & 0 & 0 \\ 2 & -3{\cdot}236 & 2 & 0 \\ 0 & 2 & -3{\cdot}236 & 2 \\ 0 & 0 & 2 & -3{\cdot}236 \end{bmatrix}.$$

Decomposition into LU form gives (on our four decimal digit floating point computer)

$$L = \begin{bmatrix} 1 & 0 & 0 & 0 \\ -0{\cdot}618 & 1 & 0 & 0 \\ 0 & -1 & 1 & 0 \\ 0 & 0 & -1{\cdot}618 & 1 \end{bmatrix} \qquad U = \begin{bmatrix} -3{\cdot}236 & 2 & 0 & 0 \\ 0 & -2 & 2 & 0 \\ 0 & 0 & -1{\cdot}236 & 2 \\ 0 & 0 & 0 & 0 \end{bmatrix}.$$

We next solve $U\mathbf{y}_1 = \mathbf{y}_0$ where $\mathbf{y}_0 = (1, 1, 1, 1)'$ and, since in this (rare) case we obtain u_{44} exactly zero, we approximate to the fourth component of $\mathbf{y}_1$ by some large number, say 10^8. We then obtain $\mathbf{y}_1 = (10^8, 1{\cdot}618 \times 10^8, 1{\cdot}618 \times 10^8, 10^8)'$ which, on normalization so that the largest element becomes 1, gives

$$\mathbf{y}_1 = (0{\cdot}6180, 1, 1, 0{\cdot}6180)'$$

as the approximation to the eigenvector. This is correct to four decimal places and we have no need to find $\mathbf{y}_2$ in this case.

9.8 Householder's method for determining eigenvalues and eigenvectors of a real symmetric matrix

The method is as follows. We let $A = A_0$, and form in turn the matrices

$$\left.\begin{aligned} A_r &= P_r A_{r-1} P_r, \qquad r = 1, 2, \ldots, n-2 \\ \text{where} \qquad P_r &= I - 2\mathbf{w}_r\mathbf{w}_r' \end{aligned}\right] \tag{9.32}$$

and $\mathbf{w}_r$ is a vector such that $\mathbf{w}_r'\mathbf{w}_r = 1$.

P_r is clearly symmetric since $I' = I$ and $(\mathbf{w}_r\mathbf{w}_r')' = \mathbf{w}_r\mathbf{w}_r'$. Further, P_r is orthogonal since

$$\begin{aligned} P_r' P_r &= (I - 2\mathbf{w}_r\mathbf{w}_r')(I - 2\mathbf{w}_r\mathbf{w}_r') \\ &= I - 4\mathbf{w}_r\mathbf{w}_r' + 4\mathbf{w}_r(\mathbf{w}_r'\mathbf{w}_r)\mathbf{w}_r' = I, \end{aligned}$$

using the normalization property of the vector $\mathbf{w}_r$. The elements of $\mathbf{w}_1$ are chosen so that $w_{11} = 0$ and the elements $a_{13}^{(1)}, a_{14}^{(1)}, \ldots, a_{1n}^{(1)}$ of the transformed matrix A_1 are zero. The elements of $\mathbf{w}_2$ are chosen so that $w_{12} = w_{22} = 0$ and the elements $a_{23}^{(2)}, a_{24}^{(2)}, \ldots, a_{2n}^{(2)}$ of the transformed matrix A_2 are zero, and so on.

Thus for a 4×4 real symmetric matrix $A = (a_{ij})$ we take

$$\mathbf{w}_1' = (0, w_{21}, w_{31}, w_{41}) \qquad \text{where } w_{21}^2 + w_{31}^2 + w_{41}^2 = 1$$

and hence

$$P_1 = \begin{bmatrix} 1 & 0 & 0 & 0 \\ 0 & 1 - 2w_{21}^2 & -2w_{21}w_{31} & -2w_{21}w_{41} \\ 0 & -2w_{21}w_{31} & 1 - 2w_{31}^2 & -2w_{31}w_{41} \\ 0 & -2w_{21}w_{41} & -2w_{31}w_{41} & 1 - 2w_{41}^2 \end{bmatrix}.$$

Evaluation of the matrix product $P_1 A P_1 = A_1$ shows that the elements $a_{11}^{(1)}, a_{12}^{(1)}, a_{13}^{(1)}$ in the top row of A_1 have values given by

$$\left.\begin{aligned} a_{11}^{(1)} &= a_{11}, \\ a_{12}^{(1)} &= a_{12} - 2w_{21}p_1, \\ a_{13}^{(1)} &= a_{13} - 2w_{31}p_1, \\ a_{14}^{(1)} &= a_{14} - 2w_{41}p_1, \\ \text{where} \qquad p_1 &= a_{12}w_{21} + a_{13}w_{31} + a_{14}w_{41} \end{aligned}\right]. \tag{9.33}$$

Now one of the properties of this type of orthogonal transformation is that the sum of the squares of the elements in any row remains invariant (exercise 12). We thus have from (9.33),

$$a_{12}^{(1)^2} + a_{13}^{(1)^2} + a_{14}^{(1)^2} = a_{12}^2 + a_{13}^2 + a_{14}^2. \tag{9.34}$$

Further, since we require $a_{13}^{(1)}$ and $a_{14}^{(1)}$ to be zero, we have

$$\left.\begin{aligned} a_{12}^{(1)} &= a_{12} - 2p_1w_{21} = \pm(a_{12}^2 + a_{13}^2 + a_{14}^2)^{\frac{1}{2}} = \pm s_1^{\frac{1}{2}}, \text{ say} \\ a_{13}^{(1)} &= a_{13} - 2p_1w_{31} = 0 \\ a_{14}^{(1)} &= a_{14} - 2p_1w_{41} = 0 \end{aligned}\right]. \tag{9.35}$$

Multiplication of the equations of (9.35) by w_{21}, w_{31}, w_{41} respectively gives (after addition)

$$p_1 = \mp w_{21}s_1^{\frac{1}{2}},$$

so that elimination of p_1 in (9.35) gives the equations for evaluating w_{21}, w_{31}, w_{41}, viz.:

$$\left.\begin{aligned} s_1 &= a_{12}^2 + a_{13}^2 + a_{14}^2 \\ w_{21}^2 &= \tfrac{1}{2}(1 \mp a_{12}/s_1^{\frac{1}{2}}) \\ w_{31} &= \mp a_{13}/2w_{21}s_1^{\frac{1}{2}} \\ w_{41} &= \mp a_{14}/2w_{21}s_1^{\frac{1}{2}} \end{aligned}\right]. \tag{9.36}$$

It is important that we choose the sign in the second equation of (9.36) so that w_{21} is as large as possible since this reduces the error in the division by w_{21} in the remaining equations. Similarly we take

$$\mathbf{w}_2' = (0, 0, w_{32}, w_{42}), \qquad w_{32}^2 + w_{42}^2 = 1$$

and

$$\left.\begin{aligned} s_2 &= a_{23}^{(1)^2} + a_{24}^{(1)^2} \\ w_{32}^2 &= \tfrac{1}{2}(1 \mp a_{23}^{(1)}/s_2^{\frac{1}{2}}) \\ w_{42} &= \mp a_{24}^{(1)}/2w_{32}s_2^{\frac{1}{2}} \end{aligned}\right] \tag{9.37}$$

again with appropriate choice of sign.

The equations (9.36) and (9.37) enable P_1 and P_2 to be determined and hence A_1 and A_2. (A slight extension enables $A_1, A_2, \ldots, A_{n-2}$ to be found in the general case.) The matrix A_2 (A_{n-2} in the general case) turns out to be tridiagonal and its eigenvalues and the corresponding eigenvectors, can be determined by the methods of section 9.6 and 9.7. Now, since P_r is symmetric and orthogonal, (9.32) represents a similarity transformation. It thus follows from Theorem 9-1 that if $\mathbf{y}$ is an eigenvector of A_2, the corresponding eigenvector of A is

$$P_1 P_2 \mathbf{y}$$

(in the general case if $\mathbf{y}$ is an eigenvector of A_{n-2}, $P_1 P_2 \ldots P_{n-2}\mathbf{y}$ is the corresponding eigenvector of A). The matrix multiplication might seem formidable but it is considerably simplified in practice by noting that for each r,

$$P_r \mathbf{y} = \mathbf{y} - 2(\mathbf{w}'_r \mathbf{y})\,\mathbf{w}_r. \tag{9.38}$$

In fact, it is not really necessary to calculate P_r at any stage of the calculation (why?) but it is essential that one keeps a record of the $\mathbf{w}_r$ and the final matrix A_{n-2} although, even here, it is only necessary to store the diagonal elements $a_{ii}^{(n-2)}$ and the off-diagonal elements $a_{i,i+1}^{(n-2)}$.

We illustrate the Householder process for the matrix

$$A = A_0 = \begin{bmatrix} 4 & 3 & 2 & 1 \\ 3 & 4 & 1 & 2 \\ 2 & 1 & 4 & 3 \\ 1 & 2 & 3 & 4 \end{bmatrix}$$

using a four decimal digit floating point computer (*vide* chapter 1). From (9.36)

$$s_1^{\frac{1}{2}} = 3{\cdot}742 = -a_{12}^{(1)}$$

$$w_{21}^2 = 0{\cdot}9010, \qquad w_{21} = 0{\cdot}9492,$$

$$w_{31} = 0{\cdot}2815, \qquad w_{41} = 0{\cdot}1408.$$

Hence $\qquad w'_1 = (0,\ 0{\cdot}9492,\ 0{\cdot}2815,\ 0{\cdot}1408),$

$$P_1 = \begin{bmatrix} 1 & 0 & 0 & 0 \\ 0 & -0{\cdot}8020 & -0{\cdot}5344 & -0{\cdot}2672 \\ 0 & -0{\cdot}5344 & 0{\cdot}8410 & -0{\cdot}079\,27 \\ 0 & -0{\cdot}2672 & -0{\cdot}07927 & 0{\cdot}9604 \end{bmatrix}$$

and

$$A_1 = P_1 A_0 P_1 = \begin{bmatrix} 4 & -3{\cdot}742 & 0 & 0 \\ -3{\cdot}742 & 6{\cdot}572 & -0{\cdot}5214 & -2{\cdot}667 \\ 0 & -0{\cdot}5214 & 2{\cdot}867 & 1{\cdot}275 \\ 0 & -2{\cdot}667 & 1{\cdot}275 & 2{\cdot}559 \end{bmatrix}.$$

Note that, because of rounding errors, the elements $a_{13}^{(1)}, a_{14}^{(1)}$, etc., will not be zero although we take them to be zero. Next

$$s_2^{\frac{1}{2}} = \{(-0{\cdot}5214)^2 + (-2{\cdot}667)^2\}^{\frac{1}{2}} = -2{\cdot}718 = -a_{23}^{(2)}$$

(we take the negative square root here to make w_{32}^2 as large as possible).

$$w_{32}^2 = 0{\cdot}5960, \qquad w_{32} = 0{\cdot}7720,$$

$$w_{42} = 0{\cdot}6355.$$

Hence

$$\mathbf{w}_2' = (0,\ 0,\ 0{\cdot}7720,\ 0{\cdot}6355),$$

$$P_2 = \begin{bmatrix} 1 & 0 & 0 & 0 \\ 0 & 1 & 0 & 0 \\ 0 & 0 & -0{\cdot}1920 & -0{\cdot}9812 \\ 0 & 0 & -0{\cdot}9812 & 0{\cdot}1920 \end{bmatrix}$$

and

$$A_2 = P_2 A_1 P_2 = \begin{bmatrix} 4 & -3{\cdot}742 & 0 & 0 \\ -3{\cdot}742 & 6{\cdot}572 & 2{\cdot}718 & 0 \\ 0 & 2{\cdot}718 & 3{\cdot}050 & 1{\cdot}238 \\ 0 & 0 & 1{\cdot}238 & 2{\cdot}374 \end{bmatrix}.$$

The result (9.19) of Theorem 9-3 provides us with a very good check that our calculations have been performed correctly up to this stage. For the sum of the eigenvalues of the matrix A is the sum of its diagonal elements which is 16 exactly. Since A_2 is a similar matrix to A, the sum of its diagonal elements should also add up to 16. This is found to be the case if the summation of the diagonal elements is performed in the correct computational order (*vide* section 1.4).

Continuing our calculations we find by the methods of section 9.6 and 9.7 our eigenvalues of A_2 are

$$\lambda_1 = 10{\cdot}00, \qquad \lambda_2 = 3{\cdot}996,$$

$$\lambda_3 = 2{\cdot}000, \qquad \lambda_4 = -0{\cdot}000\,329\,3.$$

The corresponding eigenvectors are

$$\mathbf{y}_1 = (0{\cdot}5001,\ -0{\cdot}8018,\ -0{\cdot}3229,\ -0{\cdot}05242)'$$

$$\mathbf{y}_2 = (0{\cdot}5003,\ 0{\cdot}000\,469\,6,\ 0{\cdot}6884,\ 0{\cdot}5252)'$$

$$\mathbf{y}_3 = (0{\cdot}4996,\ 0{\cdot}2671,\ 0{\cdot}2385,\ -0{\cdot}7888)'$$

$$\mathbf{y}_4 = (0{\cdot}5000,\ 0{\cdot}5345,\ -0{\cdot}6042,\ 0{\cdot}3150)'.$$

The eigenvector of A corresponding to λ_1 is

$$\mathbf{x}_1 = P_1 P_2 \mathbf{y}_1 = P_1(\mathbf{y}_1 - 2(\mathbf{w}'_2\mathbf{y}_1)\mathbf{w}_2), \qquad \text{from (9.38)}$$

$$= P_1 \mathbf{y}, \quad \text{say,}$$

where $\qquad \mathbf{y} = (0{\cdot}5001,\ -0{\cdot}8018,\ 0{\cdot}1134,\ 0{\cdot}3068)'.$

$$P_1\mathbf{y} = \mathbf{y} - 2(\mathbf{w}'_1\mathbf{y})\mathbf{w}_1 = (0{\cdot}5001,\ 0{\cdot}5000,\ 0{\cdot}4996,\ 0{\cdot}5000)'$$

i.e., $\qquad \mathbf{x}_1 = (0{\cdot}5001,\ 0{\cdot}5000,\ 0{\cdot}4996,\ 0{\cdot}5000)'.$

The remaining eigenvectors of A are determined in the same way. We can improve our approximation to an eigenvalue, say to λ_2, by evaluating the Rayleigh quotient $\mathbf{x}'_2 A\mathbf{x}_2/\mathbf{x}'_2\mathbf{x}_2$. With the improved approximation we can obtain an improved eigenvector by reducing the matrix $A - \lambda_2 I$ to upper triangular form U and solving in succession

$$U\mathbf{y}_1 = \mathbf{y}_0, \qquad LU\mathbf{y}_{r+1} = \mathbf{y}_r \qquad (r = 1, 2, \ldots)$$

where $\mathbf{y}'_0 = (1, 1, \ldots, 1)$ (as in section 9.7).

Numerical results can also be improved if double-length arithmetic is available for accumulating the inner products which occur in matrix multiplication. The exact eigenvalues of A are 10, 4, 2, 0 and the corresponding exact eigenvectors are $(1, 1, 1, 1)'$, $(1, 1, -1, -1)'$, $(1, -1, 1, -1)'$, $(1, -1, -1, 1)'$.

9.9 Householder's method for real unsymmetric matrices. The Hessenberg matrix

The method proceeds in exactly the same way, and with exactly the same formulae, as for symmetric matrices. The end product after $(n-2)$ stages, however, turns out to be a lower Hessenberg matrix. That is, a matrix for which $a_{ij} = 0$ if $j > i + 1$. Thus for a general 4 x 4 matrix A we finish with the Hessenberg matrix

$$H = \begin{bmatrix} a_{11}^{(2)} & a_{12}^{(2)} & 0 & 0 \\ a_{21}^{(2)} & a_{22}^{(2)} & a_{23}^{(2)} & 0 \\ a_{31}^{(2)} & a_{32}^{(2)} & a_{33}^{(2)} & a_{34}^{(2)} \\ a_{41}^{(2)} & a_{42}^{(2)} & a_{43}^{(2)} & a_{44}^{(2)} \end{bmatrix}.$$

The characteristic polynomial for H is obtained by evaluating $p_4(\lambda) = \det(H - \lambda I)$.

One method of performing the calculation is due to Hyman [2] and effectively consists of expanding the determinant along its last row. We have (with obvious extension to the general case for a $n \times n$ Hessenberg matrix)

$$p_4(\lambda) = \det(H - \lambda I) = (-1)^{4-1} a_{12}^{(2)} a_{23}^{(2)} a_{34}^{(2)} s_4(\lambda), \tag{9.39}$$

where

$$s_4(\lambda) = a_{41}^{(2)} x_1 + a_{42}^{(2)} x_2 + a_{43}^{(2)} x_3 + (a_{44}^{(2)} - \lambda) x_4, \tag{9.40}$$

and x_1, x_2, x_3, x_4 are obtained by solving successively the equations

$$\left.\begin{aligned} &x_1 = 1 \\ &(a_{11}^{(2)} - \lambda)x_1 + a_{12}^{(2)} x_2 = 0 \\ &a_{21}^{(2)} x_1 + (a_{22}^{(2)} - \lambda) x_2 + a_{23}^{(2)} x_3 = 0 \\ &a_{31}^{(2)} x_1 + a_{32}^{(2)} x_2 + (a_{33}^{(2)} - \lambda) x_3 + a_{34}^{(2)} x_4 = 0. \end{aligned}\right] \tag{9.41}$$

The eigenvalues are obtained from (9.39) as follows. First we use Gerschgorin's theorem to deduce upper and lower bounds for the largest eigenvalue, call them b and $-b$ respectively ($b > 0$). We then examine the interval $(-b, b)$ for changes of sign in the polynomial $s_4(\lambda)$ and hence also $p_4(\lambda)$. When a change of sign is encountered we home on to the root by the method of bisection (for example). This approach will determine all real eigenvalues. A convenient method of finding the remaining complex eigenvalues is that of Müller described in exercise 22, chapter 2. Complex arithmetic will, of course, be necessary to determine the complex eigenvalues. The eigenvectors are clearly obtained simultaneously with the eigenvalues (whether real or complex) the eigenvector being the value of $(x_1, x_2, x_3, x_4)'$ for which $p_4(\lambda) = 0$. Finally, we remark that the elements above the diagonal have all been assumed non-zero. If such an element does happen to be zero then the original Hessenberg matrix can be treated as a direct sum of lower Hessenberg matrices (although these are not now necessarily square) and each Hessenberg matrix can be treated separately. Note that we can use inverse iteration to improve the accuracy of the eigenvectors.

9.10 Error analysis of the method of bisection for finding eigenvalues of symmetric tridiagonal matrices

In this section we intend proving that the computed eigenvalue $\bar{\lambda}$ of a symmetric tridiagonal matrix A obtained (on the four decimal digit floating point computer described in chapter 1) by the method of bisection, differs by a small amount from the actual value λ.

We can assume, without loss of generality (or accuracy) that the matrix A is scaled so that $|a_r| \leqslant 0{\cdot}1$, $|b_r| \leqslant 0{\cdot}1$. From (9.23) we have $|\lambda| \leqslant \|A\|_\infty \leqslant 0{\cdot}3$ and, from (9.22), $|a_r - \lambda| \leqslant 0{\cdot}2$. To evaluate, for given λ, each $p_{r+1}(\lambda)$ in (9·25) we must form

$$A_{r+1} = fl\{(a_{r+1} - \lambda)\bar{p}_r(\lambda)\}$$
$$B_{r+1} = fl\{b_{r+1}^2 \bar{p}_{r-1}(\lambda)\}$$

where $\bar{p}_r(\lambda)$ and $\bar{p}_{r-1}(\lambda)$ are the computed values of $p_r(\lambda)$ and $p_{r-1}(\lambda)$. From section 1.10 we have

$$\left.\begin{aligned} A_{r+1} &= (a_{r+1} - \lambda)\bar{p}_r(\lambda)(1 + \epsilon_1) \\ B_{r+1} &= b_{r+1}^2 \bar{p}_{r-1}(\lambda)(1 + \epsilon_2) \end{aligned}\right] \tag{9.42}$$

where, neglecting second order terms, we have

$$|\epsilon_1| \leqslant 1{\cdot}05 \times 10^{-3},$$
$$|\epsilon_2| \leqslant 1{\cdot}0 \times 10^{-3}.$$

The computed value of $p_{r+1}(\lambda)$ is obtained from

$$\bar{p}_{r+1}(\lambda) = fl(A_{r+1} - B_{r+1}) = (A_{r+1} - B_{r+1})(1 + \epsilon_3), \tag{9.43}$$

where $$|\epsilon_3| \leqslant 0{\cdot}55 \times 10^{-3}.$$

Combining (9.42) and (9.43) we have

$$\bar{p}_{r+1}(\lambda) = (a_{r+1} - \lambda)\bar{p}_r(\lambda)(1 + \epsilon_1)(1 + \epsilon_3) - b_{r+1}^2 \bar{p}_{r-2}(\lambda)(1 + \epsilon_2)(1 + \epsilon_3). \tag{9.44}$$

Equation (9.44) shows that the computed values $\bar{p}_{r+1}$ are the exact values that would be obtained from a symmetric tridiagonal matrix A with diagonal entries $\hat{a}_{r+1}$ and off-diagonal entries $\hat{b}_{r+1}$ where

$$\hat{a}_{r+1} - \lambda = (a_{r+1} - \lambda)(1 + \epsilon_1)(1 + \epsilon_3),$$
$$\hat{b}_{r+1}^2 \qquad = b_{r+1}^2 (1 + \epsilon_2)(1 + \epsilon_3).$$

It follows that

$$\hat{a}_{r+1} - a_{r+1} = (a_{r+1} - \lambda)[(1 + \epsilon_1)(1 + \epsilon_3) - 1],$$
$$\hat{b}_{r+1} - b_{r+1} = b_{r+1} [(1 + \epsilon_2)^{\frac{1}{2}}(1 + \epsilon_3)^{\frac{1}{2}} - 1],$$

and hence, using the bounds for $|a_{r+1} - \lambda|$, ϵ_1, ϵ_2, ϵ_3, and $|b_{r+1}|$,

$$|\hat{a}_{r+1} - a_{r+1}| \leqslant 0{\cdot}2\, [(1 + 1{\cdot}05 \times 10^{-3})(1 + 0{\cdot}55 \times 10^{-3}) - 1]$$
$$\leqslant 0{\cdot}32 \times 10^{-3} \qquad \text{(neglecting second order terms)}$$

and

$$|\hat{b}_{r+1} - b_{r+1}| \leqslant 0{\cdot}1\, [(1 + 10^{-3})^{\frac{1}{2}}(1 + 0{\cdot}55 \times 10^{-3})^{\frac{1}{2}} - 1]$$
$$\leqslant 0{\cdot}0775 \times 10^{-3} \leqslant 0{\cdot}08 \times 10^{-3}.$$

Thus, at each stage of the process we can regard the calculation as being exact for a matrix $\hat{A} = A + E$, where E is different at each stage, but always satisfies.

$$|E| \leqslant 0{\cdot}08 \times 10^{-3} \begin{bmatrix} 4 & 1 & & & \\ 1 & 4 & 1 & & \\ & 1 & 4 & 1 & \\ & & \ddots & \ddots & 1 \\ & & & 1 & 4 \end{bmatrix}.$$

In the next section, (9.47), we show that the perturbation E does not change any eigenvalue of the symmetric matrix A by more than $\|E\|_2$. Further,

$$\|E\|_2 \leqslant \{\|E\|_1 \, \|E\|_\infty\}^{\frac{1}{2}}$$
$$\leqslant 0{\cdot}08 \times 10^{-3} \times 6 = 0{\cdot}48 \times 10^{-3}.$$

Now, initially, our eigenvalues were all contained in an interval of width 0·6. After t stages with the bisection process (section 2.3) we might find a particular eigenvalue λ isolated in an interval of width $0{\cdot}6 \times 2^{-t}$. If we take the computed eigenvalue $\bar{\lambda}$ to be at the centre of this interval the error analysis shows that

$$|\bar{\lambda} - \lambda| \leqslant 0{\cdot}3 \times 2^{-t} + 0{\cdot}48 \times 10^{-3}.$$

For $t = 10$, for example,

$$|\bar{\lambda} - \lambda| \leqslant 0{\cdot}78 \times 10^{-3},$$

a very satisfactory result, particularly since it is independent of the order of the matrix and the separation of the eigenvalues.

9.11 Sensitivity of the eigenvalues

Suppose that A is similar to a diagonal matrix, i.e.,

$$P^{-1}AP = \text{diag}(\lambda_1, \lambda_2, \ldots, \lambda_n) = \text{diag}(\lambda_i), \text{ say,} \tag{9.45}$$

where P is either the modal matrix X, or XD where D is a diagonal matrix.

We now investigate how close the eigenvalues of $A + E$ are to those of A. Suppose $\lambda \neq \lambda_i$ is an eigenvalue of $A + E$. Then, for some non-zero $\mathbf{x}$ we have

$$(A + E)\mathbf{x} = \lambda\mathbf{x}$$

i.e.,

$$(\lambda I - A)\mathbf{x} = E\mathbf{x}$$

which, after premultiplication by P^{-1}, may be written as

$$P^{-1}(\lambda I - A)P(P^{-1}\mathbf{x}) = (P^{-1}EP)P^{-1}\mathbf{x},$$

or, from (9.45),

$$[\text{diag}\,(\lambda - \lambda_i)]\,P^{-1}\mathbf{x} = (P^{-1}EP)P^{-1}\mathbf{x}.$$

Because of our assumption $\lambda \neq \lambda_i$, $\text{diag}(\lambda - \lambda_i)$ is non-singular and hence

$$P^{-1}\mathbf{x} = [\text{diag}(\lambda - \lambda_i)^{-1}](P^{-1}EP)P^{-1}\mathbf{x}.$$

Taking the norms of both sides we have

$$\|P^{-1}\mathbf{x}\| \leqslant \|\text{diag}(\lambda - \lambda_i)^{-1}\| \, \|P^{-1}EP\| \, \|P^{-1}\mathbf{x}\|,$$

and thus

$$1 \leqslant \|\text{diag}(\lambda - \lambda_i)^{-1}\| \, \|P^{-1}EP\|.$$

For all the norms defined in chapter 8,

$$\|\text{diag}(\lambda - \lambda_i)^{-1}\| = \max \frac{1}{|\lambda - \lambda_i|},$$

and thus

$$\min |\lambda - \lambda_i| \leqslant \| P^{-1}EP\| \leqslant \|P^{-1}\| \|E\| \, \|P\|. \tag{9.46}$$

If $\lambda = \lambda_i$, for some i, then

$$\min |\lambda - \lambda_i| = 0$$

and it is clear that (9.46) is always satisfied. This means that every eigenvalue of $A + E$ lies in at least one of the circular discs (9.46).

Since $\|P^{-1}\| \, \|P\|$, the condition number, is always greater than or equal to 1, this is the best possible condition number for the eigenvalue problem. If A is a symmetric matrix we can take P to be orthogonal (see (9.12)). In this case, since $\|P\|_2 = 1$, it follows that

$$\min |\lambda - \lambda_i| \leqslant \|E\|_2, \tag{9.47}$$

and the determination of the eigenvalues of a symmetric matrix is always a well-conditioned problem however poorly separated the eigenvalues may be.

The result (9.46) gives overall information about the sensitivity of eigenvalues. If $\mathbf{x}_i$ is the ith column of P and $\mathbf{y}_i$ is the ith column of P^{-1} we can normalize these vectors so that

$$\|\mathbf{x}_i\|_2 = \|\mathbf{y}_i\|_2 = 1,$$

then it is shown (Wilkinson [3]) that the individual sensitivity of eigenvectors depends on $1/|\cos\theta_i|$, where

$$\cos\theta_i = \mathbf{y}'_i\mathbf{x}_i.$$

If A is symmetric $\mathbf{x}_i = \mathbf{y}_i$ and $\cos\theta_i = 1$ showing again that the eigenvalues are well-conditioned. If A is transformed into $R^{-1}AR$ where R is orthogonal then $\mathbf{x}_i$ and $\mathbf{y}_i$ become $R\mathbf{x}_i$ and $R\mathbf{y}_i$ respectively and we have

$$(R\mathbf{y}_i)'(R\mathbf{x}_i) = \mathbf{y}'_iR'R\mathbf{x}_i = \mathbf{y}'_i\mathbf{x}_i$$

showing that the sensitivity of individual eigenvectors is invariant with respect to orthogonal similarity transformations.

9.12 Iterative techniques for finding the largest and smallest eigenvalues of a matrix

In many problems we are not interested in finding all the eigenvalues (and the corresponding eigenvectors) and we require only the largest or smallest eigenvalue in modulus (and the corresponding eigenvector). Since the smallest eigenvalue in modulus of A is the largest eigenvalue in modulus of A^{-1} we can restrict our attention to the former problem. We shall assume that the eigenvalues are distinct and ordered so that

$$|\lambda_1| > |\lambda_2| > \ldots > |\lambda_n| > 0.$$

Consider the iterative scheme given by

$$\mathbf{y}_k = A\mathbf{y}_{k-1} \tag{9.48}$$

where $\mathbf{y}_0$ is arbitrary. (To find eigenvalue of smallest modulus we would solve $A\mathbf{y}_k = \mathbf{y}_{k-1}$.) Equation (9.48) gives

$$\begin{aligned}\mathbf{y}_k &= A\mathbf{y}_{k-1} = A^2\mathbf{y}_{k-2} = \ldots = A^k\mathbf{y}_0 \\ &= \sum_{r=1}^{n} \alpha_r\lambda_r^k \mathbf{x}_r,\end{aligned} \tag{9.49}$$

since $\mathbf{y}_0$ can be expressed in the form

$$\mathbf{y}_0 = \sum_{r=1}^{n} \alpha_r \mathbf{x}_r$$

where $\mathbf{x}_r$ is an eigenvector and $A\mathbf{x}_r = \lambda_r\mathbf{x}_r$, for all r. Since λ_1, the root of largest modulus, is real, (9.49) can be rewritten in the form

$$\mathbf{y}_k = \lambda_1^k\{\alpha_1\mathbf{x}_1 + \sum_{r=2}^{n} \alpha_r(\lambda_r/\lambda_1)^k\mathbf{x}_r\}$$

or, if α_1 is not zero and k is sufficiently large,

$$\mathbf{y}_k = \lambda_1^k(\alpha_1\mathbf{x}_1 + \boldsymbol{\epsilon}_k), \tag{9.50}$$

$\boldsymbol{\epsilon}_k$ being a vector with very small components. When k is so large that $\boldsymbol{\epsilon}_k$ is negligible (to the required precision) it follows that $\mathbf{y}_k$ is the unnormalized vector $\mathbf{x}_1$ corresponding to the root λ_1. Further, ratios of the successive vectors $\mathbf{y}_k$ and $\mathbf{y}_{k-1}$ have the constant value λ_1. In practice it is preferable to evaluate z_k, which is the vector $\mathbf{y}_k$ normalized so that its largest element is 1, and to replace (9.48) by

$$\mathbf{y}_k = Az_{k-1}.$$

The largest element $\mathbf{y}_k$ at each stage then corresponds to the current estimate for λ_1. We consider this iterative procedure for the matrix

$$A = \begin{bmatrix} 2 & 4 & 6 \\ 3 & 9 & 15 \\ 4 & 16 & 36 \end{bmatrix}.$$

The results, using a four decimal digit floating point computer, are set out in Table 9.2. A good estimate for λ_1 is seen to be $\lambda_1 = 43{\cdot}88$. The corresponding estimate for the eigenvector is $(0{\cdot}1859,\ 0{\cdot}4460,\ 1{\cdot}000)'$. This method is known as the power method.

Table 9.2 Iterative method for finding the largest eigenvalue of a matrix

k	$\mathbf{y}'_k$			$\mathbf{z}'_k$			*Estimate for* λ_1
0	1	1	1	1	1	1	–
1	12	27	56	0·2143	0·4821	1·000	56·00
2	8·356	19·98	44·57	0·1875	0·4483	1·000	44·57
3	8·168	19·60	43·92	0·1860	0·4463	1·000	43·92
4	8·157	19·57	43·88	0·1859	0·4460	1·000	43·88
5	8·156	19·57	43·88	0·1859	0·4460	1·000	43·88

It is worthwhile to note that even when α_1 is zero initially, we ultimately converge to the required eigenvalue and eigenvector. This is so because at some stage of the calculation we obtain from $\mathbf{z}_{k-1}$ the vector $\mathbf{z}_k + \eta$, where η is a vector formed from the rounding errors in the calculation. η will almost certainly contain some non-zero multiple of $\mathbf{x}_1$, i.e., the computed value of $\mathbf{z}_k$ now contains a non-zero multiple of $\mathbf{x}_1$ and the earlier analysis applies.

The rate of convergence is clearly dependent on the ratios $|\lambda_2/\lambda_1|$, $|\lambda_3/\lambda_1|, \ldots, |\lambda_n/\lambda_1|$. The smaller these ratios the faster our rate of convergence. The rate of convergence can often be improved by working with the matrix $B = A - pI$. The eigenvalues of B are $\lambda'_1, \lambda'_2, \ldots, \lambda'_n$ where $\lambda'_1 = \lambda_1 - p, \ldots, \lambda'_n = \lambda_n - p$ and, if we assume $|\lambda'_1| > |\lambda'_k|$ for $k = 2, 3, \ldots, n$, it may be possible to obtain, for suitable choice of p, $\max_k |\lambda'_k/\lambda'_1| < |\lambda_2/\lambda_1|$ (exercise 23).

Difficulties arise when the largest eigenvalues are equal (or nearly equal) in modulus. We mention here the following special cases. The reader should consult Clint **[11]** for recent developments in this field (relating to symmetric matrices).

Case 1 $\lambda_1 = \lambda_2$; $|\lambda_1| > |\lambda_3|$.

This does not present any bother for, if $\mathbf{x}_1$ and $\mathbf{x}_2$ are independent eigenvectors corresponding to the eigenvalues λ_1 and λ_2 respectively, (9.50) becomes

$$\mathbf{y}_k = \lambda_1^{\,k}(\alpha_1\mathbf{x}_1 + \alpha_2\mathbf{x}_2 + \epsilon_k), \qquad \epsilon_k \to \mho \text{ for large } k.$$

and the ratios of the largest components of successive vectors still give λ_1. The vector $\mathbf{y}_k$, now a linear combination of $\mathbf{x}_1$ and $\mathbf{x}_2$, is an eigenvector (Why?).

Case 2 $\lambda_1 = -\lambda_2$; $|\lambda_1| > |\lambda_3|$.

Here we notice that the ratios of the largest components of $\mathbf{y}_{k+2}$ and $\mathbf{y}_k$ converge to λ_1^2 but we have no direct information regarding the eigenvectors. Let $\mathbf{x}_1$ and $\mathbf{x}_2$ be the eigenvectors corresponding to λ_1, λ_2. For large k,

$$\mathbf{y}_{k+1} \doteq \alpha\mathbf{x}_1 + \beta\mathbf{x}_2,$$

and

$$\mathbf{y}_{k+2} \doteq \lambda_1\alpha\mathbf{x}_1 - \lambda_1\beta\mathbf{x}_2.$$

We have from the above equations

$$2\alpha\mathbf{x}_1 \doteq \mathbf{y}_{k+1} + (1/\lambda_1)\mathbf{y}_{k+2},$$

$$2\beta\mathbf{x}_2 \doteq \mathbf{y}_{k+1} - (1/\lambda_1)\mathbf{y}_{k+2}$$

giving $\mathbf{x}_1$ and $\mathbf{x}_2$ by normalization.

Case 3 λ_1, λ_2 conjugate complex numbers; $|\lambda_1| > |\lambda_3|$.

There is now no consistent trend in the vectors $\mathbf{y}_k$, which are expressible, for large k, in the form

$$\mathbf{y}_k = \alpha_1\lambda_1{}^k\mathbf{x}_1 + \bar{\alpha}_1\bar{\lambda}_1{}^k\bar{\mathbf{x}}_1 + \boldsymbol{\epsilon}_k \qquad (\boldsymbol{\epsilon}_k \text{ small}),$$

where bars denote complex conjugates. If $\lambda_1, \bar{\lambda}_1$ are the roots of

$$\lambda^2 + a\lambda + b = 0, \tag{9.51}$$

we have, neglecting $\boldsymbol{\epsilon}_k, \boldsymbol{\epsilon}_{k+1}, \boldsymbol{\epsilon}_{k+2}$,

$$\mathbf{y}_{k+2} + a\mathbf{y}_{k+1} + b\mathbf{y}_k \doteq \mathbf{0}.$$

We can estimate a and b by considering the components of the vectors and hence, from (9.51), λ_1 and $\bar{\lambda}_1$. If we let

$$\lambda_1 = \mu + i\upsilon, \qquad \bar{\lambda}_1 = \mu - i\upsilon,$$

and

$$\alpha_1\lambda_1{}^k\mathbf{x}_1 = \boldsymbol{\xi} + i\boldsymbol{\eta}, \qquad \bar{\alpha}_1\bar{\lambda}_1{}^k\mathbf{x}_1 = \boldsymbol{\xi} - i\boldsymbol{\eta},$$

then

$$\mathbf{y}_k \doteq 2\boldsymbol{\xi},$$

and

$$\mathbf{y}_{k+1} \doteq 2(\mu\boldsymbol{\xi} - \upsilon\boldsymbol{\eta}),$$

i.e.,

$$\boldsymbol{\xi} \doteq \tfrac{1}{2}\mathbf{y}_k, \qquad \boldsymbol{\eta} \doteq (\mu\mathbf{y}_k - \mathbf{y}_{k+1})/2\upsilon.$$

$\mathbf{x}_1$ is obtained by normalizing $\boldsymbol{\xi} + i\boldsymbol{\eta}$ and $\mathbf{x}_2$ is the complex conjugate of $\mathbf{x}_1$.

9.13 Deflation

In those problems where we require the largest eigenvalue (in modulus) we sometimes want, in addition, the next largest eigenvalue, λ_2. This can be obtained by methods of deflation, akin to those for roots of polynomials, and often having similar catastrophic properties, which remove the largest eigenvalue. One such method is due to Hotelling [4].

Let $\mathbf{x}_1$ be the eigencolumn and $\mathbf{y}'_1$ the eigenrow of the matrix A corresponding to the eigenvalue λ_1, i.e.,

$$A\mathbf{x}_1 = \lambda_1 \mathbf{x}_1, \qquad \mathbf{y}'_1 A = \lambda_1 \mathbf{y}'_1,$$

where we suppose $\mathbf{y}'_1 \mathbf{x}_1 = 1$. Hotelling deflation is based on the following theorem.

THEOREM 9-6 *The matrix $H = A - \lambda_1 \mathbf{x}_1 \mathbf{y}'_1$ has the same eigenrows and eigencolumns as A and, apart from the eigenvalue λ_1 which has been replaced by zero, has the same eigenvalues as A.*

Proof. We have

$$\begin{aligned} H\mathbf{x}_1 &= A\mathbf{x}_1 - \lambda_1(\mathbf{x}_1\mathbf{y}'_1)\mathbf{x}_1 = A\mathbf{x}_1 - \lambda_1\mathbf{x}_1(\mathbf{y}'_1\mathbf{x}_1), \\ &= A\mathbf{x}_1 - \lambda_1\mathbf{x}_1 = \boldsymbol{\sigma}, \end{aligned}$$

i.e., $\mathbf{x}_1$ is an eigencolumn of H with corresponding eigenvalue 0. Further, if $k \neq 1$,

$$\begin{aligned} H\mathbf{x}_k &= A\mathbf{x}_k - \lambda_1(\mathbf{x}_1\mathbf{y}'_1)\mathbf{x}_k \\ &= A\mathbf{x}_k - \lambda_1\mathbf{x}_1(\mathbf{y}'_1\mathbf{x}_k) \\ &= A\mathbf{x}_k, \qquad \text{(from (9.10))}, \\ &= \lambda_k\mathbf{x}_k, \end{aligned}$$

which shows $\mathbf{x}_k$ is an eigencolumn of H with corresponding eigenvalue λ_k. A similar argument shows that $\mathbf{y}'_1$ is an eigenrow of H with corresponding eigenvalue 0 and $\mathbf{y}'_k$ $(k \neq 1)$ is an eigenrow of H with corresponding eigenvalue λ_k. This proves the theorem.

It follows from the theorem that if our eigenvalues are ordered as in section 9.12, the largest eigenvalue of H (in modulus) will be the second largest eigenvalue of A (in modulus). Hence, if H is used as the iteration matrix, λ_2 can be determined by the methods of the last section. Finally, the method is particularly well suited to symmetric matrices, since $\mathbf{y}_k = \mathbf{x}_k$ for these matrices.

An alternative method, applicable to both symmetric and non-symmetric matrices, consists of reducing the order of the matrix by one. Let the $n \times n$ matrix A be partitioned in the form

$$\begin{bmatrix} a_{11} & \mathbf{r}' \\ \mathbf{c} & X \end{bmatrix}$$

where $X = (a_{ij})$, $2 \leqslant i, j \leqslant n$, $\mathbf{r}' = (a_{1j})$, $2 \leqslant j \leqslant n$, $\mathbf{c} = (a_{i1})$, $2 \leqslant i \leqslant n$ and let λ_1, $\mathbf{x}_1$ be an eigenvalue and the corresponding eigenvector of A, where $\mathbf{x}'_1 = (1, \mathbf{y}'_1)$. It is left as an exercise for the reader to show that the $(n-1) \times (n-1)$ matrix $X - \mathbf{y}_1\mathbf{r}'$ has the eigenvalues λ_r $(r = 2, \ldots, n)$ and the (unnormalized) eigenvectors $\mathbf{y}_r - \mathbf{y}_1$ $(r = 2, \ldots, n$

REFERENCES

[1] Wilkinson, J. H., *The Algebraic Eigenvalue Problem*, Oxford University Press, London (1965).

[2] Hyman, M. A., 'Eigenvalues and Eigenvectors of General Matrices', Twelfth National Meeting A.C.M., Houston, Texas (1959).

[3] Wilkinson, J. H., Rounding Errors in Algebraic Processes, *N.P.L. Notes on Applied Science*, No. 32, H.M.S.O., London (1963).

[4] Hotelling, H., 'Analysis of a Complex of Statistical Variables into Principal Components', *J. Educ. Psychol.*, **24** (1933) 417–41, 498–520.

[5] Bodewig, E., *Matrix Calculus*, 2nd edn., North-Holland Publ. Co., Amsterdam (1959).

[6] Rall, L. B. (ed.), *Error in Digital Computation*, I and II, J. Wiley & Sons, New York (1965).

[7] Fox, L., *An Introduction to Numerical Linear Algebra*, Oxford University Press, London (1964).

[8] Walsh, J. (ed.), *Numerical Analysis: An Introduction*, Academic Press, London (1966).

[9] Nef, W., *Linear Algebra*, McGraw-Hill Publ. Co. Ltd., Maidenhead (1967).

[10] Birkhoff, G. and MacLane, S., *A Survey of Modern Algebra*, rev. edn., Macmillan Co., New York (1953).

[11] Clint, M. and Jennings, A., 'The Evaluation of Eigenvalues and Eigenvectors of Real Symmetric Matrices by Simultaneous Iteration', *Computer J.*, **13** (1970), 76–80.

Exercises

1. If the eigenvalues of a $n \times n$ matrix A are all distinct show that the eigenvectors are distinct and are linearly independent, i.e., there does not exist a relation of the form

$$\sum_{r=1}^{n} \alpha_r \mathbf{x}_r = \mathbf{0},$$

unless every $\alpha_r = 0$. (Hint: Let $\sum_{1}^{n} \alpha_r \mathbf{x}_r = \mathbf{0}$ and consider $A^p \sum_{1}^{n} \alpha_r \mathbf{x}_r = \mathbf{0}$ for $p = 0, 1, \ldots, n-1$. Show that this implies that a certain determinant is zero and this cannot possibly be true.)

2. Show that the system of differential equations

$$\mathrm{d}x/\mathrm{d}t = ax + by,$$
$$\mathrm{d}y/\mathrm{d}t = cx + dy, \qquad (a, b, c, d \text{ constants})$$

has the solution

$$x = \xi_1 e^{\lambda_1 t} + \xi_2 e^{\lambda_2 t},$$
$$y = \eta_1 e^{\lambda_1 t} + \eta_2 e^{\lambda_2 t},$$

where λ_1, λ_2 are the eigenvalues of the matrix

$$\begin{bmatrix} a & b \\ c & d \end{bmatrix}$$

and $(\xi_1, \eta_1)'$, $(\xi_2, \eta_2)'$ are the corresponding eigenvectors.

3. Show that, if H is a Hermitian matrix, i.e., one which satisfies $\tilde{H} = (\tilde{h}_{ij}) = (h_{ji}) = H'$ where the tilde denotes complex conjugate, the eigenvalues of H are real.

4. Show that if A and B are both $n \times n$ matrices the eigenvalues of AB are the same as those of BA. How are the eigenvectors of AB and BA connected?

5. Let A and B be $n \times n$ non-singular matrices. Show that if ABB' is symmetric then so is $B^{-1}AB$ and vice versa. Use this result to show that if A has a complete system of independent eigenvectors then there exists a matrix P such that AP is symmetric. Find P for the matrix A in exercise 23.

6. (Cayley–Hamilton theorem.) Show that the matrix A satisfies its characteristic equation. Use this theorem to show that the eigenvalues of A^{-1} are the reciprocal of the eigenvalues of A. Prove this result also using (9.5).

Given that the characteristic equation of the matrix

$$A = \begin{bmatrix} 2 & 4 & 1 \\ -5 & 1 & 4 \\ 1 & 3 & -3 \end{bmatrix}$$

is $\lambda^3 + 90 = 0$, show that

$$A^3 + 90I = O \quad \text{(the zero matrix)}$$

and deduce A^{-1} from this result.

7. If $\mathbf{x}_r, \mathbf{y}_r$ $(r = 1, 2, \ldots, n)$ denote the eigencolumns and eigenrows of an unsymmetric matrix A and if

$$\mathbf{x} = \sum_{r=1}^{n} \alpha_r \mathbf{x}_r, \qquad \mathbf{y} = \sum_{r=1}^{n} \beta_r \mathbf{y}_r$$

show that

$$\mathbf{y}'A\mathbf{x}/\mathbf{y}'\mathbf{x} = \sum_{r=1}^{n} \alpha_r\beta_r\lambda_r \Big/ \sum_{r=1}^{n} \alpha_r\beta_r.$$

Hence deduce that, if $\mathbf{x}$ and $\mathbf{y}$ are approximations to $\mathbf{x}_k$ and $\mathbf{y}_k$, $\mathbf{y}'A\mathbf{x}/\mathbf{y}'\mathbf{x}$ gives a better approximation to λ_k.

8. Let $B = D^{-1}A$ where $D = \text{diag}(r_1, \ldots, r_n)$ and

$$r_i = \sqrt{\left\{\sum_{j=1}^{n} |a_{ij}|^2\right\}},$$

show, by using Schur's theorem $\sum |\lambda_j|^2 \leqslant \sum\sum |a_{ij}|^2$ (Bodewig [5]), Hadamard's theorem

$$|\det A| \leqslant \prod_{i=1}^{n} r_i.$$

(Hint: Consider $|\det B|^2$ and use (9.20), the theorem of arithmetic and geometric means and Schur's theorem.)

9. If A is the 2×2 matrix

$$A = \begin{bmatrix} a & b \\ c & d \end{bmatrix}$$

with eigenvalues λ_1, λ_2 show that

$$\max(|\lambda_1|, |\lambda_2|) \leqslant \begin{cases} \tfrac{1}{2}(2|a| + |b| + |c| + 2|d|), & a \neq d \\ \tfrac{1}{2}(|a| + |b| + |c| + |d|), & a = d. \end{cases}$$

In which cases does this give a better result than Gerschgorin's theorem? Can this result be generalized for a $n \times n$ matrix?

10. Prove the result (9.25). (Hint: Expand the determinant $p_{r+1}(\lambda)$ along its bottom row.)

11. Show that the exact eigenvalues of the tridiagonal matrix

$$A = \begin{bmatrix} 2 & 1 & 0 \\ 1 & 2 & 1 \\ 0 & 1 & 2 \end{bmatrix}$$

are $2 - \sqrt{2}, 2, 2 + \sqrt{2}$. Assuming that the eigenvalue 2 is obtained by calculation as $2 - \epsilon$ show using (9.31) that the calculated eigenvector is approximately $(1 - 3\epsilon/2 + 1/2\epsilon, 3/2, -1/2\epsilon - \epsilon_4)'$ and, by normalization and letting $\epsilon \to 0$, find the ex eigenvector. Taking $\sqrt{2}$ as 1·414 obtain estimates for the remaining eigenvectors using the same method.

12. Prove (9.34) for a symmetric 4×4 matrix and generalize the result for a $n \times n$ matrix.

13. Obtain the eigenvectors corresponding to the calculated values $\lambda_2 = 3{\cdot}996$, $\lambda_3 = 2{\cdot}000$, $\lambda_4 = -0{\cdot}000\,329\,3$ of the matrix

$$\begin{bmatrix} 4 & 3 & 2 & 1 \\ 3 & 4 & 1 & 2 \\ 2 & 1 & 4 & 3 \\ 1 & 2 & 3 & 4 \end{bmatrix}$$

of section 9.8.

14. Find the eigenvalues and eigenvectors of the Hessenberg matrix

$$\begin{bmatrix} 3 & 4 & 0 & 0 \\ 2 & 3 & 4 & 0 \\ 1 & 2 & 3 & 4 \\ 0 & 1 & 2 & 3 \end{bmatrix}$$

by Hyman's method.

15. Complete an error analysis of Hyman's method for finding the eigenvalues and eigenvectors of a Hessenberg matrix and show that Hyman's method is numerically stable (Rall [6]).

16. Let S be a skew symmetric matrix, i.e., $S' = -S$. Show that the eigenvalues of S are purely imaginary (i.e., of the form $\pm i\lambda$, $i = \sqrt{-1}$). Hence show that S^2 is symmetric and has repeated eigenvalues $-\lambda^2, -\lambda^2$. Moreover if $\mathbf{x}_1, \mathbf{x}_2$ are the eigenvectors of S^2 corresponding to the repeated eigenvalues $-\lambda^2, -\lambda^2$ show that the eigenvectors of S corresponding to $\pm i\lambda$ are $\mathbf{x}_1 \pm i\mathbf{x}_2$.

17. Let $H = A + iB$ be a Hermitian matrix and let $H\mathbf{x} = \lambda\mathbf{x}$, where $\mathbf{x} = \mathbf{u} + i\mathbf{v}$. Show that A is a symmetric matrix, B is a skew symmetric matrix and

$$\begin{bmatrix} A & -B \\ B & A \end{bmatrix} \begin{bmatrix} \mathbf{u} \\ \mathbf{v} \end{bmatrix} = \lambda \begin{bmatrix} \mathbf{u} \\ \mathbf{v} \end{bmatrix}.$$

Hence deduce that the $2n \times 2n$ symmetric matrix

$$\begin{bmatrix} A & -B \\ B & A \end{bmatrix}$$

has the repeated eigenvalues $\lambda_1, \lambda_1, \lambda_2, \lambda_2, \ldots, \lambda_n, \lambda_n$ where $\lambda_1, \ldots, \lambda_n$ are the eigenvalues of H. What are the corresponding eigenvectors?

18. Use the result of the previous example to find the eigenvalues and eigenvectors of the Hermitian matrix

$$\begin{bmatrix} 2 & 1+i \\ 1-i & 2 \end{bmatrix}.$$

19. (Householder's method for a Hermitian matrix.) Let $A = A_0$ be a Hermitian matrix and form

$$A_r = P_r A_{r-1} P_r \quad (r = 1, 2, \ldots, n-2)$$

where P_r is a unitary matrix defined by

$$P_r = I - 2\mathbf{w}_r\tilde{\mathbf{w}}_r = I - \mathbf{u}_r\tilde{\mathbf{u}}_r/2K_r^2$$

where

$$u_{ir} = 0 \qquad (i = 1, 2, \ldots, r)$$

$$u_{r+1,r} = a_{r,r+1}^{(r-1)}\,(1 + S_r^2/T_r), \qquad u_{ir} = a_{ri}^{(r-1)} \qquad (i = r+2, \ldots, n)$$

$$S_r^2 = \sum_{i=r+1}^{n} \left|a_{ri}^{(r-1)}\right|^2, \qquad T_r = \left(\left|a_{r,r+1}^{(r-1)}\right|^2 S_r^2\right)^{\frac{1}{2}}$$

$$2K_r^2 = S_r^2 + T_r.$$

Show that A_{n-2} is a complex tridiagonal Hermitian matrix.

20. Find the eigenvalues and eigenvectors of the tridiagonal Hermitian matrix

$$\begin{bmatrix} 3 & 2+i & 0 \\ 2-i & 2 & 1+i \\ 0 & 1-i & 2 \end{bmatrix}.$$

21. In the Householder method for determining the eigenvalues of a matrix let $\bar{A}_r$ denote the computed value of A_r, P_r be the exact orthogonal matrix for $\bar{A}_r$ and $\bar{P}_r$ the computed value of P_r. Show that on a floating point t binary digit computer

$$\|\bar{P}_r - P_r\| \leqslant \alpha 2^{-t}, \qquad \alpha = O(1).$$

Further, if

$$\bar{A}_{r+1} = fl(\bar{P}'_r \bar{A}_r \bar{P}_r) = \bar{P}'_r \bar{A}_r \bar{P}_r + G_r,$$

show that

$$\|G_r\| \leqslant \beta 2^{-t} \|\bar{A}_r\|, \qquad \beta = O(1).$$

Hence deduce that

$$(1 - \gamma 2^{-t})^{r+1} \|A_0\| \leqslant \|\bar{A}_{r+1}\| \leqslant (1 + \gamma 2^{-t})^{r+1} \|A_0\|,$$

(where norms refer to Euclidean norms and $\gamma = \beta + 2\alpha + \alpha^2 2^{-t}$) and $\bar{A}_{r+1}$ is the exact similarity transformation of $A_0 + E$ where

$$\|E\| \leqslant (r+1)\gamma 2^{-t} \|A_0\| (1 + \gamma 2^{-t})^r.$$

(This shows that Householder's method is stable – see Rall [**6**].)

22. Find the two largest eigenvalues (and their corresponding eigenvectors) of the matrix

$$\begin{bmatrix} 2 & 0 & 0 \\ 2 & -1 & 1 \\ 1 & 1 & -1 \end{bmatrix}$$

by an iterative method. Check your result by Hyman's method.

23. Find the largest eigenvalue of the matrix

$$A = \begin{bmatrix} 2 & 0 & 0 \\ 2 & 2 & 1 \\ 1 & 1 & 2 \end{bmatrix}$$

and its corresponding eigenvector by using the iterative method of section 9.12 applied to the matrix $A - I$. By applying the method to $A - 3I$ find the smallest eigenvalue and its corresponding eigenvector.

24. What is the eigenvector corresponding to λ_2 in Case 1 of section 9.12?

25. Find the two largest eigenvalues (and corresponding eigenvectors) of the matrix

$$\begin{bmatrix} 3 & 1 & 1 \\ 2 & 1 & 1 \\ -1 & -1 & 3 \end{bmatrix}$$

using the methods of section 9.12.

26. Find the eigenvalues and eigenvectors of the following matrices by the most suitable method.

$$(a) \begin{bmatrix} 10 & 9 & 7 & 5 \\ 9 & 10 & 8 & 6 \\ 7 & 8 & 10 & 7 \\ 5 & 6 & 7 & 5 \end{bmatrix} \qquad (b) \begin{bmatrix} 6 & 4 & 4 & 1 \\ 4 & 6 & 1 & 4 \\ 4 & 1 & 6 & 4 \\ 1 & 4 & 4 & 6 \end{bmatrix}$$

$$(c) \begin{bmatrix} 5 & 4 & 1 & 1 \\ 4 & 5 & 1 & 1 \\ 1 & 1 & 4 & 2 \\ 1 & 1 & 2 & 4 \end{bmatrix} \qquad (d) \begin{bmatrix} 4 & -5 & 0 & 3 \\ 0 & 4 & -3 & -5 \\ 5 & -3 & 4 & 0 \\ 3 & 0 & 5 & 4 \end{bmatrix}$$

$$(e) \begin{bmatrix} 2 & 1 & 3 & 4 \\ 1 & -3 & 1 & 5 \\ 3 & 1 & 6 & -2 \\ 4 & 5 & -2 & -1 \end{bmatrix}.$$

These are test matrices used by Rutishauser (see reference, exercise 32).

27. Use the iterative method of section 9.12 to obtain the largest eigenvalue of (*e*) of the previous exercise. Find the second largest eigenvalue by Hotelling's method.

28. If Hotelling's method is used to find all the eigenvalues and eigenvectors of a matrix would you feel confident about the accuracy of the smallest eigenvalues?

29. Let $$A = (a_{ij}) = \frac{(i+j-2)!}{(i-1)!(j-1)!} \qquad (i, j = 1, 2, \ldots, n).$$

Show that A and A^{-1} have the same eigenvalues. (Hint: Show there exists a matrix C with $A = CC'$, $A^{-1} = C'C$.) Use this result to find all the eigenvalues of the matrix

$$\begin{bmatrix} 1 & 1 & 1 & 1 & 1 & 1 \\ 1 & 2 & 3 & 4 & 5 & 6 \\ 1 & 3 & 6 & 10 & 15 & 21 \\ 1 & 4 & 10 & 20 & 35 & 56 \\ 1 & 5 & 15 & 35 & 70 & 126 \\ 1 & 6 & 21 & 56 & 126 & 252 \end{bmatrix}.$$

30. (Method of Jacobi for symmetric matrices – Wilkinson [1].) Let $A = A_0$ be a symmetric matrix. We construct the orthogonal similarity transformation

$$A_r = Y'_r A_{r-1} Y_r$$

where $y_{pp}^{(r)} = c = \cos\theta$, $y_{qq}^{(r)} = c$; $y_{pq}^{(r)} = s = \sin\theta$, $y_{qp}^{(r)} = -s$; $y_{ii}^{(r)} = 1\ (i \neq p, q)$ $y_{ij}^{(r)} = 0$ elsewhere and

$$c^2 = \tfrac{1}{2} + 1/2\sqrt{(\alpha^2 + 1)}, \qquad s^2 = \tfrac{1}{2} - 1/2\sqrt{(\alpha^2 + 1)},$$

$$\alpha = 2a_{pq}^{(r-1)}\Big/\left(a_{pp}^{(r-1)} - a_{qq}^{(r-1)}\right).$$

The effect of the transformation is to make $a_{pq}^{(r)} = a_{qp}^{(r)} = 0$. At each stage of the calculation the numerically largest off-diagonal element is chosen to be zero and eventually the sequence A_r converges to a diagonal matrix whose elements are the eigenvalues of A. How would you decide where to terminate the sequence for a given precision?

31. Apply the method of Jacobi to finding the eigenvalues and eigenvectors of the matrix of exercise 11.

32. (Rutishauser's L–R method – *Nat. Bur. of Standards Applied Math. Series*, **49** (1958).) Let L and U denote lower and upper triangular matrices obtained by triangular decomposition (section 8.3) and consider the process

$$A = A_0 = L_0 U_0, \qquad A_1 = U_0 L_0 = L_1 U_1, \ldots, A_k = U_{k-1} L_{k-1} = L_k U_k, \ldots$$

show that A and A_k have the same eigenvalues. Rutishauser shows that if $B_k = L_1 L_2 \ldots L_k$ converges as $k \to \infty$ then A_k converges to an upper triangular matrix having the same eigenvalues as A. Convergence is guaranteed if A is symmetric and positive definite. If some of the eigenvalues are complex then under certain conditions some of the diagonal elements of A_k converge to the values of the real eigenvalues and the lower part of $A_k \to 0$ except for some elements on the sloping line below the diagonal. Suppose $a_{q,q-1}^{(k)}$ does not converge to zero then the eigenvalues of the 2 x 2 matrix

$$\begin{bmatrix} a_{q-1,q-1}^{(k)} & a_{q-1,q}^{(k)} \\ a_{q,q-1}^{(k)} & a_{qq}^{(k)} \end{bmatrix}$$

converge to a complex pair of eigenvalues of A. The method is particularly well suited to band matrices (chapter 8, exercise 11).

33. Find by the L–R method the eigenvalues of the matrices (a) and (d) of exercise 26. How would you determine the eigenvectors?

34. Suppose B is a positive definite real symmetric matrix and P is an orthogonal matrix such that

$$P'BP = D^2,$$

where D is a diagonal matrix. Show that if A is a real symmetric matrix then the values of λ satisfying

$$A\mathbf{x} = \lambda B\mathbf{x}$$

satisfy

$$Q\mathbf{z} = \lambda \mathbf{z}$$

where

$$Q = D^{-1}P'APD^{-1} \qquad \text{and} \qquad \mathbf{z} = DP'\mathbf{x}$$

(see chapter 11, exercise 37 for an application).

35. If, as in exercise 34, B is a positive definite real symmetric matrix and A is a real symmetric matrix show that if $B = LL'$ then the equation

$$AB\mathbf{x} = \lambda\mathbf{x}$$

is equivalent to

$$(L'AL)L'\mathbf{x} = \lambda L'\mathbf{x}$$

i.e., if λ is an eigenvalue and $\mathbf{y}$ an eigenvector of $L'AL$ then λ is an eigenvalue and $(L')^{-1}\mathbf{y}$ an eigenvector of AB.

10 Iterative methods for the solution of simultaneous linear equations

10.1 The general iterative process

When numerical methods are used to solve partial differential equations (see chapter 12) it is usual to approximate to the partial differential equation by a system of simultaneous equations, in which a large number of the coefficients are zero. The matrix of the coefficients in such cases is called a *sparse* matrix. If the methods of chapter 8 are used to solve such a set of equations, because the coefficients are repeatedly changed, the advantage of having zero coefficients is lost. This is not the case with the iterative methods which we now describe.

Let us assume that the system of linear simultaneous equations to be solved is

$$A\mathbf{x} = \mathbf{b}. \tag{10.1}$$

We deduce from (10.1) an iterative formula of the form

$$\mathbf{x}_{r+1} = B\mathbf{x}_r + \mathbf{c}; \tag{10.2}$$

and, by choosing an initial value $\mathbf{x}_0$, we can use (10.2) to generate $\mathbf{x}_1$, then $\mathbf{x}_2$, etc. Provided the process is convergent, successive values of $\mathbf{x}_r$ will give a closer approximation to the actual solution $\mathbf{x}$. As (10.2) will be deduced from (10.1) the advantage of having zero coefficients will not be lost, as the matrices B and $\mathbf{c}$ are the same for all steps of the procedure.

We now describe the simplest type of iterative method, the Jacobi method.

10.2 The Jacobi method

From the set of equations $A\mathbf{x} = \mathbf{b}$, that is from

$$\begin{aligned}
a_{11}x_1 + a_{12}x_2 + \ldots + a_{1n}x_n &= b_1 \\
a_{21}x_1 + a_{22}x_2 + \ldots + a_{2n}x_n &= b_2 \\
\cdot \quad \cdot \quad \cdot \quad \cdot \quad &\cdot \quad \cdot \\
a_{n1}x_1 + a_{n2}x_2 + \ldots + a_{nn}x_n &= b_n
\end{aligned}$$

we generate the iterative equations;

$$x_1^{(r+1)} = \{b_1 - a_{12}x_2^{(r)} - a_{13}x_3^{(r)} - \ldots - a_{1n}x_n^{(r)}\}/a_{11}$$

$$x_2^{(r+1)} = \{b_2 - a_{21}x_1^{(r)} - a_{23}x_3^{(r)} - \ldots - a_{2n}x_n^{(r)}\}/a_{22}$$

$$\cdot \quad \cdot \quad \cdot \quad \cdot \quad \cdot \quad \cdot \quad \cdot \quad \cdot \quad \cdot \quad \cdot$$

$$x_n^{(r+1)} = \{b_n - a_{n1}x_1^{(r)} - a_{n2}x_2^{(r)} - \ldots - a_{n,n-1}x_{n-1}^{(r)}\}/a_{nn}$$

Writing these in matrix form we have,

$$D\mathbf{x}_{r+1} = \mathbf{b} - (L + U)\mathbf{x}_r \tag{10.3}$$

where

$$L \equiv \begin{bmatrix} 0 & 0 & 0 & \ldots & 0 & 0 \\ a_{21} & 0 & 0 & \ldots & 0 & 0 \\ a_{31} & a_{32} & 0 & \ldots & 0 & 0 \\ \cdot & \cdot & \cdot & \ldots & \cdot & \cdot \\ a_{n1} & a_{n2} & a_{n3} & \ldots & a_{n,n-1} & 0 \end{bmatrix}, \qquad U \equiv \begin{bmatrix} 0 & a_{12} & a_{13} & \ldots & a_{1n} \\ 0 & 0 & a_{23} & \ldots & a_{2n} \\ \cdot & \cdot & \cdot & \ldots & \cdot \\ 0 & 0 & 0 & \ldots & a_{n-1,n} \\ 0 & 0 & 0 & \ldots & 0 \end{bmatrix}$$

and $D \equiv \mathrm{diag}(a_{11}, a_{22}, \ldots, a_{nn})$.

From these definitions of L, U, and D it is clear that our original matrix A satisfies

$$A = L + D + U. \tag{10.4}$$

Further, if D^{-1} exists (which is always the case for the applications in which we are interested) we have

$$\mathbf{x}_{r+1} = -D^{-1}(L + U)\mathbf{x}_r + D^{-1}\mathbf{b}.$$

This is then in the form of equation (10.2) with

$$\left.\begin{aligned} B &= -D^{-1}(L + U) \\ \text{and} \qquad c &= D^{-1}\mathbf{b} \end{aligned}\right] \tag{10.5}$$

We shall, for convenience, always take $\mathbf{x}_0 = D^{-1}\mathbf{b}$, although any vector could be used (see section 10.4).

To illustrate the Jacobi method we solve the equations

$$\left.\begin{aligned} 8x_1 - x_2 + x_3 &= 1 \\ 2x_1 + 10x_2 - x_3 &= 4 \\ x_1 + x_2 - 5x_3 &= 3. \end{aligned}\right] \tag{10.6}$$

We rewrite these in their iterative form

$$\left.\begin{aligned} x_1^{(r+1)} &= \tfrac{1}{8}(1 + x_2^{(r)} - x_3^{(r)}) = 0{\cdot}125 + 0{\cdot}125x_2^{(r)} - 0{\cdot}125x_3^{(r)} \\ x_2^{(r+1)} &= \tfrac{1}{10}(4 - 2x_1^{(r)} + x_3^{(r)}) = 0{\cdot}4 - 0{\cdot}2x_1^{(r)} + 0{\cdot}1x_3^{(r)} \\ x_3^{(r+1)} &= \tfrac{1}{5}(-3 + x_1^{(r)} + x_2^{(r)}) = -\,0{\cdot}6 + 0{\cdot}2x_1^{(r)} + 0{\cdot}2x_2^{(r)}. \end{aligned}\right] \tag{10.7}$$

The results of successive applications of the equation (10.7) are seen in Table 10.1. The process is continued until there is no variation in adjacent values of the x_i's or until a specified accuracy is obtained.

Table 10.1 Example of the Jacobi iteration process

r	$x_1^{(r)}$	$x_2^{(r)}$	$x_3^{(r)}$
0	0·1250	0·4000	−0·6000
1	0·2500	0·3150	−0·4950
2	0·2263	0·3005	−0·4870
3	0·2235	0·3060	−0·4946
4	0·2251	0·3058	−0·4941
5	0·2250	0·3056	−0·4938
6	0·2249	0·3056	−0·4939
7	0·2249	0·3056	−0·4939

10.3 The Gauss–Seidel method

In Table 10.1 we used $x_1^{(0)} = 0{\cdot}1250$ in the calculation of $x_2^{(1)}$ although $x_1^{(1)} = 0{\cdot}2500$ was available, in fact throughout the Jacobi process $x_i^{(r)}$ is used when $x_i^{(r+1)}$ has already been calculated. The Gauss–Seidel process makes use of this latest available information and is expressed in the form

$$x_1^{(r+1)} = \{b_1 - a_{12}x_2^{(r)} - a_{13}x_3^{(r)} - \ldots - a_{1n}x_n^{(r)}\}/a_{11}$$

$$x_2^{(r+1)} = \{b_2 - a_{21}x_1^{(r+1)} - a_{23}x_3^{(r)} - \ldots - a_{2n}x_n^{(r)}\}/a_{22}$$

$$x_3^{(r+1)} = \{b_3 - a_{31}x_1^{(r+1)} - a_{32}x_2^{(r+1)} - \ldots - a_{3n}x_n^{(r)}\}/a_{33}$$

$$\cdot \quad \cdot \quad \cdot \quad \cdot \quad \cdot \quad \cdot \quad \cdot \quad \cdot \quad \cdot$$

$$x_n^{(r+1)} = \{b_n - a_{n1}x_1^{(r+1)} - a_{n2}x_2^{(r+1)} - \ldots - a_{n,n-1}x_{n-1}^{(r+1)}\}/a_{nn}$$

With L, D, and U, as defined before, this is the same as the matrix equation

$$(L + D)\mathbf{x}_{r+1} = \mathbf{b} - U\mathbf{x}_r$$

or, assuming $(L + D)^{-1}$ exists

$$\mathbf{x}_{r+1} = -(L + D)^{-1}U\mathbf{x}_r + (L + D)^{-1}\mathbf{b}.$$

This is equation (10.2) with

$$\left.\begin{aligned} B &= -(L + D)^{-1}U \\ \text{and} \qquad \mathbf{c} &= (L + D)^{-1}\mathbf{b}. \end{aligned}\right] \qquad (10.8)$$

Using this method to solve the example given by (10.6), the iterative equations are:

$$x_1^{(r+1)} = 0{\cdot}125 + 0{\cdot}125x_2^{(r)} - 0{\cdot}125x_3^{(r)},$$
$$x_2^{(r+1)} = 0{\cdot}4 - 0{\cdot}2x_1^{(r+1)} + 0{\cdot}1x_3^{(r)},$$
$$x_3^{(r+1)} = -0{\cdot}6 + 0{\cdot}2x_1^{(r+1)} + 0{\cdot}2x_2^{(r+1)},$$

which produces the following table.

Table 10.2 Example of the Gauss–Seidel iteration process

r	$x_1^{(r)}$	$x_2^{(r)}$	$x_3^{(r)}$
0	0·1250	0·4000	−0·6000
1	0·2500	0·2900	−0·4920
2	0·2228	0·3062	−0·4942
3	0·2251	0·3056	−0·4939
4	0·2249	0·3056	−0·4939
5	0·2249	0·3056	−0·4939

It would appear, comparing Tables 10.1 and 10.2 that the Gauss–Seidel method converges faster than the Jacobi. This is not true in general but can be shown to be the case for certain types of matrices (see section 10.5). In some cases, for example, the equation

$$\left.\begin{aligned} x_1 + 2x_2 - 2x_3 &= 1 \\ x_1 + x_2 + x_3 &= 1 \\ 2x_1 + 2x_2 + x_3 &= 1 \end{aligned}\right] \qquad (10.9)$$

it can be shown that the Jacobi process converges while the Gauss–Seidel process does not. It is thus worthwhile to examine the conditions under which convergence is possible.

10.4 The convergence of an iterative process

A measure of the error at the rth stage of an iterative process defined by equation (10.2) is

$$\boldsymbol{\epsilon}_r = \mathbf{x}_r - \mathbf{x},$$

where $\mathbf{x}$ is the true solution of equation (10.1). It should be noted that this is the error of the process, assuming no rounding errors are incurred. Now by (10.2),

$$\begin{aligned} \boldsymbol{\epsilon}_r &= B\mathbf{x}_{r-1} + \mathbf{c} - \mathbf{x}, \\ &= B(\mathbf{x}_{r-1} - \mathbf{x}) + \mathbf{c} + (B - I)\mathbf{x}. \end{aligned}$$

As $\mathbf{x}$ must be the final stage of the iterative process it must satisfy equation (10.2) and so

$$\mathbf{x} = B\mathbf{x} + \mathbf{c},$$

or

$$\mathbf{c} + (B - I)\mathbf{x} = \boldsymbol{\mho},$$

and so

$$\epsilon_r = B(\mathbf{x}_{r-1} - \mathbf{x}) = B\epsilon_{r-1}. \tag{10.10}$$

By repeated application of (10.10) we get

$$\epsilon_r = B^r \epsilon_0. \tag{10.11}$$

If the iterative process is to converge we require that $\epsilon_r \to \boldsymbol{\mho}$ as $r \to \infty$, which would require, by (10.11), that

$$B^r \to \text{the zero matrix as } r \to \infty.$$

Note, it follows from (10.11), that if we have convergence this is independent of ϵ_0 and hence on the initial value of $\mathbf{x}_0$ we choose. Consequently, starting values other than $\mathbf{x}_0 = D^{-1}\mathbf{b}$ can be taken.

Now suppose Y is the modal matrix of B and Λ the spectral matrix. That is. from (9.4),

$$BY = Y\Lambda$$

i.e.,

$$B = Y\Lambda Y^{-1}.$$

It follows that $B^r = Y\Lambda^r Y^{-1}$, and for B^r to tend to the zero matrix, we must have, as Y and Y^{-1} are fixed, Λ^r tending to the zero matrix. But

$$\Lambda^r = \text{diag}(\lambda_1{}^r, \lambda_2{}^r, \ldots, \lambda_n{}^r),$$

and the diagonal elements will only tend to zero if

$$|\lambda_i| < 1 \qquad (i = 1, \ldots, n).$$

The largest value of $|\lambda_i|$ for a given matrix B is called the *spectral radius* and is denoted by $\rho(B)$. The necessary condition for an iterative process, defined by (10.2), to converge is thus $\rho(B) < 1$. It also follows immediately that if two systems converge, that system with the smaller spectral radius will converge the quicker.

From equations (10.5) and (10.8) we see that, for the Jacobi and Gauss–Seidel processes, we require the spectral radii ρ_J of $-D^{-1}(L + U)$ and ρ_G of $-(L + D)^{-1}U$ to be less than one.

In the example given by (10.6)

$$-D^{-1}(L + U) = -\begin{bmatrix} \frac{1}{8} & 0 & 0 \\ 0 & \frac{1}{10} & 0 \\ 0 & 0 & -\frac{1}{5} \end{bmatrix}\begin{bmatrix} 0 & -1 & 1 \\ 2 & 0 & -1 \\ 1 & 1 & 0 \end{bmatrix} = \begin{bmatrix} 0 & \frac{1}{8} & -\frac{1}{8} \\ -\frac{1}{5} & 0 & \frac{1}{10} \\ \frac{1}{5} & \frac{1}{5} & 0 \end{bmatrix}$$

which has eigenvalues 0·146, –0·0725 ± 0·214i. Thus the spectral radius $\rho_J \doteqdot 0{\cdot}22$.

Also

$$-(L+D)^{-1}U = -\begin{bmatrix} 8 & 0 & 0 \\ 2 & 10 & 0 \\ 1 & 1 & -5 \end{bmatrix}^{-1} \begin{bmatrix} 0 & -1 & 1 \\ 0 & 0 & -1 \\ 0 & 0 & 0 \end{bmatrix} = \begin{bmatrix} 0 & \frac{1}{8} & -\frac{1}{8} \\ 0 & -\frac{1}{40} & \frac{1}{8} \\ 0 & \frac{1}{50} & 0 \end{bmatrix}$$

has eigenvalues 0, −0·063, −0·038 and so $\rho_G \doteqdot 0{\cdot}063$. Thus both the Jacobi and Gauss–Seidel systems converge but the latter is the faster. This confirms our earlier practical experience.

For the example given by equation (10.9), $\rho_J = 0$ and $\rho_G = 2$, so only the Jacobi process converges.

10.5 Tests of convergence

The final test of whether an iterative process converges or not is that the spectral radius of the iterative matrix must be less than one. To evaluate this can be quite a long process. However, in certain cases, we can make a decision on convergence or compare rates of convergence – without actually evaluating the eigenvalues – by means of the following tests.

(*a*) If any of the diagonal elements of A are zero the Jacobi and Gauss–Seidel methods cannot be used, as D and $(D+L)$ will be singular.

(*b*) By Gerschgorin's theorem, Theorem 9·4, we have

$$\rho(B) \leqslant \min\left(\|B\|_1, \|B\|_\infty\right).$$

Thus if the maximum sum of the moduli of elements of the rows or columns of B is less than 1, then the iterative process will converge.

In the example given by (10.6) for the Jacobi method the relevant totals are 0·25, 0·3, 0·4, 0·4, 0·325, and 0·225. Thus $\rho_J \leqslant 0{\cdot}4$ and hence the process converges. Similarly $\rho_G \leqslant 0{\cdot}25$ and the Gauss–Seidel process also converges. However, no decision can be made from this test as to the rate of convergence and even if a row or column total is greater than 1 the process can still converge.

(*c*) It may be shown (see exercise 2), that in the case of the Jacobi method, rule (*b*) requires that for convergence

$$|a_{i1}| + |a_{i2}| + \ldots + |a_{i,i-1}| + |a_{i,i+1}| + \ldots + |a_{in}| < |a_{ii}|$$

for all i. A similar inequality holds for the columns of A. This means that if the elements of A, defined by equation (10.1), are such that the sum of the moduli of the off-diagonal terms is less than the modulus of the diagonal term, then the Jacobi method will converge.

When such a condition exists we call the matrix A, *diagonally dominant*. By looking at our example of equation (10.6) we can see immediately that A is diagonally dominant and the Jacobi method converges.

(d) If we divide each equation by its diagonal element a_{ij}, the Gauss–Seidel iterative matrix becomes $-(I+L)^{-1}U$ and in this form we may apply the Stein–Rosenberg theorem (Varga **[1]**) which states:

THEOREM 10-1 (Stein–Rosenberg) *If $B=-(L+U)$ is a non-negative matrix ($b_{ij}\geqslant 0$), then either*

$$(a)\ \rho_J=\rho_G=0,\qquad (b)\ \rho_J=\rho_G=1,\qquad (c)\ 0<\rho_G<\rho_J<1$$

$$\text{or}\qquad (d)\ 1<\rho_J<\rho_G.$$

It follows immediately from this theorem that if the Jacobi iterative matrix is non-negative then the Jacobi and Gauss–Seidel methods both converge or diverge together and in the case of convergence the Gauss–Seidel method will always be at least as fast as the Jacobi and in most cases faster.

(e) If A is real, symmetric and positive definite (section 8.1), then the Gauss–Seidel iterative matrix $-(D+L)^{-1}L'$, has spectral radius less than 1. This is proved as follows.

If $-\lambda$ is an eigenvalue of this matrix and $\mathbf{y}$ an eigenvector we have from (9.1)

$$(D+L)^{-1}L'\mathbf{y}=\lambda\mathbf{y},$$

and hence

$$\mathbf{y}'L'\mathbf{y}=\lambda(\mathbf{y}'D\mathbf{y}+\mathbf{y}'L\mathbf{y}). \tag{10.12}$$

Transposition of (10.12) gives

$$\begin{aligned}\mathbf{y}'L\mathbf{y}&=\lambda(\mathbf{y}'D'\mathbf{y}+\mathbf{y}'L'\mathbf{y}),\\&=\lambda(\mathbf{y}'D\mathbf{y}+\mathbf{y}'L'\mathbf{y}),\\&=\lambda(\mathbf{y}'D\mathbf{y}+\lambda\mathbf{y}'D\mathbf{y}+\lambda\mathbf{y}'L\mathbf{y})\qquad\text{from (10.12)}\end{aligned}$$

and thus, rearranging,

$$(1-\lambda^2)\mathbf{y}'L\mathbf{y}=\lambda(1+\lambda)\mathbf{y}'D\mathbf{y}. \tag{10.13}$$

Since from (10.4), $A=L+D+L'$, it follows from (10.13) that

$$\begin{aligned}(1-\lambda^2)\mathbf{y}'A\mathbf{y}&=\lambda(1+\lambda)\mathbf{y}'D\mathbf{y}+(1-\lambda^2)\mathbf{y}'D\mathbf{y}+(1-\lambda^2)\mathbf{y}'L'\mathbf{y}\\&=(1+\lambda)\mathbf{y}'D\mathbf{y}+(1-\lambda^2)\lambda(\mathbf{y}'D\mathbf{y}+\mathbf{y}'L\mathbf{y}),\qquad\text{from (10.12)}\\&=(1+\lambda)\mathbf{y}'D\mathbf{y}+\lambda\{(1-\lambda^2)+\lambda(1+\lambda)\}\mathbf{y}'D\mathbf{y},\qquad\text{from (10.13)}\\&=(1+\lambda)^2\mathbf{y}'D\mathbf{y}.\end{aligned} \tag{10.14}$$

Since A and D are positive definite $\mathbf{y}'A\mathbf{y}>0$ and $\mathbf{y}'D\mathbf{y}>0$, that is we must have in (10.14)

$$1-\lambda^2>0$$

$$\text{or}\qquad |\lambda|<1,$$

and (e) is proved.

All these points (*a*) to (*e*) help to select the iterative method to be used. If, in a particular case, none of these is applicable, or if a test when applied is inconclusive, then the true behaviour of the method can only be found by evaluating the eigenvalues of the iterative matrix. The difficulty of case (*a*) can sometimes be removed by changing the order of the equations thereby removing the zero elements from the diagonal position. Ordering of the equations, by moving large elements to the diagonal, can also bring about a situation where (*c*) can be used (see exercise 4).

As an illustration of the application of these rules, consider the set of equations found in section 12.6. If for convenience we multiply each equation by −1, the matrix of the coefficients A is

$$\begin{bmatrix} 4 & -1 & 0 & 0 & 0 \\ -1 & 4 & -1 & 0 & 0 \\ 0 & -1 & 4 & -1 & 0 \\ 0 & 0 & -1 & 4 & -1 \\ 0 & 0 & 0 & -1 & 4 \end{bmatrix}$$

which is a typical sparse matrix. Application of the above rules shows that

(*a*) A is diagonally dominant so that the Jacobi method will converge,

(*b*) the Gauss–Seidel method converges, because A is real, symmetric and positive definite

(*c*)

$$-D^{-1}(L+U) = \begin{bmatrix} 0 & \frac{1}{4} & 0 & 0 & 0 \\ \frac{1}{4} & 0 & \frac{1}{4} & 0 & 0 \\ 0 & \frac{1}{4} & 0 & \frac{1}{4} & 0 \\ 0 & 0 & \frac{1}{4} & 0 & \frac{1}{4} \\ 0 & 0 & 0 & \frac{1}{4} & 0 \end{bmatrix}$$

is a non-negative matrix and thus the Gauss–Seidel method converges faster than the Jacobi.

The actual Gauss–Seidel calculation converges to the solution $u_1 = 0{\cdot}076\,923$, $u_2 = 0{\cdot}120\,193$, $u_3 = 0{\cdot}153\,846$, $u_4 = 0{\cdot}120\,193$, $u_5 = 0{\cdot}076\,923$ in nine iterations, whereas the Jacobi method requires 15. The respective spectral radii are

$$\rho_J = \frac{1}{2\sqrt{2}} \doteq 0{\cdot}353\,55 \qquad \text{and} \qquad \rho_G = 3/16 = 0{\cdot}1875.$$

10.6 Accelerated convergence

A measure of the rate of convergence of a particular method can be obtained by taking the difference of two successive iterative values, that is

$$\mathbf{x}_{r+1} - \mathbf{x}_r.$$

Consider the Jacobi case. By (10.3),

$$\begin{aligned} D\{\mathbf{x}_{r+1} - \mathbf{x}_r\} &= b - (L + U)\mathbf{x}_r - D\mathbf{x}_r, \\ &= b - (L + D + U)\mathbf{x}_r, \\ &= b - A\mathbf{x}_r. \end{aligned}$$

Thus the change in $\mathbf{x}$, between two successive stages of the calculation, is dependent on the residual vector of the previous step. Bearing this in mind we might anticipate the action of the process by taking a different proportion of this quantity to accelerate the convergence. That is

$$D\{\mathbf{x}_{r+1} - \mathbf{x}_r\} = w\{\mathbf{b} - A\mathbf{x}_r\}, \tag{10.15}$$

where w is some selected weighting factor.

From (10.15) we deduce the iterative relationship

$$\mathbf{x}_{r+1} = (I - wD^{-1}A)\mathbf{x}_r + wD^{-1}\mathbf{b}.$$

This defines the *extrapolated Jacobi* method and the rate of convergence is governed by the spectral radius of $(I - wD^{-1}A)$ as compared with the Jacobi case of

$$-D^{-1}(L + U) = -D^{-1}(A - D) = I - D^{-1}A.$$

From this it would seem reasonable to suppose that we can choose w so that the largest eigenvalue λ' of $(I - wD^{-1}A)$ is smaller than λ, that of $(I - D^{-1}A)$, the two sets of eigenvalues being linked by the relationship

$$\lambda'_r = 1 - w(1 - \lambda_r).$$

If we carry out a similar exercise in the Gauss–Seidel case we have

$$\begin{aligned} D\{\mathbf{x}_{r+1} - \mathbf{x}_r\} &= \mathbf{b} - U\mathbf{x}_r - D\mathbf{x}_r - L\mathbf{x}_{r+1} \\ &= \mathbf{b} - A\mathbf{x}_r - L\{\mathbf{x}_{r+1} - \mathbf{x}_r\}. \end{aligned} \tag{10.16}$$

To get the accelerated *extrapolated Gauss–Seidel* system we now multiply the right-hand side by the weighting function w to give

$$D\{\mathbf{x}_{r+1} - \mathbf{x}_r\} = w\{\mathbf{b} - A\mathbf{x}_r\} - wL\{\mathbf{x}_{r+1} - \mathbf{x}_r\}.$$

From which we deduce the iterative equation

$$\mathbf{x}_{r+1} = \{I - (D + wL)^{-1}A\}\mathbf{x}_r + w(D + wL)^{-1}\mathbf{b}.$$

The relevant iterative matrices for the Gauss–Seidel method and its extrapolated form are

$$-(D + L)^{-1}U = -(D + L)^{-1}(A - D - L) = I - (D + L)^{-1}A$$

and $I - (D + wL)^{-1}A$.

By comparing these matrices with the equivalent ones for the Jacobi method it will be seen that the choice of w for the Gauss–Seidel case is a much more complicated problem. A lot of work has been done on such problems but it is outside the scope of this chapter (see Ralston and Wilf [2] and the bibliography given in Varga [1]). The problem of accelerating the Gauss–Seidel method can be made less complicated if we return to (10.16), re-write it as

$$(D+L)(\mathbf{x}_{r+1}-\mathbf{x}_r)=\mathbf{b}-A\mathbf{x}_r$$

and then multiply the right-hand side by the weighting function w. This then gives the iterative form

$$\mathbf{x}_{r+1}=\{I-w(D+L)^{-1}A\}\mathbf{x}_r+(D+L)^{-1}w\mathbf{b}.$$

The choice of w then becomes the less complicated one of comparing the eigenvalues of $\{I-(D+L)^{-1}A\}$ and $\{I-w(D+L)^{-1}A\}$, which is similar to the work required in the Jacobi case. This method does introduce extra terms equivalent to $-(w-1)D^{-1}L\mathbf{x}^{(r)}$, into the calculation but this is bearable in view of the reduction in complexity in the choice of w.

As an example we re-consider the example described in section 12.6, which has been previously mentioned in section 10.5. For the Gauss–Seidel case the eigenvalues of $-(L+D)^{-1}U$ that is $\{I-(D+L)^{-1}A\}$ are 1/16, 3/16, and 0 repeated twice. The eigenvalues of $\{I-w(D+L)^{-1}A\}$ will be $(1-\frac{15}{16}w)$, $(1-\frac{13}{16}w)$, and $1-w$ and by taking $w=1{\cdot}1$ these become $-0{\cdot}1$, $-0{\cdot}031\,25$, and $0{\cdot}106\,25$. With this choice of w the spectral radius is reduced from 0·1875 to 0·106 25. The actual iterative equations are:

$$u_1^{(r+1)}=0{\cdot}051\,562\,5-0{\cdot}1u_1^{(r)}+0{\cdot}275u_2^{(r)}$$

$$u_2^{(r+1)}=0{\cdot}068\,75+0{\cdot}025u_1^{(r)}-0{\cdot}1u_2^{(r)}+0{\cdot}275u_3^{(r)}+0{\cdot}25u_1^{(r+1)}$$

$$u_3^{(r+1)}=0{\cdot}103\,125+0{\cdot}025u_2^{(r)}-0{\cdot}1u_3^{(r)}+0{\cdot}275u_4^{(r)}+0{\cdot}25u_2^{(r+1)}$$

$$u_4^{(r+1)}=0{\cdot}068\,75+0{\cdot}025u_3^{(r)}-0{\cdot}1u_4^{(r)}+0{\cdot}275u_5^{(r)}+0{\cdot}25u_3^{(r+1)}$$

$$u_5^{(r+1)}=0{\cdot}051\,562\,5+0{\cdot}025u_4^{(r)}-0{\cdot}1u_5^{(r)}+0{\cdot}25u_4^{(r+1)}$$

With these equations we converge to the solution, given in section 10.5, in eight iterations. Comparing this with the Gauss–Seidel method we see that only one iteration is saved, but it must be remembered that we are trying to improve on a good system. This method of acceleration would be more impressive on a slowly convergent system.

10.7 Rounding errors

Unless all arithmetic is exact, there is bound to be a small round-off error at each stage of the calculation. Thus in calculating $\mathbf{x}_i^{(r+1)}$ there will be a small round-off error η_{ir} where

$$|\eta_{ir}|\leqslant\eta. \tag{10.17}$$

The actual magnitude of η depends on B and $\mathbf{c}$ and the computer being used and can be deduced from section 8.4. Taking rounding errors into account, equation (10.2) becomes

$$\tilde{\mathbf{x}}_{r+1} = B\tilde{\mathbf{x}}_r + \mathbf{c} + \boldsymbol{\eta}_r \tag{10.18}$$

where $\tilde{\mathbf{x}}_r$ is the approximation to the rth iterative value $\mathbf{x}_r$, and $\boldsymbol{\eta}_r$ is the column vector of the errors η_{ir}. The error at each stage of the calculation, using (10.18), is now

$$\begin{aligned}\boldsymbol{\xi}_{r+1} &= \tilde{\mathbf{x}}_{r+1} - \mathbf{x}\\ &= B\tilde{\mathbf{x}}_r + \mathbf{C} + \boldsymbol{\eta}_r - \mathbf{x}\\ &= B\{\tilde{\mathbf{x}}_r - \mathbf{x}\} + \boldsymbol{\eta}_r\\ &= B\boldsymbol{\xi}_r + \boldsymbol{\eta}_r \end{aligned} \tag{10.19}$$

which is to be compared to the true error given by (10.10). By repeated application of (10.19) we get

$$\boldsymbol{\xi}_{r+1} = B^{r+1}\boldsymbol{\xi}_0 + \boldsymbol{\eta}_r + B\boldsymbol{\eta}_{r-1} + B^2\boldsymbol{\eta}_{r-2} + \ldots + B^r\boldsymbol{\eta}_0.$$

As

$$\boldsymbol{\xi}_0 = \tilde{\mathbf{x}}_0 - \mathbf{x} = \mathbf{x}_0 - \mathbf{x} = \boldsymbol{\epsilon}_0$$

$$B^{r+1}\boldsymbol{\xi}_0 = \boldsymbol{\epsilon}_{r+1}$$

and so

$$\boldsymbol{\xi}_{r+1} = \boldsymbol{\epsilon}_{r+1} + \boldsymbol{\eta}_r + B\boldsymbol{\eta}_{r-1} + \ldots + B^r\boldsymbol{\eta}_0.$$

Taking norms

$$\begin{aligned}\|\boldsymbol{\xi}_{r+1}\| &\leqslant \|\boldsymbol{\epsilon}_{r+1}\| + \|\boldsymbol{\eta}_r\| + \|B\boldsymbol{\eta}_{r-1}\| + \ldots + \|B^r\boldsymbol{\eta}_0\|\\ &\leqslant \|\boldsymbol{\epsilon}_{r+1}\| + \|\boldsymbol{\eta}_r\| + \|B\|\,\|\boldsymbol{\eta}_{r-1}\| + \ldots + \|B^r\|\,\|\boldsymbol{\eta}_0\|. \end{aligned} \tag{10.20}$$

If the method used is convergent then $B^r \to O$ and $\boldsymbol{\epsilon}_{r+1} \to \boldsymbol{\sigma}$ as $r \to \infty$, and it follows from this and equations (10.17) and (10.20) that the build up of rounding errors is of little consequence, provided the criterion we apply to terminate the iterative process does not conflict with error considerations. However, if $\rho(B)$ is very close to 1, powers of B will be quite large in the initial stages and so quite a number of terms in (10.20) will contribute. Thus if $\rho(B)$ is near to 1 round-off can cause trouble. It may easily be shown that in the case of Jacobi and Gauss–Seidel if $\rho(B) = 1$ then we are in fact dealing with the case when $\det A = 0$, that is, singular equations.

10.8 Summary and notes

The advantage of iterative methods are (a) the method is repetitive, (b) the matrix B is the same at each stage, (c) round-off error is in general no problem. The first of these means that these methods are ideally suited for digital computers, whilst the second ensures that if they are used with sparse matrices a greater number of equations can be dealt with than with the methods described in chapter 8, as less storage space is required.

Iterative methods will converge provided the spectral radius of the iterative matrix is less than 1. In section 10.5 we have deduced a set of rules which enable us to decide which of the two main methods, Gauss–Seidel or Jacobi, it is best to use. This is usually the Gauss–Seidel but not always.

We have also shown that for particular equations we may accelerate the rate of convergence by intelligent use of a weighting function w. To use these methods we must have knowledge of the eigenvalues of B and we select w accordingly. This varies according to the equations and we have dealt with the two main methods and a simplification of one of these. However, the optimum choice of w is in general outside the range of this chapter, but quite a lot of useful work has been done on this problem (Ralston and Wilf [2] and the bibliography given in Varga [1]). We have only dealt with w being the same at each stage. This is not essential, in fact it is shown (Fox [3]) that in the case of the extrapolated Gauss–Seidel method the problem of selecting w's at each stage is the problem of choosing w_r's such that

$$P_k(t) = \prod_{r=1}^{k} (1 - w_r t)$$

is as small as possible over the whole range of the eigenvalues ($a < \lambda_i < b$) of B. This is a problem to which Chebyshev polynomials can be applied and in fact we choose w_r's to be the reciprocals of the zeros of

$$P_k(t) = T_k\left(\frac{b-a-2t}{b-a}\right) \Big/ T_k\left(\frac{b+a}{b-a}\right)$$

An alternative type of iterative method, using the techniques of chapter 13, is given in exercise 14.

REFERENCES

[1] Varga, R., *Matrix Iterative Analysis*, Prentice-Hall Inc., New Jersey (1962).

[2] Ralston, A. and Wilf, H. S., *Mathematical Methods for Digital Computers*, J. Wiley & Sons, New York (1960).

[3] Fox, L., *An Introduction to Numerical Linear Algebra*, Oxford University Press, London (1964).

Exercises

1. Solve the following set of simultaneous equations by the Jacobi and Gauss–Seidel methods giving your answer correct to five decimal places:

$$\begin{aligned} 20x + 2y + 3z &= 24 \\ x + 8y + z &= 12 \\ 2x - 3y + 15z &= 30. \end{aligned}$$

2. If the Jacobi method is to be used to solve the set of simultaneous equations $A\mathbf{x} = \mathbf{b}$, show that the elements of the iterative matrix $B = \{b_{ij}\}$ are given by

$$b_{ij} = \begin{cases} 0, & i = j \\ -a_{ij}/a_{ii}, & i \neq j. \end{cases}$$

Use Gerschgorin's theorem to show that the convergence of this method is guaranteed if the matrix A is diagonally dominant.

3. Explain why at least one eigenvalue of the Gauss–Seidel iterative matrix must be zero.

4. Deduce that the Jacobi method of solution cannot be used to solve the equations

$$\begin{aligned} x + 2y + 10z &= 59 \\ 2x + 8y + z &= -4 \\ 20x - y + 2z &= 74 \end{aligned}$$

in their given order. Suggest a re-ordering of the equations which would yield a convergent solution and find this solution.

5. Obtain a solution of the following equations to six decimal places by means of an iterative method

$$\begin{aligned} 1{\cdot}72x - 2{\cdot}97y + 0{\cdot}21z &= 6{\cdot}21 \\ 5{\cdot}21x + 1{\cdot}52y - 2{\cdot}37z &= 1{\cdot}68 \\ 2{\cdot}01x + 0{\cdot}92y + 3{\cdot}89z &= 7{\cdot}76. \end{aligned}$$

6. Verify that for the equations (10.9) we have, with the usual notation $\rho_J = 0$, $\rho_G = 2$. Solve the equations by the convergent system.

7. Solve the equations

$$\begin{aligned} 16x_1 + 2x_2 + 3x_3 + x_4 &= 15 \\ x_1 + 10x_2 + 4x_3 + 3x_4 &= 1 \\ 3x_1 + x_2 + 15x_3 + 2x_4 &= -40 \\ x_1 + 2x_2 + 4x_3 + 18x_4 &= 61 \end{aligned}$$

by the quickest basic iterative method, giving the reasons for the choice of method.

8. The set of equations

$$\begin{aligned} x - y \qquad\quad &= 0{\cdot}7 \\ -0{\cdot}25x + y - 0{\cdot}5z &= 0{\cdot}8 \\ -0{\cdot}5y + z &= 0{\cdot}9 \end{aligned}$$

is to be solved by an iterative method. Show that the Gauss–Seidel method converges faster than the Jacobi method. Show further that if the extrapolated Jacobi method is used the best value of w is 2/3. Show that even with this choice of w the extrapolated Jacobi method converges slower than the Gauss–Seidel method.

9. Let $A = D + L + U$ be a tridiagonal matrix. If $A_1 = D + \lambda L + U, A_2 = D + \lambda^{\frac{1}{2}} L + \lambda^{\frac{1}{2}} U$ show that

$$\det A_1 = \det A_2.$$

Does this result extend to other types of matrices (see Young, D. M., *Trans. Amer. Math. Soc.*, **76** (1954) 92–111)?

10. Use the result of exercise 9 to determine the best value of w in the extrapolated Jacobi process for solving $A\mathbf{x} = \mathbf{b}$, where A is a tridiagonal matrix.

11. For a symmetric matrix A of the form $A = I - L - U$, where $L = U'$, Aitken proposed the iterative scheme

$$(I - L)\tilde{\mathbf{x}}_{n+1} = U\mathbf{x}_n + \mathbf{b},$$
$$\mathbf{x}_{n+1} = L\tilde{\mathbf{x}}_{n+1} + U\mathbf{x}_{n+1} + \mathbf{b}.$$

Show that the error vector ϵ_{n+1} satisfies

$$\epsilon_{n+1} = (I - U)^{-1} L (I - L)^{-1} U\epsilon_n = B\epsilon_n.$$

Further, if A is the matrix associated with elliptic partial differential equations, show that B is non-negative definite and all its eigenvalues are less than 1.

12. Evans, D. J. (*Computer J.*, 1963) has suggested the modified Aitken iterative method. He notes that B can be written in the form $(I - U)^{-1}(I - L)^{-1}LU$ and he develops the extrapolated modified Aitken iteration

$$(I - wL)(I - wU)\mathbf{x}_{n+1} = [w^2 LU + (1 - w)I]\,\mathbf{x}_n + w\mathbf{b}.$$

Show that ϵ_n satisfies

$$\epsilon_{n+1} = B_w \epsilon_n$$

where

$$B_w = I - w(I - wU)^{-1}(I - wL)^{-1}A.$$

If

$$T_w = (I - wU)B_w(I - wU)^{-1}.$$

show that $I - T_w$ and $I - B_w$ are positive definite for $w > 0$.

13. Let $\{x_k^{(r)}\}$ be the sequence of successive iterates of the variable x_k in the Jacobi or Gauss–Seidel methods. Show that, if the method is convergent, then the sequence of values $\{y_k^{(r)}\}$ defined by Aitken's δ^2-extrapolation formula

$$y_k^{(r+1)} = x_k^{(r+1)} - \frac{(x_k^{(r+1)} - x_k^{(r)})^2}{x_k^{(r+1)} - 2x_k^{(r)} + x_k^{(r-1)}},$$

converges to the exact solution faster than the sequence $\{x_k^{(r)}\}$. Hence, from Table 10.2, show that $y_1^{(3)} = x_1^{(4)}$ (to four decimal places).

14. Solve the linear equations (10.6) by minimizing

$$(8x_1 - x_2 + x_3 - 1)^2 + (2x_1 + 10x_2 - x_3 - 4)^2 + (x_1 + x_2 - 5x_3 - 3)^2$$

using the methods of chapter 13. (Take as initial approximations $x_1^{(0)} = x_2^{(0)} = 0$, $x_3^{(0)} = -1$.)

11 Ordinary differential equations

11.1 Introduction

The general form of an ordinary differential equation of order n is

$$y^{(n)} = \frac{d^n y}{dx^n} = f(x, y, y^{(1)}, y^{(2)}, \ldots, y^{(n-1)}), \tag{11.1}$$

with associated conditions equal in number to the order of the differential equation. If these conditions are given for one value of x (which can usually be arranged to be $x = 0$ by a suitable change of variable) then we have an 'initial value' system. Otherwise we have a 'boundary value' system. We shall consider the problem of determining numerically the value of the function $y(x)$ at points other than those specified in the associated conditions.

The simplest type of differential equation system is the first order linear equation

$$\left.\begin{aligned} y' &= \frac{dy}{dx} = f(x, y), \\ \text{with} \quad y(x_0) &= y_0. \end{aligned}\right] \tag{11.2}$$

We shall assume that the 'analytic' solution of (11.2) is unique, continuous and differentiable in some finite region $x_0 \leqslant x \leqslant x_N$. That is, we assume that $f(x, y)$ is defined and continuous in the strip $x_0 \leqslant x \leqslant x_N$, $-\infty < y < \infty$, and satisfies the Lipschitz condition

$$|f(x, y) - f(x, Y)| \leqslant L|y - Y|$$

where y and Y are any two numbers and L is a constant (see Goursat [**1**]).

Equations of second and higher order can be reduced to systems of first order equations, e.g., $y'' = y^{(2)} = f(x, y, y')$ can be written as

$$\left.\begin{aligned} y' &= z, \\ z' &= f(x, y, z) \end{aligned}\right] \tag{11.3}$$

a pair of simultaneous first order equations. Thus we shall, on the whole, restrict our attention to methods for solving first order equations. However, for certain problems, it is advisable to use a special technique which does not involve reduction of the original differential equation to a system of simultaneous first order equations (exercise 6).

We now give some numerical methods for solving the equation (11.2). Some methods exhibit instability and a discussion of this is given in section 11.7.

11.2 Euler's method

One fairly obvious approach is to replace the derivative $\mathrm{d}y/\mathrm{d}x$ in (11.2) by a finite-difference approximation. We obtain from formula (3.33),

$$\frac{y(x+h)-y(x)}{h}=f(x, y(x))+O(h). \tag{11.4}$$

Since we know the solution y_0 at $x = x_0$ we can find $y_1 = y(x_0 + h)$ by re-arranging (11.4). Thus

$$y_1 = y(x_0+h) = y(x_0) + hf(x_0, y(x_0)) + O(h^2),$$

i.e., by ignoring the error term of order h^2, we have

$$y_1 \mathrel{\dot{=}} y_0 + hf(x_0, y_0). \tag{11.5}$$

With the usual difference notation of $x_k = x_0 + kh$ and $y_k = y(x_k)$ (section 3.4) we obtain the recurrence relation

$$y_{k+1} \mathrel{\dot{=}} y_k + hf(x_k, y_k), \qquad k = 0, 1, 2, \ldots \tag{11.6}$$

with y_0 given by the initial condition.

This simple procedure is known as Euler's method. To illustrate its use we estimate $y(0{\cdot}1)$, $y(0{\cdot}2)$, $y(0{\cdot}3)$ where $y(x)$ satisfies

$$y' = y, \; y(0) = 1. \tag{11.7}$$

In this instance $f(x, y) = y$ and we shall take $h = 0{\cdot}1$. Successive use of (11.6) gives

$$y(0{\cdot}1) = y_1 = 1 + 0{\cdot}1(1) = 1{\cdot}1,$$
$$y(0{\cdot}2) = y_2 = 1{\cdot}1 + 0{\cdot}1(1{\cdot}1) = 1{\cdot}21,$$
$$y(0{\cdot}3) = y_3 = 1{\cdot}21 = 0{\cdot}1(1{\cdot}21) = 1{\cdot}331,$$

etc. (The exact solution of (11.7) is $y = \mathrm{e}^x$.)

Note that the truncation error in each step is $O(h^2)$ and this is carried through at each stage of the computation so that the accumulated truncation error in y_k after k steps is proportional to $k(h^2) = h(kh) = hx_k$ (for large k) and is, therefore, of order h. The accumulated round-off should also be taken into account (exercise 1).

11.3 Difference equations

Equation (11.6), and many other approximations derived in this chapter, belong to a class of equations known as difference equations. One type of difference equation frequently encountered is a linear homogeneous equation with constant coefficients of the form

$$y_{r+n} + a_1 y_{r+n-1} + \ldots + a_{n-1} y_{r+1} + a_n y_r = 0 \tag{11.8}$$

where the a_i $(i = 1, 2, \ldots, n)$ are constants. As this equation connects y_{r+n} with the

n previous values in the sequence it is said to be of order n. The method of solution of (11.8) can be found by a generalization of the technique which we now employ to solve the equation

$$y_{r+2} - 5y_{r+1} + 6y_r = 0. \tag{11.9}$$

If E is the shift operator, defined in section 3.4, $y_{r+1} = Ey_r$, $y_{r+2} = E^2y_r$, so that (11.9) can be expressed in the form:

$$(E^2 - 5E + 6)y_r = 0,$$

or

$$(E-2)(E-3)y_r = 0. \tag{11.10}$$

Let

$$(E-3)y_r = z_r, \tag{11.11}$$

then we have from (11.10)

$$(E-2)z_r = 0,$$

i.e.,

$$z_{r+1} - 2z_r = 0,$$

or

$$\frac{z_{r+1}}{2^{r+1}} - \frac{z_r}{2^r} = 0. \tag{11.12}$$

This last equation, (11.12), implies

$$\frac{z_{r+1}}{2^{r+1}} = \frac{z_r}{2^r} = \frac{z_{r-1}}{2^{r-1}} = \ldots = \frac{z_1}{2^1} = \frac{z_0}{2^0} = \text{a constant, } C, \text{ say;}$$

that is $z_r = C.2^r$.

Substituting in (11.11) we have

$$(E-3)y_r = y_{r+1} - 3y_r = C.2^r$$

or

$$\frac{y_{r+1}}{2^{r+1}} - \frac{3y_r}{2^{r+1}} = \frac{C}{2}. \tag{11.13}$$

If we let $w_r = y_r/2^r$ then (11.13) becomes

$$w_{r+1} - \frac{3}{2}w_r = \frac{C}{2}.$$

If we make the further substitution $w_r = u_r + \alpha$, where α is a constant, we have

$$(u_{r+1} + \alpha) - \frac{3}{2}(u_r + \alpha) = \frac{C}{2},$$

i.e.,

$$u_{r+1} - \frac{3}{2}u_r = \frac{C}{2} + \frac{1}{2}\alpha. \tag{11.14}$$

Now choose $\alpha = -C$, then (11.14) becomes

$$u_{r+1} - \frac{3}{2}u_r = 0,$$

which yields the solution

$$u_r = k(3/2)^r, \qquad \text{where } k \text{ is a constant.}$$

Hence,

$$w_r = k(3/2)^r - C,$$

and

$$y_r = k3^r - C2^r,$$

i.e., the solution has the form

$$y_r = A_1 2^r + A_2 3^r,$$

where A_1 and A_2 are arbitrary constants which can be determined from a knowledge of y_0, y_1. This method can easily be extended to the general equation (11.8) of order n and we have

$$y_r = A_1\mu_1{}^r + A_2\mu_2{}^r + \ldots + A_n\mu_n{}^r,$$

where $\mu_1, \mu_2, \ldots, \mu_n$ are the distinct roots of the polynomial

$$\mu^n + a_1\mu^{n-1} + \ldots + a_{n-1}\mu + a_n = 0. \qquad (11.15)$$

A slightly different result holds when the roots of (11.15) are not distinct (exercise 3).

11.4 The Taylor series method

This method is based on calculating the coefficients of the Taylor series

$$y(x+h) = y + hy' + \frac{h^2}{2!}y'' + \ldots . \qquad (11.16)$$

If we know y_r and x_r then we can calculate y_{r+1} at $x_{r+1} = x_r + h$, since the derivatives $y'_r, y''_r, \ldots$, can be obtained from (11.2). It is usual to keep h constant although there are advantages in varying h. For example, changing the sign of h in (11.16) estimates $y(x-h)$, the value of y at the previous step, and this provides a useful check that the value of h taken is not too large. Another advantage of this method is that the coefficients of the series decrease faster than those of most finite difference formulae. Thus we can use a large interval h with many terms or small h with few terms.

The disadvantages of this method lies in the calculation of the derivatives, which can be rather tiresome if $f(x, y)$ is an unwieldly function. However, if Leibnitz's rule for successive differentiation of a product is easy to apply, this drawback vanishes.

The application of the Taylor series method to the solution of first order equations is fairly straightforward and we shall consider application of the method to the second order equation

$$y'' = xy, \qquad y(0) = 1, \qquad y'(0) = 0. \tag{11.17}$$

For this equation Leibnitz's rule gives

$$y^{(n+2)} = xy^{(n)} + ny^{(n-1)}, \qquad n \geqslant 1. \tag{11.18}$$

Thus if we take $h = 0{\cdot}1$, in (11.16), we have

$$y(0{\cdot}1) = y(0) + (0{\cdot}1)y'(0) + \frac{(0{\cdot}1)^2}{2!}y''(0) + \frac{(0{\cdot}1)^3}{3!}y'''(0) + \ldots . \tag{11.19}$$

Equation (11.17) gives $y(0) = 1, y'(0) = 0, y''(0) = 0$. The recurrence relations (11.18) gives

$$y^{(3)}(0) = y(0) = 1, \qquad y^{(4)}(0) = 2y'(0) = 0,$$
$$y^{(5)}(0) = 3y^{(2)}(0) = 0, \qquad y^{(6)}(0) = 4y^{(3)}(0) = 4.$$

The next non-zero value will be $y^{(9)}(0) = 7y^{(6)}(0) = 28$. Equation (11.19) thus gives

$$y(0{\cdot}1) = 1 + \frac{(0{\cdot}1)^3}{3!} \times 1 + \frac{(0{\cdot}1)^6}{6!} \times 4 + \frac{(0{\cdot}1)^9}{9!} \times 28 + \ldots,$$
$$= 1{\cdot}0002 \text{ (correct to four decimal places).}$$

$y(0{\cdot}2)$ can be computed directly from $y(0)$ by taking $h = 0{\cdot}2$ in (11.16), that is,

$$y(0{\cdot}2) = 1 + \frac{(0{\cdot}2)^3}{3!} + \frac{(0{\cdot}2)^6}{6!} \times 4 + \frac{(0{\cdot}1)^9}{9!} \times 28 + \ldots,$$
$$= 1{\cdot}0013.$$

Alternatively we can estimate $y(0{\cdot}2)$ from the computed value of $y(0{\cdot}1)$. Taking $h = 0{\cdot}1$, (11.16) implies

$$y(0{\cdot}2) = y(0{\cdot}1) + (0{\cdot}1)y'(0{\cdot}1) + \frac{(0{\cdot}1)^2}{2!}y''(0{\cdot}1) + \ldots . \tag{11.20}$$

In this example we do not have an analytic expression for $y'(x)$ and so we have to compute $y'(0{\cdot}1)$ from the Taylor series at the same time as we compute $y(0{\cdot}1)$. Since

$$y'(h) = y'_0 + hy''_0 + \frac{h^2}{2!}y'''_0 + \frac{h^3}{3!}y_0^{(4)} + \ldots, \tag{11.21}$$

$$v'(0{\cdot}1) = y'(0) + (0{\cdot}1)y''(0) + \frac{(0{\cdot}1)^2}{2!}y'''(0) + \frac{(0{\cdot}1)^3}{3!}y^{(4)}(0) + \ldots,$$
$$= \frac{(0{\cdot}1)^2}{2!} + \frac{(0{\cdot}1)^5}{5!} \times 4 + \frac{(0{\cdot}1)^8}{8!} \times 28 + \ldots,$$
$$= 0{\cdot}0050 \text{ (correct to four decimal places).}$$

From equation (11.17) we have

$$y''(0{\cdot}1) = (0{\cdot}1)(1{\cdot}0002) = 0{\cdot}100\,02,$$

and from (11.18),

$$y'''(0{\cdot}1) = (0{\cdot}1)(0{\cdot}0050) + 1(1{\cdot}0002) = 1{\cdot}0007,$$

$$y^{(4)}(0{\cdot}1) = (0{\cdot}1)(0{\cdot}100\,02) + 2(0{\cdot}0050) = 0{\cdot}0200.$$

Substituting these values in (11.20) we have

$$\begin{aligned} y(0{\cdot}2) &= 1{\cdot}0002 + 0{\cdot}000\,50 + (0{\cdot}01)(0{\cdot}050\,02) + (0{\cdot}0002) \\ &= 1{\cdot}0014. \end{aligned}$$

Notice that $y'''(0{\cdot}1)$ is greater than $y''(0{\cdot}1)$ and therefore it is advisable to compute a few more derivatives in order to avoid truncating at an accidentally negligible value. The difference in the two values for $y(0{\cdot}2)$ is due to round-off to four places of decimal.

Many variations of the Taylor's series method have been suggested usually with the aim of easing the differentiation process. Gibbons [2] considers the general nth order equation (11.1)

$$y^{(n)} = f(x, y, y', \ldots, y^{(n-1)}),$$

with values of $y, y', \ldots, y^{(n-1)}$ given as initial conditions at $x_0 = 0$. The quantity y and its first $(n-1)$ derivatives are then expressed as power series

$$\left.\begin{aligned} y &= a_0 + a_1 x + a_2 x^2 + \ldots, \\ y^{(k)} &= a_0^{(k)} + a_1^{(k)} x + a_2^{(k)} x^2 + \ldots, k = 1, \ldots, n-1 \end{aligned}\right] \qquad (11.22)$$

The initial conditions imply that at least the first coefficient of each of the series (11.22) is known and thus we can find the first n coefficients $a_0, a_1, a_2, \ldots, a_{n-1}$ of the series for y. By using the differential equation and comparing coefficients of x^0 we can find $a_0^{(n)}$, and consequently (by integrating n times) we have a_n. By repeating the process (comparing coefficients of x^i) we can find as many coefficients $a_i^{(n)}$ as are required.

To illustrate the process consider the example

$$(1 + x^2)y'' + y = 0, \qquad y(0) = y'(0) = 1. \qquad (11.23)$$

We write

$$\begin{aligned} y &= a_0 + a_1 x + a_2 x^2 + \ldots \\ y' &= a'_0 + a'_1 x + a'_2 x^2 + \ldots \\ y'' &= a''_0 + a''_1 x + a''_2 x^2 + \ldots. \end{aligned}$$

The initial conditions imply $a_0 = 1$ and $a'_0 = 1$. But $a'_0 = a_1$. Thus $a_1 = 1$. From (11.23)

$$(1 + x^2)(a''_0 + a''_1 x + a''_2 x^2 + \ldots) + (a_0 + a_1 x + a_2 x^2 + \ldots) = 0.$$

By comparing coefficients we have

$$x^0 : a''_0 + a_0 = 0, \qquad \text{i.e.,}\ a''_0 = -1.$$

But $a''_0 = a'_1 = 2a_2$, so $a'_1 = -1$, $a_2 = -\frac{1}{2}$.

$$x : a''_1 + a_1 = 0, \qquad \text{i.e.,}\ a''_1 = -1.$$

Again $$a''_1 = 2a'_2 = 2(3a_3).$$

Thus $$a'_2 = -\tfrac{1}{2} \qquad \text{and} \qquad a_3 = -\tfrac{1}{6}.$$

We can tabulate our results as follows:

Table 11.1 Scheme for coefficients in Gibbons's method

i	0	1	2	3
a_i	1	1	$-\frac{1}{2}$	$-\frac{1}{6}$
a'_i	1	-1	$-\frac{1}{2}$	
a''_i	-1	-1		

Proceeding in this way we can obtain any coefficient that is required.

In his paper Gibbons gives recurrence relations for obtaining the coefficients when the function 'f' in (11.1) contains such operations as addition, multiplication, reciprocation, exponentiation, logarithm, square root, sine, and cosine. Gibbons claims that large truncation errors and destructive cancellation are minimized when h is 'small', and he gives a method for determining a practical radius of convergence.

11.5 Predictor—corrector methods

We have already seen in section 11.2 how derivatives can be replaced by a finite-difference approximation. Alternatively, the first-order differential equation (11.1) can be written as the integral equation

$$y(x) - y(x_0) = \int_{x_0}^{x} f(x, y)\, \mathrm{d}x. \tag{11.24}$$

To obtain the solution of (11.24) we integrate over small intervals (x_k, x_{k+1}) and add individual results, that is,

$$y_{k+1} = y_k + \int_{x_k}^{x_{k+1}} f(x, y)\, \mathrm{d}x, \qquad k = 0, 1, \ldots . \tag{11.25}$$

The integration of (11.25) can be accomplished by means of the finite-difference quadrature formulae given in chapter 5. For example we could use the trapezium rule

$$y_{k+1} = y_k + \frac{h}{2}(f_k + f_{k+1}) \tag{11.26}$$

where $f_k = f(x_k, y_k)$ and $h = x_{k+1} - x_k$. However, f_{k+1} on the right-hand side of (11.26) depends on y_{k+1} and thus the equation, as it stands, cannot in general be solved without resorting to techniques for the solution of non-linear equations. This can be overcome by, for instance, using Euler's formula (11.6) to estimate or predict a value of y_{k+1} (denoted by y_{k+1}^P) and using y_{k+1}^P to compute $f_{k+1}^P = f(x_{k+1}, y_{k+1}^P)$. Equation (11.26) can now be used to improve or correct the value of y_{k+1}, that is we calculate

$$y_{k+1}^C = y_k + \frac{h}{2}(f_k + f_{k+1}^P).$$

We now have a predictor–corrector system

$$\text{Predictor:}\quad y_{k+1}^P = y_k + hf_k, \qquad k = 0, 1, 2, \ldots$$

$$\text{Corrector:}\quad y_{k+1}^C = y_k + \tfrac{1}{2}h(f_k + f_{k+1}^P), \qquad k = 0, 1, 2, \ldots.$$

The corrector is used as many times as is necessary to achieve convergence, that is, until successive y_{k+1}^C correspond within given limits.

To illustrate the method consider again the system (11.7), i.e.,

$$y' = y, \qquad y(0) = 1.$$

Euler's predictor gives $y_1{}^P = 1{\cdot}1$. The trapezium rule corrects this to

$$y_1{}^{C_1} = 1 + \frac{0{\cdot}1}{2}(1 + 1{\cdot}1) = 1{\cdot}105$$

$$y_1{}^{C_2} = 1 + \frac{0{\cdot}1}{2}(1 + 1{\cdot}105) = 1{\cdot}105\,25$$

$$y_1{}^{C_3} = 1 + \frac{0{\cdot}1}{2}(1 + 1{\cdot}105\,25) = 1{\cdot}105\,262\,5.$$

The question that immediately arises concerns the number of figures that are accurate in each approximation. Now, from chapter 5, the error in the trapezium rule is approximately $-(h^3/12)f''_k$ (cf. with Euler formula error of $O(h^2)$), so that in this example, $y' = y$, i.e., $f = y, f'' = y$, thus $f''_0 = y_0 = 1$ and the error is about $-(0{\cdot}1)^3/12 \doteqdot -10^{-4}$, so that we can only expect an accuracy of about three or four decimal places. It would thus be appropriate to stop at $y_1{}^{C_3}$, rounding our answer to four decimal places.

The Euler–trapezium system is a special case of the more general Adams–Bashforth predictor–corrector system (derived in section 5.3) viz.:

$$\text{Predictor:}\quad y_{k+1} = y_k + h\left(1 + \frac{1}{2}\nabla + \frac{5}{12}\nabla^2 + \frac{3}{8}\nabla^3 + \frac{251}{720}\nabla^4 + \ldots\right)f_k \qquad (11.27)$$

$$\text{Corrector:}\quad y_{k+1} = y_k + h\left(1 - \frac{1}{2}\nabla - \frac{1}{12}\nabla^2 - \frac{1}{24}\nabla^3 - \frac{19}{720}\nabla^4 - \ldots\right)f_{k+1}. \qquad (11.28)$$

Equations (11.27) and (11.28) can be truncated at some particular difference. ((11.27) with no differences gives Euler's method and (11.28) with only the first difference term gives the trapezium rule). Suppose, for instance, the first difference in (11.27) is retained, then we obtain the predictor

$$y_{k+1} = y_k + \tfrac{1}{2}h(3f_k - f_{k-1}), \qquad k = 1, 2, \ldots, n-1.$$

Clearly, this expression requires a knowledge of y_1 before it is of use in evaluating y_2. Further, when p differences are retained in (11.27) the first $p + 1$ values of y must be known in order to proceed with the integration of the original differential equation. These first $p + 1$ values can be calculated by a self-starting method such as Taylor series or Runge–Kutta (section 11.6). The classic Adams–Bashforth methods are obtained by truncating equations (11.27) and (11.28) after second or third differences. For example, using all terms up to and including second differences we have

$$\left.\begin{aligned}\text{Predictor:}\quad & y^P_{k+1} = y_k + \frac{h}{12}(23f_k - 16f_{k-1} + 5f_{k-2})\\ \text{Corrector:}\quad & y^C_{k+1} = y_k + \frac{h}{12}(5f^P_{k+1} + 8f_k - f_{k-1})\end{aligned}\right] \qquad (11.29)$$

with truncation errors of $\frac{3}{8}h\nabla^3 f_k$ or $\frac{3}{8}h^4 y^{(4)}(\xi_1)$ and $-\frac{1}{24}h\nabla^3 f_{k+1}$ or $-\frac{1}{24}h^4 y^{(4)}(\xi_2)$ where ξ_1 and ξ_2 lie somewhere in the range bounded by the extreme pivotal points occurring in the formula. For this formula $x_{k-2} < \xi_1 < x_{k+1}$ and $x_{k-1} < \xi_2 < x_{k+1}$. Note that the coefficient in the truncation error is smaller in the corrector formula than it is in the predictor formula and is of opposite sign.

Another popular predictor–corrector formula is that due to Milne (see Ralston [3], for example):

$$\left.\begin{aligned}\text{Predictor:}\quad & y^P_{k+1} = y_{k-3} + \frac{4h}{3}(2f_k - f_{k-1} + 2f_{k-2}) + \frac{14}{45}h^5 y^{(5)}(\eta_1)\\ \text{Corrector:}\quad & y^C_{k+1} = y_{k-1} + \frac{h}{3}(f^P_{k+1} + 4f_k + f_{k-1}) - \frac{1}{90}h^5 y^{(5)}(\eta_2)\end{aligned}\right] \qquad (11.30)$$

where $x_{k-3} < \eta_1 < x_{k+1}, x_{k-1} < \eta_2 < x_{k+1}$ (see exercise 12).

The corrector is Simpson's rule and the predictor is also a Newton–Cotes type formula

Before using any iterative method, such as these predictor–correctors, we must investigate the convergence of the iterations. For the trapezium corrector we have

$$y_{k+1}^{[r+1]} = \frac{h}{2} f_{k+1}^{[r]} + y_k + \frac{h}{2} f_k = \frac{h}{2} f_{k+1}^{[r]} + \phi_k \qquad (r = 0, 1, \ldots)$$

where ϕ_k denotes all the terms not involving y_{k+1}, and the index $[r]$ indicates the rth estimate of the corrected value in the sequence. The exact solution of this corrector equation is

$$Y_{k+1} = (h/2) f(x_{k+1}, Y_{k+1}) + \phi_k,$$

whereas the result after r iterations is

$$y_{k+1}^{[r+1]} = (h/2) f(x_{k+1}, y_{k+1}^{[r]}) + \phi_k.$$

Thus

$$\left|Y_{k+1} - y_{k+1}^{[r+1]}\right| = \tfrac{1}{2} h \left| f(x_{k+1}, Y_{k+1}) - f(x_{k+1}, y_{k+1}^{[r]})\right|.$$

The function $f(x, y)$ must satisfy a Lipschitz condition (section 11.1) and so

$$|Y_{k+1} - y_{k+1}^{[r+1]}| \leqslant (h/2) L \, |Y_{k+1} - y_{k+1}^{[r]}|.$$

By reduction on r,

$$\left|Y_{k+1} - y_{k+1}^{[r+1]}\right| \leqslant (hL/2)^{r+1} \left| Y_{k+1} - y_{k+1}^{[0]}\right|$$

Thus, if $(hL/2) < 1$,

$$\lim_{r \to \infty} |Y_{k+1} - y_{k+1}^{[r+1]}| = 0.$$

By choosing h sufficiently small we can ensure that $hL < 2$ and obtain convergence. For the Adams–Bashforth formulae with error $O(h^4)$ (equation (11.28)) we must have $hL < 2{\cdot}4$, and with Milne's equation (11.30) we must have $hL < 3$ for convergence.

11.6 The method of Runge–Kutta

This method applies to first order differential equations or to sets of first order equations and can, therefore, be used to solve ordinary differential equations of any order. The derivation of this method in the general case is fairly lengthy and we shall just consider the second order process (i.e., error $O(h^3)$). The derivation of the fourth order process is sketched in the exercises – for fuller details see Ralston [3] and Kopal [4]. We define the quantities

$$\left.\begin{aligned} k_1 &= hf(x_r, y_r) = hf_r \\ k_2 &= hf(x_r + \alpha h,\ y_r + \beta k_1) \end{aligned}\right] \qquad (11.31)$$

and form

$$y_{r+1} = y_r + w_1 k_1 + w_2 k_2 \tag{11.32}$$

so that the expansion (in powers of h) of y_{r+1} given by (11.32) agrees with the solution of the differential equation to three terms of the Taylor series. Now the Taylor series expansion of the solution of the differential equation gives

$$y_{r+1} = y_r + h y'_r + \frac{1}{2} h^2 y''_r + O(h^3)$$

$$= y_r + hf_r + \frac{h^2}{2}\left(\frac{\partial f_r}{\partial x} + f_r \frac{\partial f_r}{\partial y}\right) + O(h^3) \tag{11.33}$$

since $$\left[\frac{\mathrm{d}}{\mathrm{d}x}(f(x,y))\right]_r = \left[\frac{\partial f}{\partial x} + \frac{\mathrm{d}y}{\mathrm{d}x}\frac{\partial f}{\partial y}\right]_r = \frac{\partial f_r}{\partial x} + f_r \frac{\partial f_r}{\partial y}.$$

Expanding (11.31) in the same manner and substituting into (11.32) gives

$$y_{r+1} = y_r + w_1 h f_r + w_2 \left[h f_r + h^2\left(\alpha \frac{\partial f_r}{\partial x} + \beta f_r \frac{\partial f_r}{\partial y}\right)\right] + O(h^3). \tag{11.34}$$

Comparing coefficients of powers of h in (11.33) and (11.34) we have

$$w_1 + w_2 = 1, \qquad w_2 \alpha = \tfrac{1}{2}, \qquad w_2 \beta = \tfrac{1}{2},$$

i.e., three equations in four unknowns. Equation (11.31) suggests $\alpha = 1$ might be an appropriate value and so we have as a particular solution

$$\alpha = \beta = 1, \qquad w_1 = w_2 = \tfrac{1}{2}.$$

The truncation error can be obtained by examining the next term of the Taylor series. We have

$$T_2 = -h^3 w_2 \left(\frac{1}{2}\alpha^2 \frac{\partial^2 f}{\partial x^2} + \alpha\beta f \frac{\partial^2 f}{\partial x \partial y} + \frac{1}{2}\beta^2 f^2 \frac{\partial^2 f}{\partial y^2}\right) + \frac{1}{6} h^3 \left(\frac{\partial^2 f}{\partial x^2} + f \frac{\partial^2 f}{\partial x \partial y} + f^2 \frac{\partial^2 f}{\partial y^2}\right) + $$
$$+ \frac{1}{6} h^3 \frac{\partial f}{\partial y}\left(\frac{\partial f}{\partial x} + f \frac{\partial f}{\partial y}\right),$$

$$= -\frac{1}{12} h^3 \left(\frac{\partial^2 f}{\partial x^2} + 2f \frac{\partial^2 f}{\partial x \partial y} + f^2 \frac{\partial^2 f}{\partial y^2}\right) + \frac{1}{6} h^3 \frac{\partial f}{\partial y}\left(\frac{\partial f}{\partial x} + f \frac{\partial f}{\partial y}\right).$$

For the general nth order method we have

$$\begin{aligned}
k_1 &= hf(x_r, y_r),\\
k_2 &= hf(x_r + \alpha_2 h,\ y_r + \beta_{21} k_1),\\
k_3 &= hf(x_r + \alpha_3 h,\ y_r + \beta_{31} k_1 + \beta_{32} k_2),\\
k_4 &= hf(x_r + \alpha_4 h,\ y_r + \beta_{41} k_1 + \beta_{42} k_2 + \beta_{43} k_3),
\end{aligned}$$

etc., and

$$y_{r+1} = y_r + w_1 k_1 + w_2 k_2 + \ldots + w_n k_n.$$

The most common Runge–Kutta method is the fourth order process (error $O(h^5)$) where

$$\alpha_2 = \beta_{21}$$
$$\alpha_3 = \beta_{31} + \beta_{32}$$
$$\alpha_4 = \beta_{41} + \beta_{42} + \beta_{43}$$
$$w_1 + w_2 + w_3 + w_4 = 1$$
$$w_2\alpha_2 + w_3\alpha_3 + w_4\alpha_4 = \tfrac{1}{2}$$
$$w_2\alpha_2^2 + w_3\alpha_3^2 + w_4\alpha_4^2 = \tfrac{1}{3}$$
$$w_2\alpha_2^3 + w_3\alpha_3^3 + w_4\alpha_4^3 = \tfrac{1}{4}$$
$$w_3\alpha_2\beta_{32} + w_4(\alpha_2\beta_{42} + \alpha_3\beta_{43}) = \tfrac{1}{6}$$
$$w_3\alpha_2^2\beta_{32} + w_4(\alpha_2^2\beta_{42} + \alpha_3^2\beta_{43}) = \tfrac{1}{12}$$
$$w_3\alpha_2\alpha_3\beta_{32} + w_4(\alpha_2\beta_{42} + \alpha_3\beta_{43})\alpha_4 = \tfrac{1}{8}$$
$$w_4\alpha_2\beta_{32}\beta_{43} = \tfrac{1}{24}$$

i.e., we have 11 equations in 13 unknowns. Two of the constants, e.g., α_2 and α_3 may be given arbitrary values and the other equations can be solved in terms of these parameters. For convenient applications there are obvious restrictions on these values. A number of methods have been proposed (see Fox [5]) each using different values of the parameters to gain some advantage. Two of the most common methods in use are the methods originally developed by Runge which we shall designate the standard Runge–Kutta method and a process due to Gill [6]. The value used for the parameters are given in Table 11.2.

Table 11.2 Parameter values in the standard Runge–Kutta method and Gill's method

	Standard Runge–Kutta method	*Gill's method*
α_2	$\frac{1}{2}$	$\frac{1}{2}$
α_3	$\frac{1}{2}$	$\frac{1}{2}$
α_4	1	1
β_{21}	$\frac{1}{2}$	$\frac{1}{2}$
β_{31}	0	$\frac{1}{2}(2^{\frac{1}{2}} - 1)$
β_{32}	$\frac{1}{2}$	$1 - 2^{\frac{1}{2}}$
β_{41}	0	0
β_{42}	0	$-2^{-\frac{1}{2}}$
β_{43}	1	$1 + 2^{-\frac{1}{2}}$
w_1	$\frac{1}{6}$	$\frac{1}{6}$
w_2	$\frac{1}{3}$	$\frac{1}{3}(1 - 2^{-\frac{1}{2}})$
w_3	$\frac{1}{3}$	$\frac{1}{3}(1 + 2^{-\frac{1}{2}})$
w_4	$\frac{1}{6}$	$\frac{1}{6}$

We now solve $y' = 1 + y^2$, $y(0) = 0$, by the standard Runge–Kutta fourth order process. The equations to be used are

$$k_1 = hf(x_r, y_r), \qquad k_2 = hf(x_r + \tfrac{1}{2}h,\ y_r + \tfrac{1}{2}k_1),$$

$$k_3 = hf(x_r + \tfrac{1}{2}h,\ y_r + \tfrac{1}{2}k_2), \qquad k_4 = hf(x_r + h,\ y_r + k_3),$$

$$y_{r+1} = y_r + \tfrac{1}{6}(k_1 + 2k_2 + 2k_3 + k_4) + O(h^5).$$

For this example we choose $h = 0{\cdot}2$. To compute $y_1 = y(0{\cdot}2)$ from $y = y(0) = 0$ we first calculate

$$\begin{aligned}
k_1 &= 0{\cdot}2[1 + (0)^2] = 0{\cdot}2,\\
k_2 &= 0{\cdot}2\,[1 + (\tfrac{1}{2} \times 0{\cdot}2)^2] = 0{\cdot}202,\\
k_3 &= 0{\cdot}2\,[1 + (\tfrac{1}{2} \times 0{\cdot}202)^2] = 0{\cdot}202\,04,\\
k_4 &= 0{\cdot}2\,[1 + (0{\cdot}202\,04)^2] = 0{\cdot}208\,16,\\
y_1 &\doteqdot 0 + \tfrac{1}{6}(0{\cdot}2 + 0{\cdot}404 + 0{\cdot}404\,08 + 0{\cdot}208\,10)\\
&= 0{\cdot}2027.
\end{aligned}$$

We can now calculate $y_2(= y(0{\cdot}4))$ from y_1. We have

$$\begin{aligned}
k_1 &= 0{\cdot}2[1 + (0{\cdot}2027)^2] = 0{\cdot}208\,22,\\
k_2 &= 0{\cdot}2[1 + (0{\cdot}2027 + \tfrac{1}{2} \times 0{\cdot}208\,22)^2] = 0{\cdot}218\,82,\\
k_3 &= 0{\cdot}2[1 + (0{\cdot}2027 + \tfrac{1}{2} \times 0{\cdot}218\,82)^2] = 0{\cdot}219\,48,\\
k_4 &= 0{\cdot}2[1 + (0{\cdot}2027 + 0{\cdot}219\,48)^2] = 0{\cdot}235\,64,\\
y_2 &\doteqdot 0{\cdot}2027 + \tfrac{1}{6}[0{\cdot}208\,22 + 0{\cdot}437\,64 + 0{\cdot}438\,96 + 0{\cdot}235\,64]\\
&= 0{\cdot}4228.
\end{aligned}$$

(cf. with the true solution $y = \tan x$).

We have to try and estimate the accuracy of these figures. The expression for the truncation error is too complicated to be of any practical use. In most cases it is not possible to reduce the truncation error by a suitable choice of parameters. One check is to re-solve the problem using half the interval, to compare the results, and to adjust the step length if necessary. In the above example, after computing $y(0{\cdot}2)$ we would choose $h = 0{\cdot}1$, calculate $y(0{\cdot}1)$ from $y(0)$ and $y(0{\cdot}2)$ from $y(0{\cdot}1)$. If the two values for $y(0{\cdot}2)$ agree to some required accuracy then we take this value for $y(0{\cdot}2)$. If the answers disagree we take $h = 0{\cdot}05$, compute $y(0{\cdot}05)$, $y(0{\cdot}1)$, $y(0{\cdot}15)$, and $y(0{\cdot}2)$ and compare these values with the values previously computed, and so on. These intermediate results can also be used in conjunction with Richardson extrapolation to improve answers.

The Runge–Kutta fourth order process can be represented geometrically as illustrated in Fig. 11.1.

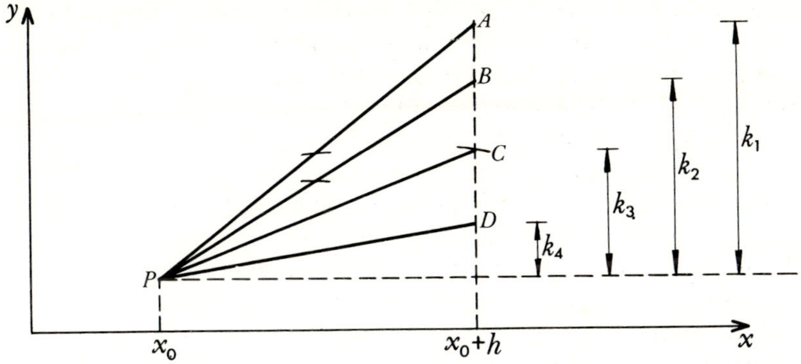

Fig. 11.1 Geometrical representation of Runge–Kutta method.

The process consists of four stages:

(*a*) Draw a straight line through $P(x_0, y_0)$ with slope $f(x_0, y_0)$ and let it pass through A. This approximation to the differential equation gives the increment k_1.
(*b*) Find the coefficients (x_1, y_1) of the mid-point of PA and draw the straight line PB through P having slope $f(x_1, y_1)$. This gives the increment k_2.
(*c*) Draw PC with slope equal to the slope at the mid-point of PB. This gives k_3.
(*d*) Draw PD with slope corresponding to the slope at C – this gives k_4.

The increment $k = \frac{1}{6}(k_1 + 2k_2 + 2k_3 + k_4)$ is the weighted mean of k_1, k_2, k_3, k_4.

11.7 Stability

One important question that must be considered is the stability of the numerical procedure. Instability arises when any error that is introduced (e.g., due to round-off) produces an unwanted solution of the difference equation which then increases more rapidly than the required solution, eventually swamping it. We are concerned that our method does not produce any spurious results that can grow without bound. We are not interested in how this unwanted solution arises but in what happens to it during further computation.

The formulae that we use give rise to a truncation error and a knowledge of this is valuable, but not sufficient, to determine the accuracy of the method. We shall just consider what is involved in instability by considering simple examples. A fuller discussion can be found in Fox [5], Ralston [3], and Dahlquist [7a,b].

Consider first the equation

$$y' = y - \tfrac{1}{3}x - \tfrac{2}{3}.$$

This has the general solution $y = Ae^x + \frac{1}{3}x + 1$ and the special condition $y(0) = 1$ makes A vanish. The solution is now a simple linear term, varying slowly in comparison with the exponential term. During computation it is more than likely that the exponential term will be introduced (why?). Even if the local multiple of e^x is small the resulting error will grow and swamp the required solution $(\frac{1}{3}x + 1)$. This phenomenon is called 'inherent instability' since a small change in the starting value will cause a large change, for large x, in the true solution. It is analogous to the 'ill-conditioning' of linear algebra (chapter 8) and is relatively independent of the numerical method of solution. However, by using higher order formulae and multilength arithmetic, greater precision can be obtained. Alternatively, it may be possible to restate the problem in such a way that the instability is overcome.

Note that differential equations such as $y' = -10y + \frac{1}{2}x + \frac{1}{4}$, whose general solution is $y = Ae^{-10x} + \frac{1}{20}x + \frac{1}{50}$, do not suffer from inherent instability since, even with the condition $y(0) = \frac{1}{50}$ which makes $A = 0$, we have, for large x, the term Ae^{-10x} tending to zero and so this term makes a negligible contribution to the computed value of y.

Another type of instability arises in the solution of $y' = -10y$, $y(0) = y_0$ by the method of Runge–Kutta. In the second order method consisting of equations (11.31) and (11.32) we obtain

$$y_{r+1} = (1 - 10h + 50h^2)y_r.$$

By reduction on r we have

$$y_{r+1} = (1 - 10h + 50h^2)^{r+1}y_0$$

or

$$y_{r+1} = \lambda^{r+1}y_0$$

where

$$\lambda = 1 - 10h + 50h^2$$

is part of the expansion of e^{-10h}.

If λ is greater than unity then this computed solution will grow unboundedly and will behave quite differently to the true solution y_0e^{-10x}. Thus, clearly, if $0 < h < 0{\cdot}2$ we have a stable system. The fourth order Runge–Kutta method gives

$$y_{r+1} = \left[1 - 10h + 50h^2 - \frac{500}{3}h^3 + \frac{1250}{3}h^4\right]^{r+1} y_0,$$

and from Fig. 11.2 we see that we have to impose the interval restrictions $0 < h < 0{\cdot}27$. In general for the system $y' = -Ky$ $(K > 0)$ the standard Runge–Kutta fourth order method has the interval restriction $Kh < 2{\cdot}7$. When there is a limiting value of h below which we have stability and above which we have instability then we say that the system exhibits 'partial instability', and this is often found in Runge–Kutta methods.

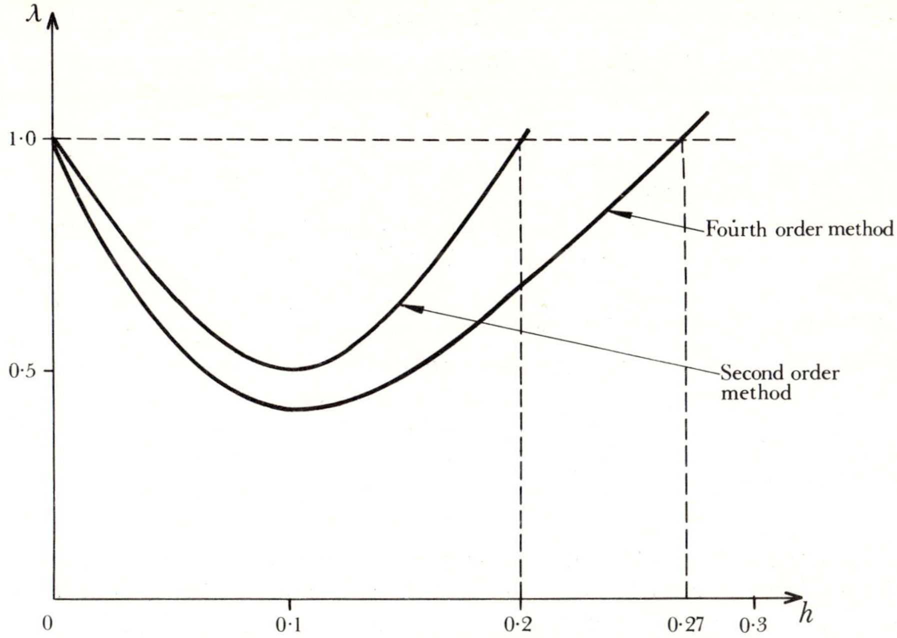

Fig. 11.2 Stability for 2nd and 4th order Runge–Kutta (Stable for $|\lambda| < 1$)

Far more catastrophic than partial instability is the 'strong instability' commonly found in predictor–corrector methods. Fox **[5]** gives as an illustration of a strongly unstable corrector, the formula

$$11y_{r+1} = -27y_r + 27y_{r-1} + 11y_{r-2} + 3h(y'_{r+1} + 9y'_r + 9y'_{r-1} + y'_{r-2}).$$

Thus for the differential equation $y' = -Ky$ we have

$$y_{r+1}(11 + 3Kh) + 27y_r(1 + Kh) - 27y_{r-1}(1 - Kh) - y_{r-2}(11 - 3Kh) = 0.$$

This will have the general solution (see section 11.3)

$$y_r = A\mu_1{}^r + B\mu_2{}^r + C\mu_3{}^r$$

where A, B, and C are constants and μ_1, μ_2, and μ_3 are the roots of the cubic

$$(11 + 3Kh)\mu^3 + 27(1 + Kh)\mu^2 - 27(1 - Kh)\mu - (11 - 3Kh) = 0.$$

For all values of Kh this cubic has a root smaller than $-1{\cdot}5$ and this gives rise to a term in the general solution which will grow rapidly and swamp the true solution $y_r = y(rh) = e^{-rKh}$. In this case a reduction of the step length h does not improve the situation and the numerical solution will eventually diverge from e^{-Kx}. With the Milne–Simpson corrector

$$y_{r+1} = y_{r-1} + \tfrac{1}{3}h(y'_{r+1} + 4y'_r + y'_{r-1})$$

we have (when $y' = Ky$) the recurrence relation

$$(1 - \tfrac{1}{3}Kh)y_{r+1} - \tfrac{4}{3}Khy_r - (1 + \tfrac{1}{3}Kh)y_{r-1} = 0$$

whose solution is

$$y_r = A\mu_1{}^r + B\mu_2{}^r$$

where μ_1 and μ_2 are given by

$$\mu_1, \mu_2 = \frac{\tfrac{2}{3}Kh \pm \sqrt{(1 + \tfrac{1}{3}K^2h^2)}}{(1 - \tfrac{1}{3}Kh)}\ .$$

Expanding in powers of Kh we find

$$\mu_1 = \{1 + O(h^5)\}e^{Kh}$$

$$\mu_2 = -\{1 - O(h^3)\}e^{-\frac{1}{3}Kh}$$

that is, the solution at $x = rh$ is of the order

$$Ae^{Kx} + Be^{-\frac{1}{3}Kx}\ .$$

The first term corresponds to the true solution while the second is spurious. If K is positive then the spurious solution decays and does not matter. If K is negative then we have instability. Dahlquist calls this 'weak instability' as it is unstable for only some types of differential equations.

For this same differential equation, $y' = Ky$, the Adams–Bashforth formula

$$y_{r+1} = y_r + \tfrac{1}{12}h(5y'_{r+1} + 8y'_r - y'_{r-1})$$

introduces one spurious solution. Here

$$\mu_1, \mu_2 = \frac{1 + \tfrac{2}{3}Kh \pm \sqrt{(1 + Kh + \tfrac{7}{12}K^2h^2)}}{2(1 - \tfrac{5}{12}Kh)}$$

For stability we must have $|\mu_1| < 1$ and $|\mu_2| < 1$. If K is real and negative then $|\mu_1| < 1$ for all h but for $|\mu_2| < 1$ we must have $|Kh| < 6$. Thus we have a restriction on K and h giving a type of partial instability.

In general partial instability depends on the differential equation as well as the method used to solve it. Strongly unstable systems are independent of the differential equation and arise in finite-difference formulae because the order of the finite-difference equation exceeds that of the differential equation. Weak instability arises in the same way as strong instability but is found only for some types of differential equations.

11.8 Boundary-value problems

We now consider briefly systems in which the given conditions are shared between two or more points (see Fox [8]). For simplicity we shall look at second order linear equations with the conditions specified at the two end points (the two point boundary-value problem), for example

$$\left.\begin{aligned} y'' + f(x)y' + g(x)y &= k(x), \\ y(x_0) = y_0, \qquad y(x_n) &= y_n. \end{aligned}\right] \tag{11.35}$$

In theory we can solve all boundary-value problems by the initial-value techniques discussed in the earlier sections. We form initial-value systems by adding sufficient guessed conditions at one end (the end involving least work) and we adjust these conditions until the required relations are satisfied at the other end (see Fox [5] and [8]).

Special techniques, however, are available for boundary-value problems, one of the most useful being the deferred-correction method of Fox and Goodwin [9]. We replace y'' and y' in equation (11.35) by the central difference formulae (see chapter 3)

$$h^2 y''_r = \delta^2 y_r - (\tfrac{1}{12}\delta^4 - \tfrac{1}{90}\delta^6 + \ldots)y_r$$

$$hy'_r = \mu\delta y_r - (\tfrac{1}{6}\mu\delta^3 - \tfrac{1}{30}\mu\delta^5 + \ldots)y_r$$

and we then have

$$(1 - \tfrac{1}{2}hf_r)y_{r-1} - (2 - h^2 g_r)y_r + (1 + \tfrac{1}{2}hf_r)y_{r+1} = h^2 k_r + C_r \tag{11.36}$$

where

$$C_r = (\tfrac{1}{12}\delta^4 - \tfrac{1}{90}\delta^6 + \ldots)y_r + hf_r(\tfrac{1}{6}\mu\delta^3 - \tfrac{1}{30}\mu\delta^5 + \ldots)y_r, \qquad r = 1, 2, \ldots, n-1.$$

If we disregard the difference correction C_r we have $n-1$ linear equations in the $n+1$ unknowns $y_0, y_1, \ldots, y_n$. The two boundary conditions supply the extra two equations necessary to solve the system. If y_0 and y_n are known (as in (11.35)) we then have

$$-(2 - h^2 g_1)y_1 + (1 + \tfrac{1}{2}hf_1)y_2 = h^2 k_1 + C_1 - (1 - \tfrac{1}{2}hf_1)y_0$$

$$(1 - \tfrac{1}{2}hf_2)y_1 - (2 - h^2 g_2)y_2 + (1 + \tfrac{1}{2}hf_2)y_3 = h^2 k_2 + C_2$$

$$\cdot \quad \cdot \quad \cdot \quad \cdot \quad \cdot \quad \cdot \quad \cdot \quad \cdot \quad \cdot$$

$$(1 - \tfrac{1}{2}hf_{n-2})y_{n-3} - (2 - h^2 g_{n-2})y_{n-2} + (1 + \tfrac{1}{2}hf_{n-2})y_{n-1} = h^2 k_{n-2} + C_{n-2}$$

$$(1 - \tfrac{1}{2}hf_{n-1})y_{n-2} - (2 - h^2 g_{n-1})y_{n-1} = h^2 k_{n-1} + C_{n-1} - (1 + \tfrac{1}{2}hf_{n-1})y_n$$

or, in matrix form,

$$A\mathbf{y} = \mathbf{d} + C(\mathbf{y}).$$

where

$$A = (a_{ij}), a_{ij} = 0, |i - j| > 1,$$

$$a_{ii} = -(2 - h^2 g_i), i = 1, \ldots, n-1;$$

$$a_{i,i+1} = 1 + \tfrac{1}{2}hf_i,\ a_{i+1,i} = 1 - \tfrac{1}{2}hf_{i+1}, i = 1, \ldots, n-2;$$

$$\mathbf{y} = (y_1, \ldots, y_{n-1})', C(\mathbf{y}) = (C_1, \ldots, C_{n-1})'$$

$$\mathbf{d} = (d_1, \ldots, d_{n-1})', d_1 = h^2 k_1 - (1 - \tfrac{1}{2}hf_1)y_0;$$

$$d_{n-1} = h^2 k_{n-1} - (1 + \tfrac{1}{2}hf_{n-1})y_n;\ d_i = h^2 k_i, i = 2, \ldots, n-2.$$

We can obtain a first approximation by solving $A\mathbf{y}^{[1]} = \mathbf{d}$, then we find the correction terms, and solve $A\mathbf{y}^{[2]} = \mathbf{d} + C(\mathbf{y}^{[1]})$, and so on. For an economic method of solution of these equations see [8] and [10].

To illustrate the method consider the example

$$y'' + xy' + y = 0,$$

with $y = 0$ at $x = 0$, and $y = 1$ at $x = 1$.

Taking $h = 0{\cdot}2$, the equations we have to solve are

$$\begin{aligned}
-1{\cdot}96y(0{\cdot}2) + 1{\cdot}02y(0{\cdot}4) \qquad\qquad\qquad\qquad\qquad &= 0 + C(0{\cdot}2)\\
0{\cdot}96y(0{\cdot}2) - 1{\cdot}96y(0{\cdot}4) + 1{\cdot}04y(0{\cdot}6) \qquad\qquad &= 0 + C(0{\cdot}4)\\
0{\cdot}94y(0{\cdot}4) - 1{\cdot}96y(0{\cdot}6) + 1{\cdot}06y(0{\cdot}8) &= 0 + C(0{\cdot}6)\\
0{\cdot}92y(0{\cdot}6) - 1{\cdot}96y(0{\cdot}8) &= -1{\cdot}08 + C(0{\cdot}8).
\end{aligned}$$

Neglecting the correction term we have

$$\mathbf{y}^{[1]} = (0,\ 0{\cdot}2729,\ 0{\cdot}5245,\ 0{\cdot}7365,\ 0{\cdot}8967,\ 1)'.$$

We must now form a difference table in order to obtain the difference corrections. To do this we must compute some extra values of y on each side of the range of integration, viz., $y(-0{\cdot}2)$, $y(-0{\cdot}4)$, $y(1{\cdot}2)$, $y(1{\cdot}4)$, etc. These can be found from the recurrence relation (11.36). The difference table is shown in Table 11.3.

Table 11.3 Difference table for the solution of a boundary value type differential equations

x	$y^{[1]}$	δ	δ^2	δ^3	δ^4	δ^5	δ^6	C
−0·4	−0·5245							
		2516						
−0·2	−0·2729		213					
		2729		−213				
0	0		0		0			
		2729		−213		30		
0·2	0·2729		−213		30		1	0·0001
		2516		−183		31		
0·4	0·5245		−396		61		−21	0·0003
		2120		−122		10		
0·6	0·7365		−518		71		−12	0·0004
		1602		−51		−2		
0·8	0·8967		−569		69		−12	0·0005
		1033		18		−14		
1·0	1·0000		−551		55			
		482		73				
1·2	1·0482		−478					
		4						
1·4	1·0486							

The correction terms C shown in the last column of Table 11.3 are computed from

$$C_r = (\tfrac{1}{12}\delta^4 - \tfrac{1}{90}\delta^6)y_r + hf_r(\tfrac{1}{6}\mu\delta^3 - \tfrac{1}{30}\mu\delta^5)y_r, \qquad r = 1, 2, 3, 4.$$

We now have to solve the revised equations

$$\begin{aligned}
-1{\cdot}96y(0{\cdot}2) + 1{\cdot}02y(0{\cdot}4) &= 0{\cdot}0001\\
0{\cdot}96y(0{\cdot}2) - 1{\cdot}96y(0{\cdot}4) + 1{\cdot}04y(0{\cdot}6) &= 0{\cdot}0003\\
0{\cdot}94y(0{\cdot}4) - 1{\cdot}96y(0{\cdot}6) + 1{\cdot}06y(0{\cdot}8) &= 0{\cdot}0004\\
0{\cdot}92y(0{\cdot}6) - 1{\cdot}96y(0{\cdot}8) &= -1{\cdot}0795
\end{aligned}$$

and we find

$$\mathbf{y}^{[2]} = (0,\ 0{\cdot}2723,\ 0{\cdot}5234,\ 0{\cdot}7354,\ 0{\cdot}8959,\ 1)'.$$

The process should be repeated to find a new difference correction and $\mathbf{y}^{[3]}$ determined, and so on until a convergent solution is obtained.

In this type of problem it is also quite common to have some of the boundary conditions involving derivatives, for example $y'_0 + ay_0 = b$ (cf. chapter 12). This means that y_0 is not known explicitly and the above method has to be modified.

If we take $r = 0$ in the general equation (11.36) we have

$$(1 - \tfrac{1}{2}hf_0)y_{-1} - (2 - h^2g_0)y_0 + (1 + \tfrac{1}{2}hf_0)y_1 = h^2k_0 + C_0 \tag{11.37}$$

and this involves y_{-1} which lies outside the range y_0 to y_n. We can eliminate this 'fictitious' value by replacing the boundary condition by its central difference equivalent

$$y_1 - y_{-1} + 2hay_0 + C_0' = 2hb \tag{11.38}$$

where C_0' is the difference corrector given by

$$C'_0 = 2(-\tfrac{1}{6}\mu\delta^3 + \tfrac{1}{30}\mu\delta^5 - \ldots)y_0.$$

From (11.37) and (11.38) we have

$$[(1 - \tfrac{1}{2}hf_0)2ha - (2 - h^2g_0)]\,y_0 + 2y_1$$
$$= h^2k_0 + 2hb(1 - \tfrac{1}{2}hf_0) + C_0 - (1 - \tfrac{1}{2}hf_0)C'_0.$$

Thus we have introduced one extra equation involving the unknown y_0. The system can now be solved by the method just described.

Another possibility is that one boundary condition is at infinity (e.g., $y \to 0$ as $x \to \infty$). In this case we replace this condition by $y = 0$ at $x = X_1$ and solve. We re-solve with $y = 0$ at $x = X_2$ $(X_2 > X_1)$. If the solutions disagree to a specified accuracy then we repeat with a value $x = X_3$ $(X_3 > X_2)$ and compare the results for $x = X_2, x = X_3$ and so on.

11.9 Chebyshev methods

There are two methods in use which take advantage of Chebyshev polynomials and both methods can be used for equations of any order both of initial-value or boundary value type. The first of these is due to Lanczos **[11]**. He approximates to the solution by a polynomial and determines the coefficients by substituting into the differential equation and solving the linear equations that arise. For example, consider the differential system

$$\left.\begin{aligned} &xy'' + xy' + y = 0, \\ &y(0) = 0, \qquad y'(0) = 1 \end{aligned}\right] \tag{11.39}$$

where we seek a solution in the range $-1 \leqslant x \leqslant 1$.

Suppose we represent y by a polynomial (of degree five say), then the left-hand side of the differential equation (11.39) will be a polynomial $P(x)$. If we write (having used the given conditions)

$$y = x + a_2x^2 + a_3x^3 + a_4x^4 + a_5x^5$$

then

$$P(x) = (2 + 2a_2)x + (3a_2 + 6a_3)x^2 + (4a_3 + 12a_4)x^3 + (5a_4 + 20a_5)x^4 + 6a_5x^5.$$

The classic approach is to take $P(x) = \tau x^5$. By comparing coefficients of powers of x, we find

$$a_2 = -1, \quad a_3 = 1/2, \quad a_4 = -1/3!, \quad a_5 = 1/4!, \quad \tau = 1/4,$$

(cf. Taylor's series expansion section 11.4). Note that the maximum error which will occur is at the ends of the range and is $\frac{1}{4}$ (in magnitude).

Lanczos suggests we replace $P(x)$ by a Chebyshev polynomial $T_n(x)$ (section 4.2). In this case we take

$$P(x) = \tau T_5(x) = \tau.(16x^5 - 20x^3 + 5x).$$

Equating coefficients we find

$$\begin{aligned}
(2 + 2a_2) &= 5\tau & \text{or} \quad a_2 &= \tfrac{5}{2}\tau - 1 \\
(3a_2 + 6a_3) &= 0 & \text{or} \quad a_3 &= -\tfrac{5}{4}\tau + \tfrac{1}{2} \\
(4a_3 + 12a_4) &= -20\tau & \text{or} \quad a_4 &= -\tfrac{5}{4}\tau - \tfrac{1}{6} \\
(5a_4 + 20a_5) &= 0 & \text{or} \quad a_5 &= \tfrac{5}{16}\tau + \tfrac{1}{24} \\
6a_5 &= 16\tau & \text{or} \quad \tau &= \tfrac{2}{113}\,.
\end{aligned}$$

The maximum error in magnitude will now be $\frac{2}{113}$, as compared with $\frac{1}{4}$ previously. Provided τ is sufficiently small then we say that we have the approximate solution of (11.39). (In fact we have the exact solution of (11.39) with the right-hand side of the differential equation replaced by $\tau T_5(x)$.)

The second method is due to Clenshaw [12]. He replaces the solution, in the range $-1 \leqslant x \leqslant 1$, by a Chebyshev series, i.e.,

$$y = \tfrac{1}{2}a_0 T_0(x) + a_1 T_1(x) + \ldots,$$

and determines $a_0, a_1, a_2, \ldots$. To do this he obtains relations between the coefficients appearing in y and y', y' and y'', etc., and he uses these recursively to produce the coefficients. Further details of this method are left to exercise 27.

If we have more equations than unknowns then we have the problem of 'over-determination' (cf. chapter 7). Assuming we have one extra equation then Lanczos introduces a perturbation of the form $\tau_0 T_n(x) + \tau_1 T_{n+1}(x)$ giving one extra unknown. Clenshaw, on the other hand, forms a residual from the extra equation, as follows. Let α_r denote estimates of the coefficients a_r in the approximation and let $R(\alpha)$ be the residual obtained on substituting the coefficients α_r into the extra equation. If we re-solve the equations using different starting conditions then we obtain the coefficients β_r and the residual $R(\beta)$. The correct solution is the linear combination

$$a_r = A\alpha_r + B\beta_r$$

where A and B are given by

$$AR(\alpha) + BR(\beta) = 0.$$

11.10 Summary and notes

In this chapter we have considered a variety of methods for solving the simplest type of ordinary linear differential equation. Ralston [3] is of the opinion that predictor–corrector methods are better than Runge–Kutta methods and are faster by a factor of between two and four. This advantage in speed is due to the fact that for a fourth order predictor–corrector the function $f(x, y)\,(=y')$ only has to be evaluated twice (certainly in Hamming's variation, exercise 13, where the corrector need only be used once) since function values at earlier steps are known. For the fourth order Runge–Kutta method the function has to be evaluated four times. Another disadvantage with the method of Runge–Kutta is the difficulty in estimating the local truncation error at each step. In practice we have to repeat the calculations using a smaller step length and to compare the results. This leads to many more evaluations of the function $f(x, y)$ and slows down the process even further. However, since Runge–Kutta methods do not require a special starting procedure (a disadvantage with predictor–corrector methods) and since its operations are largely repetitive it can be used conveniently on an automatic computer. Other advantages of Runge–Kutta over predictor–corrector methods are that the increment h need not be constant, and they are more likely to be stable.

Differential equations of order two or higher can be treated as systems of first order equations and can therefore be treated by either Runge–Kutta or predictor–corrector methods. Runge–Kutta methods have an advantage as large, complicated, and even non-linear systems can be dealt with by very little extra programming.

We mentioned, very briefly, the method of Fox and Goodwin [9] with reference to boundary-value problems (section 11.8). This method can be applied to initial-value problems, sets of equations, equations of any order, linear or non-linear. The principle is similar to that used to solve partial differential equations (chapter 12). This method has the advantage that differences in the corrections term are small, but, to take full advantage of the method, programming becomes heavy. No general theory of convergence has been produced but there are no known examples of failure to converge.

With regard to the method of Taylor series the disadvantage of finding derivatives can be overcome by using electronic computers to find analytic expressions for the derivatives but this might be time consuming. The modern trend, however, is towards the more rapidly convergent Chebyshev series. In both these methods the interval length h can be quite large and, further, h need not be kept constant. Chebyshev methods can be applied to most types of ordinary differential equations provided the range of integration can be transformed to the range

$$-1 \leqslant x \leqslant 1 \qquad \text{or} \qquad 0 \leqslant x \leqslant 1.$$

There is a good summary of available methods and literature on the subject in Fox [5] and Walsh [13]. Some classical methods are also useful and a number of these methods are sketched in the exercises (e.g., Picard's and Frobenius's). Further information about classical methods can be obtained from Buckingham [**14**] and Jeffreys and Jeffreys [**15**].

REFERENCES

[1] Goursat, E., *A Course of Mathematical Analysis.* Vol. II, Part II, *Differential Equations* (translation), Dover Publications Inc., New York (1945).

[2] Gibbons, A., 'A Program for the Automatic Integration of Differential Equations using the Method of Taylor Series'. *Computer J.*, **3** (1960) 108–11.

[3] Ralston, A., *A First Course in Numerical Analysis*, McGraw-Hill Book Co. Ltd., New York (1965).

[4] Kopal, Z., *Numerical Analysis*, J. Wiley & Son, New York (1955).

[5] Fox, L. (ed.), *Numerical Solution of Ordinary and Partial Differential Equations*, Pergamon Press, Oxford (1962).

[6] Gill, S., 'A Process for the Step-by-Step Integration of Differential Equations in an Automatic Digital Computing Machine', *Proc. Camb. Phil. Soc.*, **47** (1951) 96–108.

[7a] Dahlquist, G., 'Convergence and Stability in the Numerical Integration of Ordinary Differential Equations', *Math. Scand.*, **4** (1956) 33–53.

[7b] Dahlquist, G., 'Stability and Error Bounds in the Numerical Integration of Ordinary Differential Equations', *K. Tekn. Hogsk, Handl.*, **130** (1959).

[8] Fox, L., *The Numerical Solution of Two-Point Boundary-Value Problems in Ordinary Differential Equations*, Oxford University Press, Oxford (1957).

[9] Fox, L. and Goodwin, E. T., 'Some New Methods for the Numerical Integration of Ordinary Differential Equations'. *Proc. Camb. Phil. Soc.*, **45** (1949) 373–88.

[10] *Modern Computing Methods,* H.M.S.O., London (1961).

[11] Lanczos, C., 'Trigonometrical Interpolation of Empirical and Analytic Functions', *J. Maths. Phys.*, **17** (1938) 123–29.

[12] Clenshaw, C. W., 'The Numerical Solution of Linear Differential Equations in Chebyshev Series', *Proc. Camb. Phil. Soc.*, **53** (1957) 134–49.

[13] Walsh, J. (ed.), *Numerical Analysis; An Introduction*, Academic Press, London (1966).

[14] Buckingham, R. A., *Numerical Methods*, Pitman, London (1957).

[15] Jeffreys, H. and Jeffreys, B. S., *Methods of Mathematical Physics*, 3rd edn., Cambridge University Press, Cambridge (1956).

Exercises

1. Euler's approximate solution to the differential equation (11.7), with $h = 0{\cdot}1$, is given by

$$y_{n+1} = (1 + 0{\cdot}1)y_n, \qquad y_0 = 1.$$

Given that the arithmetic is performed on a four decimal digit floating point computer, estimate bounds for the round-off error obtained with n applications of this formula.

2. If Euler's formula is used to find the value of y_{n+1} from y_n
   (*a*) by application with step length h;
   (*b*) by two applications with step length $\frac{1}{2}h$;

and the answers are denoted respectively by $y_{n+1}(h)$ and $y_{n+1}(\frac{1}{2}h)$, show that a better approximation to y_{n+1} is, in general,

$$2y_{n+1}(\tfrac{1}{2}h) - y_n(h).$$

Extend this approach (see section 3.7).

3. Find the solution of the difference equation

$$y_{n+2} - 2y_{n+1} + y_n = 0, \qquad y_0 = 1, \qquad y_1 = 2.$$

Generalize your result to give the solution of a nth order linear equation having k roots repeated and the remainder distinct.

4. Evaluate

$$I_k(\phi) = \int_0^{\pi} \frac{\cos k\theta - \cos k\phi}{\cos \theta - \cos \theta} \mathrm{d}\theta, \qquad k \text{ an integer} \geqslant 0,$$

by showing $I_0 = 0, I_1 = \pi$,

$$I_{k+1} - 2I_k \cos \phi + I_{k-1} = 0, \qquad (k = 1, 2, \ldots)$$

and solving the recurrence relation. (This integral occurs in the Prandtl 'lifting line' theory of aerodynamics.)

5. Evaluate $\tan^{-1} x$ for $x = 0(0{\cdot}1)5$ by
   (*a*) using a series expansion for $x \leqslant 0{\cdot}5$;
   (*b*) solving the differential equation

$$y' = 1/(1 + x^2), \qquad 0{\cdot}5 < x \leqslant 2;$$

   (*c*) using a series expansion for $x > 2$.

Why might it be better to solve the second order differential equation

$$y'' + 2x/(1 + x^2)^2 = 0,$$

for the range (*b*), using the boundary values $y(0{\cdot}5)$, $y(2)$ obtained in (*a*) and (*c*)?

6. (Numerov or Gauss–Jackson method for solving $y'' = f(x, y)$.) We assume y_r and f_r and enough differences are known for $r = 0, 1, \ldots, n-1$ and we use

$$\delta^2 y_n = h^2 (f_n + \tfrac{1}{12}\delta^2 f_n - \tfrac{1}{240}\delta^4 f_n + \ldots),$$

estimating the central differences $\delta^2 f_n$, $\delta^4 f_n$ by judicious estimation from earlier differences. This gives $\delta^2 y_n$ from which y_{n+1} and hence f_{n+1} can be obtained and also a revised value of $\delta^2 f_n$ and a revised guess for $\delta^4 f_n$. We may iterate the process if necessary. Use this method to solve

$$y'' = xy^2, \qquad y(0) = 1, \qquad y'(0) = -1,$$

using the Taylor series method as a starter.

7. Using the trapezium rule

$$y_{n+1} = y_n + \tfrac{1}{2}h\,(y'_n + y'_{n+1}), \qquad h = 0{\cdot}1,$$

find the solution after n steps of

$$y' = 10y - 9, \qquad y(0) = 1,$$

and compare your result (assuming no rounding errors are incurred) with the exact result. What result would be obtained if the method of solution is to predict y_{n+1} for each n by Euler's method and to correct y_{n+1} by one application of the trapezium rule?

8. Given the differential equation

$$y'' = xy^2, \qquad y(0) = 1, \qquad y'(0) = -1,$$

calculate y and y' for $x = 0(0{\cdot}1)1$ using the predictor–corrector formulae

$$\text{Predictor:} \quad y_{n+1} = y_{n-3} + \tfrac{4}{3}h(2f_n - f_{n-1} + 2f_{n-2}),$$
$$\text{Corrector:} \quad y_{n+1} = y_{n-1} + \tfrac{1}{3}h(f_{n-1} + 4f_n + f_{n+1}).$$

(Hint: Calculate starting values y_1, y_2, y_3 and y'_1, y'_2, y'_3 by the methods of Runge–Kutta or Taylor series.)

9. The function $y(x)$ satisfies the differential equation

$$y' = -y - 2x/y, \qquad y(-4) = 3{\cdot}0.$$

Solve the differential equation numerically, tabulating

(*a*) for $x = -4(0{\cdot}8) - 1{\cdot}6$,
(*b*) for $x = -4(0{\cdot}4) - 1{\cdot}6$.

You should work to an accuracy of four decimal places. Use the two tabulations to give improved estimates for $y(-3{\cdot}2)$, $y(-2{\cdot}4)$, and $y(-1{\cdot}6)$.

10. Determine the values of a, b, c, and d, such that the truncation error of the formula

$$y_{n+1} = ay_{n-1} + h(by'_{n+1} + cy'_n + dy'_{n-1})$$

is $O(h^5)$.

11. Show

$$\int_0^{x_{2n}} f(x)\,dx = 2nh\left[1 + \frac{1}{6}n^2\delta^2 + \left(\frac{n^4}{120} - \frac{n^2}{72}\right)\delta^4 + \ldots\right]f_n.$$

12. By taking $n = 2$ and $n = 1$ in exercise 11 show that

$$\int_{x_0}^{x_4} f(x)\,dx = \frac{4}{3}h(2f_3 - f_2 + 2f_1) + \frac{14}{45}h^5 f^{(4)}(\xi), \qquad x_0 \leqslant \xi \leqslant x_4,$$

$$\int_{x_0}^{x_2} f(x)\,dx = \frac{1}{3}h(f_2 + 4f_1 + f_0) - \frac{1}{90}h^5 f^{(4)}(\eta), \qquad x_0 \leqslant \eta \leqslant x_2,$$

i.e., we have the Milne–Simpson predictor–corrector formulae used in exercise 8.

13. (Hamming's method.) Show that for the predictor–corrector system (11.29) the true value y_{k+1} is given by

$$y_{k+1} = y_{k+1}^P + \tfrac{3}{8}h^4 y^{(4)}(\xi_1) = y_{k+1}^C - \tfrac{1}{24}h^4 y^{(4)}(\xi_2).$$

Hence, assuming that $y^{(4)}(x)$ is slowly varying show that

$$y_{k+1}^C - y_{k+1}^P \mp \tfrac{5}{12}h^4 y^{(4)}(\xi) \mp y_k^C - y_k^P$$

and derive the predictor–corrector–modifier system.

$$y_{k+1}^P = y_k + \tfrac{1}{12}h(23f_k - 16f_{k-1} + 5f_{k-2}),$$

$$y_{k+1}^M = y_{k+1}^P + \tfrac{9}{10}(y_k^C - y_k^P),$$

$$y_{k+1}^C = y_k + \tfrac{1}{12}h(5f_{k+1}^M + 8f_k - f_{k-1}),$$

$$y_{k+1} = y_{k+1}^C - \tfrac{1}{10}(y_{k+1}^C - y_{k+1}^P).$$

Further details of this method can be found in Ralston [3].

14. Show that the corrector

$$y_{n+1} = (1 - a)y_n + ay_{n-1} + \tfrac{1}{12}h[(5 - a)y'_{n+1} + (8 + 8a)y'_n + (5a - 1)y'_{n-1}]$$

is convergent for $-1 \leqslant a \leqslant 1$.

15. (General formulation of Runge–Kutta methods.) If D is an operator defined by

$$D = \frac{\partial}{\partial x} + f\frac{\partial}{\partial y}$$

show that

$$y_1 - y_0 = \sum_{t=0}^{\infty} \frac{h^{t+1}}{(t+1)!}\left\{\left(\frac{\partial}{\partial x} + f\frac{\partial}{\partial y}\right)^t f\right\}_0$$

$$= hf_0 + \frac{h^2}{2!}Df_0 + \frac{h^3}{3!}(D^2f + f_yDf)_0 + \frac{h^4}{4!}(D^3f + f_yD^2f + f_y{}^2 + 3DfDf_y)_0 +$$

$$+ \frac{h^5}{5!}(D^4f + 6DfD^2f_y + D^2fDf_y + f_y{}^2D^2f + f_y{}^3Df + 3f_{yy}(Df)^2 +$$

$$+ f_yD^3f + 7f_yDfDf_y)_0 + O(h^6).$$

16. Define $D_i = \alpha_i \frac{\partial}{\partial x} + \left(\sum_{j=1}^{i-1} \beta_{ij} \right) f \frac{\partial}{\partial y}$. Hence show that, with the notation adopted in the text,

$k_1 = hf_0$

$k_2 = hf_0 + h^2 D_2 f_0 + \frac{h^3}{2!} D_2^2 f_0 + \frac{h^4}{3!} D_2^3 f_0 + \frac{h^5}{5!} D_2^4 f_0 + O(h^6)$

$$k_3 = hf_0 + h^2 D_3 f_0 + h^3 (\tfrac{1}{2} D_3^2 f + \beta_{32} f_y D_2 f)_0 + h^4 (\tfrac{1}{6} D_3^3 f + \tfrac{1}{2} \beta_{32} f_y D_2^2 f + \beta_{32} D_2 f D_3 f_y)_0$$
$$+ h^5 (\tfrac{1}{24} D_3^4 f + \tfrac{1}{6} \beta_{32} f_y D_2^3 f + \tfrac{1}{2} \beta_{32} D_2^2 f D_3 f_y + \tfrac{1}{2} \beta_{32}^2 f_{yy} (D_2 f)^2 +$$
$$+ \tfrac{1}{2} \beta_{32} D_2 f D_3^2 f_y)_0 + O(h^6).$$

What are the expansions for k_4 and k_5?

17. Using the fact that we require

$$y_1 - y_0 = \sum_{i=1}^{n} w_i k_i$$

where the $w_i, \alpha_i, \beta_{ij}$ are independent of $f(x, y)$, i.e., $D_i = \alpha_i D$, verify by means of the results of exercises 15 and 16 that the equations satisfied by the constants in the third order Runge–Kutta system are

$$\alpha_2 = \beta_{21}, \qquad \alpha_3 = \beta_{31} + \beta_{32}$$
$$w_1 + w_2 + w_3 = 1,$$
$$w_2 \alpha_2 + w_3 \alpha_3 = \tfrac{1}{2},$$
$$w_2 \alpha_2^2 + w_3 \alpha_3^2 = \tfrac{1}{3},$$
$$w_3 \beta_{32} \alpha_2 = \tfrac{1}{6}.$$

Solve these equations on the assumptions

(*a*) $\alpha_2 \neq \alpha_3$ $(\alpha_2 \neq 0,\ \alpha_3 \neq 0)$
(*b*) $\alpha_2 = \alpha_3$ $(\neq 0)$.

18. Establish the fourth order Runge–Kutta equations of section 11.6. On the assumption $\alpha_2 \neq \alpha_3 \neq \alpha_4$ $(\alpha_2 \neq 0, \alpha_3 \neq 0, \alpha_4 \neq 0)$ show that $\alpha_4 = 1$ and the remaining constants can be expressed in terms of α_2 and α_3. Find a series of solutions when $\alpha_2 = \alpha_3 \neq \alpha_4$ $(\alpha_2 \neq 0,\ \alpha_4 \neq 0)$.

19. Use the standard Runge–Kutta method to solve the simultaneous differential equations

$$y' = z,$$
$$z' = -y, \qquad y(0) = 0, \qquad z(0) = 1,$$

– these equations are equivalent to the second order system $y'' + y = 0$, $y(0) = 0$, $y'(0) = 1$. (Hint: Solve as a vector differential equation.)

20. The linear differential equation

$$y'' + \alpha y' + \beta y = 0,$$

where $\alpha > 0$ and $\beta > 0$ are constants such that $\alpha^2 - 4\beta > 0$, is to be solved numerically using the recurrence formula

$$(\delta^2 + \alpha h \mu \delta + \beta h^2) y = 0.$$

(*a*) Determine the recurrence relation in terms of ordinates when the formula is applied at the point x_r.

(*b*) Show that the method is stable if $\beta h^2 < 4$.

21. If the corrector in exercise 14 is used to solve $y' = -Ky$ $(K > 0)$ show that the roots of the 'complementary polynomial' of the difference equation lie in the unit circle if

$$Kh \leqslant (6 - 6a)/(1 + a)$$

22. Solve the differential equation $y'' - y = 0$, $y(0) = 0$, $y(1) = 1$ by taking initial approximations $y'(0) = \alpha_1$, $y'(0) = \alpha_2$ and using an initial value method to find the corresponding values of y at $x = 1$, call them y_{α_1} and y_{α_2}. Interpolate (or extrapolate) to find the value of α, corresponding to $y_\alpha = 1$. Repeat the process as many times as may be necessary until a value of α is obtained such that $|y_\alpha - 1|$ is less than a prescribed amount.

23. The deflection y of an elastic beam under a transverse load $w(x)$ is given by an equation of the form

$$y^{(4)}(x) = w(x).$$

If the beam is simply supported at the end $x = 0$ and rigidly clamped at the end $x = 1$, the boundary conditions are

$$y(0) = y''(0) = y(1) = y'(1) = 0.$$

Solve this system when $w(x) = 1$ using a step length of size 0·1 and replacing $y^{(4)}$ by its central difference approximation.

24. Simplify the method of deferred correction to deal with the second order equation $y'' = f(x)y$ and show that the correction term is now

$$C_r = \left(-\frac{1}{240}\delta^6 + \frac{13}{15\,120}\delta^8 - \ldots\right) y_r.$$

(Hint: Consider $(1 + \frac{1}{12}\delta^2)y''$). Hence solve the differential equation $y'' = xy$ over the range $x = 0(0{\cdot}2)2$, with boundary conditions $y = 0$ at $x = 0$ and $y = 1$ at $x = 2$.

25. Use a deferred correction method to obtain values of y for $x = 2(0{\cdot}2)3$, where

$$(1 - x^2)y'' + 2(3 + x)y' + 4y = 0, \qquad y(2) = 1, \qquad y(3) = -7.$$

26. Find y in the range $x = 0(0{\cdot}2)1$ given that

$$y'' + y + x = 0, \qquad y(0) = 0, \qquad y'(1) + y(1) = 1.$$

27. (The method of Clenshaw.) Establish relations between the coefficients in the Chebyshev series for y' and y and hence those between y'' and y (see chapter 4, exercise 9). Using these results find the coefficient of $T_r(x)$ in

$$xy'' + y' + 16xy$$

and hence show that the solution of the differential equation

$$xy'' + y' + 16xy = 0, \qquad y(0) = 1, \qquad y'(0) = 0.$$

at $x = 1$ is $y(1) = -0{\cdot}3972$.

28. (Picard's method of solution.) The first order differential equation $y' = f(x, y)$, $y(x_0) = y_0$ may be rewritten

$$y = y_0 + \int_{x_0}^{x} f(x, y)\,\mathrm{d}x.$$

This leads to the iterative formula for obtaining y as a function of x, viz.,

$$y^{[n+1]} = y_0 + \int_{x_0}^{x} f(x, y^{[n]})\,\mathrm{d}x, \qquad y^{[0]} = y_0.$$

If $f(x, y) = x + y^2$, $y_0 = 0$ show that Picard's iterative formula gives

$$y^{[3]} = \frac{1}{2}x^2 + \frac{1}{20}x^5 + \frac{1}{160}x^8 + \frac{1}{4400}x^{11}$$

and this provides the estimate $y(0{\cdot}3) = 0{\cdot}045\,121\,9$. Use a fourth order Runge–Kutta formula to check this value.

29. Show that, provided $f(x, y)$ is bounded and satisfies a Lipschitz condition, the sequence $y^{[n]}$ in Picard's method converges to the exact solution $y(x)$ of the differential equation.

30. (The method of Frobenius.) Let y satisfy the differential equation

$$y'' + \frac{p(x)}{x}y' + \frac{q(x)}{x^2}y = 0$$

where
$$p(x) = p_0 + p_1x + p_2x^2 + \dots,$$
$$q(x) = q_0 + q_1x + q_2x^2 + \dots, \qquad (p_0, q_0) \neq (0, 0).$$

Then a solution exists of the form

$$y = x^m(a_0 + a_1x + a_2x^2 + \dots + a_nx^n + \dots), \qquad a_0 \neq 0,$$

where m satisfies the 'indicial equation'

$$m(m-1) + p_0m + q_0 = 0$$

and the coefficients a_n can be determined from a recurrence relation. Hence find the two independent solutions and the general solution of the differential equation

$$(2x + x^3)y'' - y' - 6xy = 0.$$

What is the range of validity of the general solution? How can one compute results outside the range of validity?

31. If, in Frobenius's method, the indicial equation has two equal roots m_1, or one root is m_1 and the other is $m_2 = m_1 -$ an integer (and a coefficient is infinite), then the solutions are given by

$$y_1 = x^{m_1}(a_0 + a_1x + \ldots + a_nx^n + \ldots) \quad \text{or} \quad y_1 + x^{m_2}(a_0 + a_1x + \ldots + a_nx^n + \ldots)$$

$$\text{and } y_2 = \left(\frac{\partial y_1}{\partial m}\right)_{m=m_1} \quad \text{or} \quad y_2 = \left[\frac{\partial}{\partial m}\left\{(m - m_1)y_1\right\}\right]_{m=m_1}, \text{ respectively.}$$

Use this information to show that the two independent solutions of

$$(x - x^2)y'' + (1 - 5x)y' - 4y = 0$$

are

$$y_1 = 1^2 + 2^2x + 3^2x^2 + 4^2x^3 + \ldots,$$

$$y_2 = y_1 \ln x - 2\{1\cdot 2x + 2\cdot 3x^2 + 3\cdot 4x^3 + \ldots\}.$$

32. Find, by the method of exercise 31, the two independent solutions of Bessel's equation of order unity, viz.,

$$x^2y'' + xy' + (x^2 - 1)y = 0.$$

33. Find the solution of the differential equation

$$2x^2(x - 1)y'' - y' - 6xy = 0$$

for $x = 10(1)20$ given that $y(\infty) = 0$, $y(100) = 1$.

34. If $u(x)$ is a particular solution of the differential equation

$$y'' + a(x)y' + b(x)y = 0,$$

show that the general solution is

$$y = Au\int u^{-2} \exp\left[-\int a(x)\, dx\right] dx + Bu.$$

(Hint: Assume a second solution can be found of the form $y = uv$.) Use this result to find the second solution of the differential equation

$$(x^2 - x)y'' + (x - 3)y' - 4y = 0,$$

given that $y = 1/x^2$ is a particular solution.

35. (The method of perturbations.) We solve the equation

$$y' + p(x)y = 0, \qquad p(x) = p_0(x) + \epsilon p_1(x) + \epsilon^2 p_2(x),$$

where ϵ is small, by first obtaining the solution $y_0(x)$ of

$$y' + p_0(x)y = 0$$

and then, assuming the true solution is

$$y = y_0(x) + \epsilon y_1(x) + \epsilon^2 y_2(x) + \ldots,$$

we substitute in the original differential equation and compare coefficients of ϵ^k $(k = 1, 2, \ldots)$. This leads to equations for determining y_1, y_2, etc. Use this method to solve

$$y' + (1 - \epsilon x)y = 0, \qquad y(0) = 1, \qquad \epsilon x \ll 1.$$

36. Show that the differential equation

$$y'' + \lambda y = 0, \qquad y(0) = y(1) = 0, \qquad \lambda \neq 0$$

when suitably approximated by central differences, reduces to the matrix equation

$$A\mathbf{y} - \tfrac{1}{12}\lambda h^2 B\mathbf{y} = 0, \qquad \text{or} \qquad B^{-1}A\mathbf{y} = \tfrac{1}{12}h^2\lambda\mathbf{y}$$

where $\mathbf{y}$ is an approximation to the true solution of the differential equation.

37. The n eigenvalues of the matrix equation in exercise 36 correspond to the n smallest eigenvalues of the given differential equation. By taking h to be some convenient value find the smallest eigenvalue, (*a*) by the power method (section 9.11), (*b*) by means of Rayleigh's quotient. What is the next largest eigenvalue?

12 Partial differential equations

12.1 Introduction. First order equations and characteristics

Before commencing with methods for the solution of partial differential equations we investigate some of the basic properties of these equations and lay the groundwork for our later work. The general first order linear partial differential equation has the form

$$a\frac{\partial u}{\partial x}+b\frac{\partial u}{\partial y}=c, \tag{12.1}$$

where a, b, c are functions of x and y not involving u (which is also a function of x and y) or any partial derivatives of u. If a, b, and c are functions of u then the equation (12.1) is termed a 'quasi-linear' partial differential equation.

By analogy with ordinary differential equations it is to be expected that known values of u on some initial curve C in the x,y plane are sufficient for the unique determination of u elsewhere. Since any displacement du, tangential to the surface, $u = u(x, y)$, is given by

$$\mathrm{d}u=\frac{\partial u}{\partial x}\,\mathrm{d}x+\frac{\partial u}{\partial y}\,\mathrm{d}y, \tag{12.2}$$

we have, by the linear dependence of equations (12.1) and (12.2)

$$\frac{\mathrm{d}x}{a}=\frac{\mathrm{d}y}{b}=\frac{\mathrm{d}u}{c}=\mathrm{d}s,\ \text{say}, \tag{12.3}$$

where 's' is measured along the tangent line given by (12.3). In particular we have from (12.3),

$$\frac{\mathrm{d}y}{\mathrm{d}x}=\frac{b}{a}. \tag{12.4}$$

Equation (12.4) defines a curve, relating x and y, which is called a 'characteristic'.

To familiarize ourselves with the idea of characteristics, consider the simple equation used by Fox [1], viz.:

$$\frac{\partial u}{\partial x}+\frac{\partial u}{\partial y}=1. \tag{12.5}$$

The characteristic equation is $dy/dx = 1$ and the characteristics are the family of straight lines having slope 1 (Fig. 12.1).

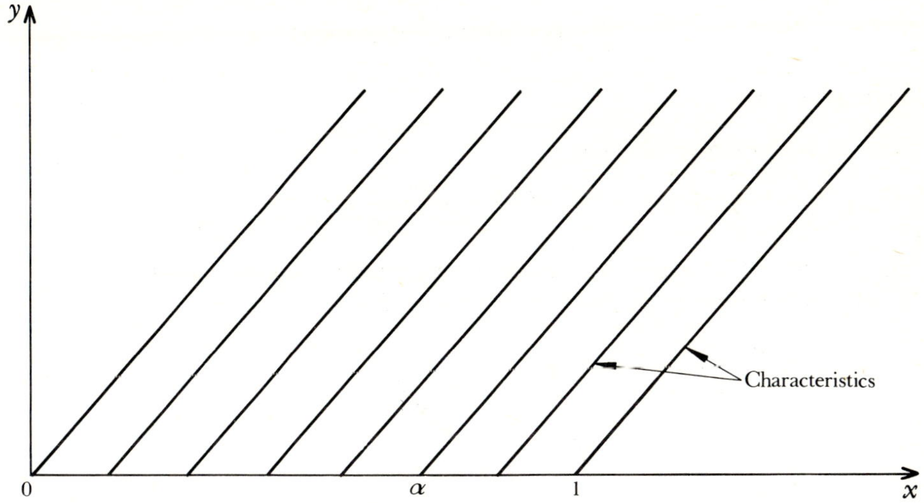

Fig. 12.1 Characteristics for $\frac{\partial u}{\partial x} + \frac{\partial u}{\partial y} = 1$

Suppose u is given on the line segment $y = 0, 0 \leqslant x \leqslant 1$ (that is, the initial curve C is the x-axis between 0 and 1), then along the characteristic $y = x - \alpha$ the solution of (12.5) is, from (12.3),

$$u = x + \beta = y + (\beta + \alpha), \tag{12.6}$$

where β is a constant of integration which can be determined from the known value of u on C. Note that the solution is unique in the region bounded by the terminal characteristics at $x = 0$ and $x = 1$.

If the initial curve C in the above example is taken to be $y = x$, that is, one of the characteristics, there is a solution only if $u = x = y$ on C.

Another possibility is that of discontinuous initial values. Suppose that on the line $y = 0$ the initial condition for the differential equation (12.5) is

$$u = \begin{cases} f_1(x), & 0 \leqslant x < x_1 \\ f_2(x), & x_1 < x \leqslant 1 \end{cases}$$

and

$$f_1(x_1) \neq f_2(x_1),$$

then the discontinuous nature of u persists all along the characteristic $y = x - x_1$.

If the characteristics depend on the solution (quasi-linear case) then two characteristics may cross. This causes a discontinuity and gives rise to 'shock waves'. Propagated discontinuities of this type restrict the use of the finite-difference methods we discuss later.

Before leaving first order partial differential equations we examine a more complicated system. Consider the equation

$$\frac{\partial u}{\partial x} - (x-y-1)\frac{\partial u}{\partial y} = x(y-x),$$

with $u = y^2$ on $x = 0$.

The characteristics are given by

$$\frac{\mathrm{d}x}{1} = \frac{-\mathrm{d}y}{(x-y-1)},$$

that is $y = x + Ce^x$, where C is a constant of integration. In this case there are three possibilities depending on the value of C (Fig. 12.2).

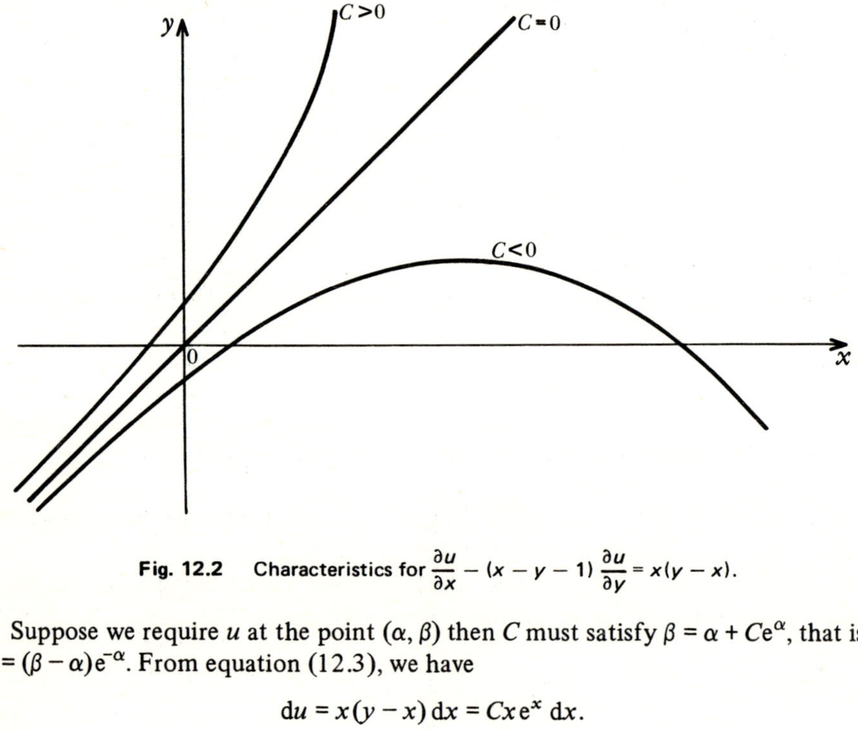

Fig. 12.2 Characteristics for $\frac{\partial u}{\partial x} - (x-y-1)\frac{\partial u}{\partial y} = x(y-x)$.

Suppose we require u at the point (α, β) then C must satisfy $\beta = \alpha + Ce^{\alpha}$, that is $C = (\beta - \alpha)e^{-\alpha}$. From equation (12.3), we have

$$\mathrm{d}u = x(y-x)\,\mathrm{d}x = Cxe^x\,\mathrm{d}x.$$

Hence

$$u = C\int xe^x\,\mathrm{d}x = Cxe^x - Ce^x + D,$$

where D is a constant.

The characteristic through (α, β) meets $x = 0$ where $y = C$, and thus, from the initial condition, $u = C^2$ at $(0, C)$. Substituting these values into the above equation for u we have

$$D = C^2 + C = e^{-\alpha}(\beta - \alpha)\,[(\beta - \alpha)e^{-\alpha} + 1].$$

In particular, at the point (α, β), u is given by

$$u = (\beta - \alpha)\,[(\alpha - 1) + (\beta - \alpha)\mathrm{e}^{-2\alpha} + \mathrm{e}^{-\alpha}].$$

The numerical values of u for $\beta = 0, 0{\cdot}1, 0{\cdot}2$ and for α ranging from 0 to 1 in steps of 0·1 can be seen in Table 12.1.

This problem, again, is rather idealized since the actual equations corresponding to equation (12.3) can be integrated analytically. When this is not the case (see exercise 2) then the integration has to be performed numerically, the method used being similar to that discussed in section 12.3.

Table 12.1 Numerical solution of $\dfrac{\partial u}{\partial x} - (x - y - 1)\dfrac{\partial u}{\partial y} = x(y - x)$

α \ β	0·0	0·1	0·2
0·0	0·0000	0·0100	0·0400
0·1	0·0077	0·0000	0·0087
0·2	0·0231	0·0048	0·0000
0·3	0·0371	0·0138	0·0014
0·4	0·0438	0·0193	0·0039
0·5	0·0387	0·0162	0·0011
0·6	0·0191	0·0009	−0·0113
0·7	−0·0168	−0·0292	−0·0366
0·8	−0·0702	−0·0756	−0·0769
0·9	−0·1420	−0·1395	−0·1336
1·0	−0·2325	−0·2215	−0·2077

12.2 Second order partial differential equations

Most of the problems involving partial differential equations encountered in physics, engineering, etc., are of second order and of the form

$$a\frac{\partial^2 u}{\partial x^2} + b\frac{\partial^2 u}{\partial x \partial y} + c\frac{\partial^2 u}{\partial y^2} = e, \qquad (12.7)$$

where a, b, c, and e are functions of $\partial u/\partial x$, $\partial u/\partial y$, u, x, and y (quasi-linear second order partial differential equation). If a, b, c, are independent of u and its first derivatives then equation (12.7) is termed a linear second order partial differential equation. The usual notation will be adopted, viz.,

$$p = \frac{\partial u}{\partial x}, \quad q = \frac{\partial u}{\partial y}, \quad r = \frac{\partial^2 u}{\partial x^2}, \quad s = \frac{\partial^2 u}{\partial x \partial y}, \quad t = \frac{\partial^2 u}{\partial y^2},$$

so that (12.7) becomes

$$ar + bs + ct = e.$$

The analysis of the first order case can be extended and we have, for a curve Γ in the x, y plane, equation (12.2) which we now write as

$$\mathrm{d}u = p\,\mathrm{d}x + q\,\mathrm{d}y. \qquad (12.8)$$

Two other equations must be formed in order that the values of u, p, and q on the curve enable r, s, and t to be determined on Γ. Namely,

$$\mathrm{d}\left(\frac{\partial u}{\partial x}\right)=\frac{\partial^2 u}{\partial x^2}\,\mathrm{d}x+\frac{\partial^2 u}{\partial y\partial x}\mathrm{d}y,$$

or

$$\mathrm{d}p = r\,\mathrm{d}x + s\,\mathrm{d}y, \tag{12.9}$$

and

$$\mathrm{d}q = s\,\mathrm{d}x + t\,\mathrm{d}y. \tag{12.10}$$

Equations (12.7), (12.9), and (12.10) form a set of three linear equations in r, s, and t and we now choose Γ so that coefficients in these equations satisfy

$$\begin{vmatrix} a & b & c \\ \mathrm{d}x & \mathrm{d}y & 0 \\ 0 & \mathrm{d}x & \mathrm{d}y \end{vmatrix} = 0,$$

i.e., $c(\mathrm{d}x)^2 - b\,\mathrm{d}x\mathrm{d}y + a(\mathrm{d}y)^2 = 0$,

or

$$a\left(\frac{\mathrm{d}y}{\mathrm{d}x}\right)^2 - b\left(\frac{\mathrm{d}y}{\mathrm{d}x}\right) + c = 0. \tag{12.11}$$

When (12.11) holds the matrix

$$\begin{bmatrix} e & a & b & c \\ \mathrm{d}p & \mathrm{d}x & \mathrm{d}y & 0 \\ \mathrm{d}q & 0 & \mathrm{d}x & \mathrm{d}y \end{bmatrix}$$

must be of rank two for a solution to be possible. This gives three relations any one of which is dependent on the other two. A typical relation is

$$e\,\mathrm{d}x\mathrm{d}y - a\,\mathrm{d}p\mathrm{d}y - c\,\mathrm{d}q\mathrm{d}x = 0. \tag{12.12}$$

Equation (12.11) defines two characteristic curves (Γ_1, Γ_2) (as it is a quadratic in $\mathrm{d}y/\mathrm{d}x$) and (12.12) specifies the conditions to be satisfied along the characteristics by the solution. If we have a domain in the x, y plane in which u is defined then, by comparison with homogeneous quadratic forms, if $b^2 > 4ac$ in that domain, then the differential equation is termed 'hyperbolic'. Similarly if $b^2 = 4ac$, the partial differential equation is termed 'parabolic', while if $b^2 < 4ac$ it is termed 'elliptic'.

12.3 Hyperbolic equations

These are characterized by the possession of real, distinct characteristics. Thus for

$$a\frac{\partial^2 u}{\partial x^2} + b\frac{\partial^2 u}{\partial x\partial y} + c\frac{\partial^2 u}{\partial y^2} = e, \qquad b^2 > 4ac,$$

there are two characteristics passing through any point, which are given by

$$c(\mathrm{d}x)^2 - b(\mathrm{d}x)(\mathrm{d}y) + a(\mathrm{d}y)^2 = 0. \tag{12.13}$$

For the solutions we write

$$\left.\begin{aligned} \mathrm{d}y &= f\,\mathrm{d}x \\ \mathrm{d}y &= g\,\mathrm{d}x \end{aligned}\right] \tag{12.14}$$

where f and g are the roots of (12.13) and will, in general, be functions of x, y, u, p, and q. For a solution to be possible we must have (cf. section 12.2)

$$e\,\mathrm{d}x\mathrm{d}y - a\,\mathrm{d}p\,\mathrm{d}y - c\,\mathrm{d}q\mathrm{d}x = 0,$$

which gives

and

$$\left.\begin{aligned} e\,\mathrm{d}y - af\,\mathrm{d}p - c\,\mathrm{d}q &= 0 \\ e\,\mathrm{d}y - ag\,\mathrm{d}p - c\,\mathrm{d}q &= 0. \end{aligned}\right] \tag{12.15}$$

If we are given u, p, and q on the initial line C, then we can draw from such points as P and Q the two families of characteristics (Fig. 12.3).

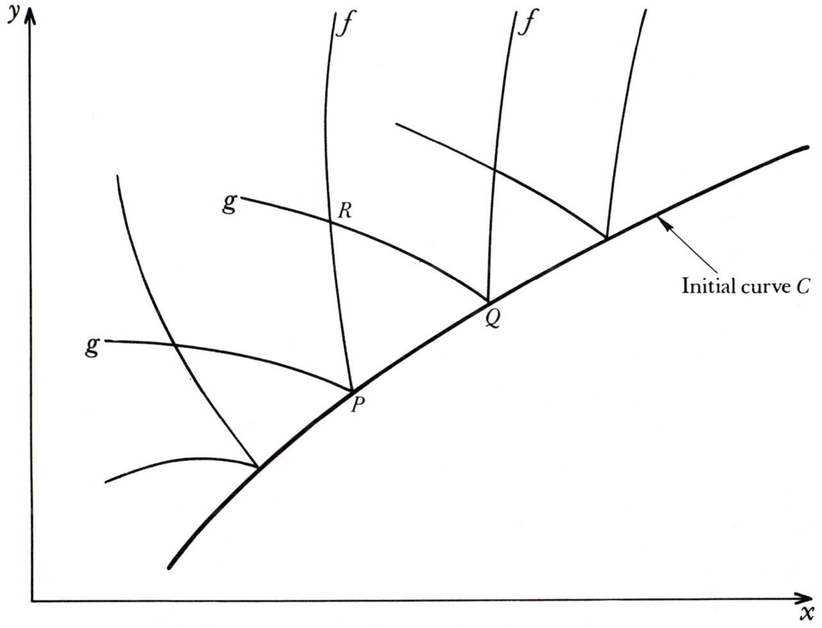

Fig. 12.3 Initial curve C with 'f' and 'g' characteristics

Consider two adjacent points P and Q and let the 'f' characteristic from P meet the 'g' characteristic from Q in R. If the characteristics depend on the solution u then the coordinates of R, as well as p and q, have to be determined at this juncture.

The coordinates can be calculated by the use of quadrature formulae such as the trapezium rule (section 5.1). For instance, integration of the first equation of (12.14) over an interval AB gives

$$y_B - y_A = \int_A^B f\,\mathrm{d}x \mp \tfrac{1}{2}(f_A + f_B)(x_B - x_A),$$

where f_A, for example, denotes evaluation of the function f at the point A. We can treat the equations (12.15) similarly. We have, for example,

$$\int_A^B e\,\mathrm{d}y = \int_A^B af\,\mathrm{d}p + \int_A^B c\,\mathrm{d}q,$$

i.e., $\tfrac{1}{2}(e_A + e_B)(y_B - y_A) \mp \tfrac{1}{2}[(af)_A + (af)_B]\,(p_B - p_A) + \tfrac{1}{2}(c_A + c_B)(q_B - q_A)$.

Thus (12.14) implies, PR and QR now being the intervals of integration,

$$\left.\begin{aligned} y_R - y_P &\mp \tfrac{1}{2}(f_R + f_P)(x_R - x_P) \\ y_R - y_Q &\mp \tfrac{1}{2}(g_R + g_Q)(x_R - x_Q) \end{aligned}\right] \qquad (12.16)$$

and (12.15) implies

$$\left.\begin{aligned} (e_R + e_P)(y_R - y_P) &\mp (a_R f_R + a_P f_P)\,(p_R - p_P) + (c_R + c_P)(q_R - q_P) \\ (e_R + e_Q)(y_R - y_Q) &\mp (a_R g_R + a_Q g_Q)\,(p_R - p_Q) + (c_R + c_Q)(q_R - q_Q). \end{aligned}\right] \qquad (12.17)$$

To determine u_R we integrate $\mathrm{d}u = p\,\mathrm{d}x + q\,\mathrm{d}y$ to obtain

$$\left.\begin{aligned} u_R - u_P &\mp \tfrac{1}{2}(p_R + p_P)(x_R - x_P) + \tfrac{1}{2}(q_R + q_P)(y_R - y_P) \\ u_R - u_Q &\mp \tfrac{1}{2}(p_R + p_Q)(x_R - x_Q) + \tfrac{1}{2}(q_R + q_Q)(y_R - y_Q). \end{aligned}\right] \qquad (12.18)$$

Equations (12.16), (12.17), and (12.18) comprise six non-linear equations for the determination of the unknowns x, y, p, q, and u at the point R. These are usually solved iteratively using the two linear equations (12.16) to determine x_R and y_R, the equations (12.17) to determine p_R and q_R, and any one of the equations (12.18) to give u_R. Only the first step of the iteration is a special case and, in the absence of any other information, f_R and g_R could be identified initially with f_P and g_P, respectively. The iterations will usually converge rapidly if the interval PQ is not too large, which is also necessary to minimize the truncation errors in the quadrature formulae.

Quite often the solution is required on a rectangular grid but the points of intersection of the two families of characteristics do not always occur at grid points; for instance, in Fig. 12.4, the characteristics through P' and Q' meet in R', which is not a grid point. Suppose our problem is as follows. The initial curve C is the line $y = 0$ and we require the solution at equidistant points alone the line $y = rh$, $r = 1, 2, \ldots$.

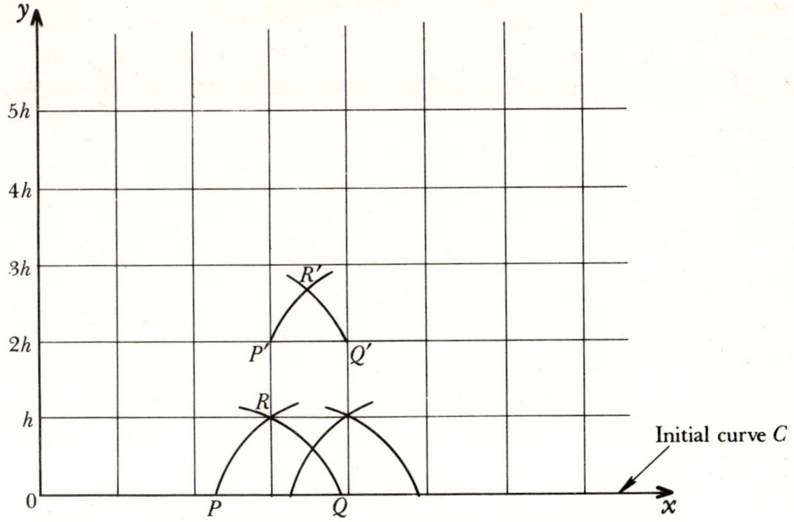

Fig. 12.4 Characteristics on a rectangular grid

Referring to Fig. 12.4 the equations relating P, Q, and R are clearly the equations (12.16), (12.17), and (12.18) but this time x_R and y_R are known and x_P and x_Q are unknown. We have just enough equations to obtain a solution and the only complication now is the necessity to interpolate to find u, p, and q at P and Q at each stage of the iteration (see Hartree [2]).

We now give an example to illustrate the solution of hyperbolic partial differential equations by the method described. We require the solution of the equation

$$\frac{\partial^2 u}{\partial x^2} + \frac{\partial^2 u}{\partial x \partial y} - 2\frac{\partial^2 u}{\partial y^2} = -1 \tag{12.19}$$

given the initial conditions

$$u = \frac{\partial u}{\partial y} = x, \qquad \text{on } y = 0, \qquad 0 \leqslant x \leqslant 1.$$

In this case the characteristic equation is

$$\left(\frac{\mathrm{d}y}{\mathrm{d}x}\right)^2 - \left(\frac{\mathrm{d}y}{\mathrm{d}x}\right) - 2 = 0,$$

giving

$$\frac{\mathrm{d}y}{\mathrm{d}x} = 2 \qquad \text{or} \qquad \frac{\mathrm{d}y}{\mathrm{d}x} = -1,$$

that is, the characteristics are a pair of straight lines having slope 2 and -1 respectively.

We divide the line $y = 0$ into N equal parts of width Δx. From Fig. 12.5 it is clear that the characteristics intersect at the same level and, referring to the figure, we see that the coordinates of R can be found directly without the use of (12.16).

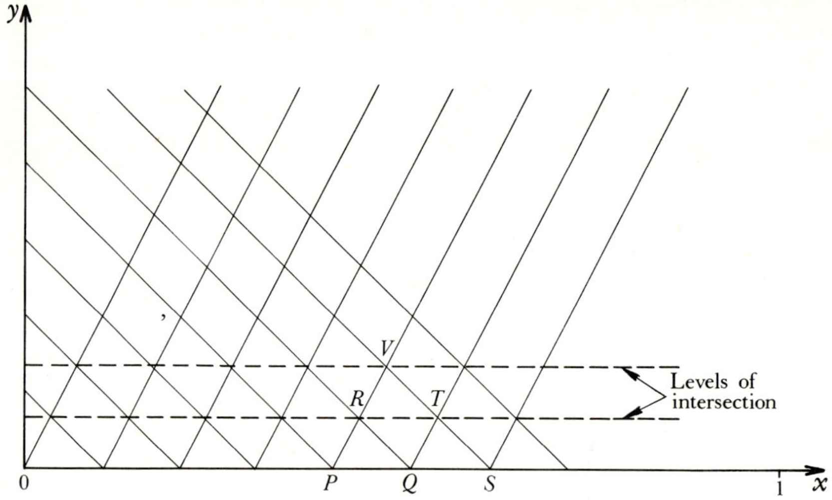

Fig. 12.5 Characteristics for $\dfrac{\partial^2 u}{\partial x^2} + \dfrac{\partial^2 u}{\partial x \partial y} - 2\dfrac{\partial^2 u}{\partial y^2} = -1$

For all points P, Q, and R we have

$$f_P = f_Q = f_R = \ \ 2,$$
$$g_P = g_Q = g_R = -1.$$

Initially $$u = x, \qquad \partial u/\partial y = q = x,$$

and hence $\partial u/\partial x = p = 1$.

For the purpose of this example we shall take $\Delta x = 0{\cdot}1$ and consider the computation for the first characteristic grid point R between $P(0{\cdot}4, 0)$ and $Q(0{\cdot}5, 0)$.

From the diagram R has coordinates $(0{\cdot}4333, 0{\cdot}066\,67)$. Equation (12.17) then yields

$$(-1-1)(0{\cdot}066\,67) = (2+2)(p_R - 1) + (-2-2)(q_R - 0{\cdot}4)$$
$$(-1-1)(0{\cdot}066\,67) = (-1-1)(p_R - 1) + (-2-2)(q_R - 0{\cdot}5)$$

that is, $p_R = 1{\cdot}067$ and $q_R = 0{\cdot}5$. Equation (12.18) now gives $u_R = 0{\cdot}4644$.

We similarly find that the characteristic grid point T between $Q(0{\cdot}5, 0)$ and $S(0{\cdot}6, 0)$ has coordinates $(0{\cdot}5333, 0{\cdot}066\,67)$. Equation (12.17) gives

$$(-1-1)(0{\cdot}0667) = (2+2)(p_T - 1) + (-2-2)(q_T - 0{\cdot}5)$$
$$(-1-1)(0{\cdot}0667) = (-1-1)(p_T - 1) + (-2-2)(q_T - 0{\cdot}6)$$

or $p_T = 1{\cdot}067$ and $q_T = 0{\cdot}6$, and equation (12.18) gives $u_T = 0{\cdot}5711$. Proceeding in this way can obtain the solution along the first level of intersection. These calculated values can be considered as given values on a new initial curve and we can proceed as before taking points such as R and T to obtain the value of u at the point V on the next level. The complete solution for the first three y-levels is shown in Table 12.2.

Table 12.2 Numerical solution of $\dfrac{\partial^2 u}{\partial x^2} + \dfrac{\partial^2 u}{\partial x \partial y} - 2\dfrac{\partial^2 u}{\partial y^2} = -1$

First y-level $x = q = u = 0(0.1)1, \quad y = 0, \quad p = 1$					
Second y-level $y = 0{\cdot}0667, \quad p = 1{\cdot}067$			*Third y-level* $y = 0{\cdot}1333, \quad p = 1{\cdot}133$		
x	q	u	x	q	u
0·0333	0·1	0·0378	0·0667	0·2	0·0844
0·1333	0·2	0·1444	0·1667	0·3	0·1978
0·2333	0·3	0·2511	0·2667	0·4	0·3111
0·3333	0·4	0·3578	0·3667	0·5	0·4244
0·4333	0·5	0·4644	0·4667	0·6	0·5378
0·5333	0·6	0·5711	0·5667	0·7	0·6511
0·6333	0·7	0·6778	0·6667	0·8	0·7644
0·7333	0·8	0·7844	0·7667	0·9	0·8778
0·8333	0·9	0·8911	0·8667	1·0	0·9911
0·9333	1·0	0·9978			

12.4 Finite-difference methods

With this type of method we obtain the solution of the differential equations over a uniform rectangular grid or network. A general grid point will be denoted by $(x_j, y_k) = (x_0 + j(\Delta x), y_0 + k(\Delta y))$ with reference to some origin (x_0, y_0), where Δx and Δy are respectively uniform intervals in the x and y directions. A function u, evaluated at the point (x_j, y_k) will be denoted by $u_{j,k}$ or u_{jk}. When the function u and its derivatives are single-valued, finite, and continuous functions of x and y we have from Taylor's theorem

$$u_{j+1,k} = u_{j,k} + (\Delta x)\left(\frac{\partial u}{\partial x}\right)_{j,k} + \frac{(\Delta x)^2}{2!}\left(\frac{\partial^2 u}{\partial x^2}\right)_{j,k} + \ldots$$

$$u_{j-1,k} = u_{j,k} - (\Delta x)\left(\frac{\partial u}{\partial x}\right)_{j,k} + \frac{(\Delta x)^2}{2!}\left(\frac{\partial^2 u}{\partial x^2}\right)_{j,k} + \ldots$$

Adding these equations and rearranging we obtain

$$\left(\frac{\partial^2 u}{\partial x^2}\right)_{j,k} = \frac{u_{j+1,k} - 2u_{jk} + u_{j-1,k}}{(\Delta x)^2} + O[(\Delta x)^2]. \tag{12.20}$$

By subtracting the Taylor's expansions we obtain

$$(\partial u/\partial x)_{j,k} = (u_{j+1,k} - u_{j-1,k})/(2\Delta x) + O\,[(\Delta x)^2]\,.$$

Consider, for example, the simplest hyperbolic system (the wave equation) defined by

$$\frac{\partial^2 u}{\partial x^2} = \frac{\partial^2 u}{\partial y^2}, \tag{12.21}$$

together with the conditions

$$\left.\begin{aligned} u &= \tfrac{1}{2}x(1-x) \\ \frac{\partial u}{\partial y} &= 0 \end{aligned}\right] \quad 0 \leqslant x \leqslant 1, \quad y = 0,$$

$$u = 0 \text{ on } x = 0 \text{ and } x = 1, \qquad y > 0.$$

In terms of finite-differences this becomes

$$\frac{u_{j+1,k} - 2u_{jk} + u_{j-1,k}}{(\Delta x)^2} = \frac{u_{j,k+1} - 2u_{j,k} + u_{j,k-1}}{(\Delta y)^2}$$

or

$$u_{j,k+1} = -u_{j,k-1} + 2(1-\rho^2)u_{j,k} + \rho^2(u_{j+1,k} + u_{j-1,k}) \tag{12.22}$$

where $\rho = (\Delta y)/(\Delta x)$.

Equation (12.22) means, in terms of Fig. 12.6, that the value of u at the point P is calculated from the known values at the points marked ×. The computation is

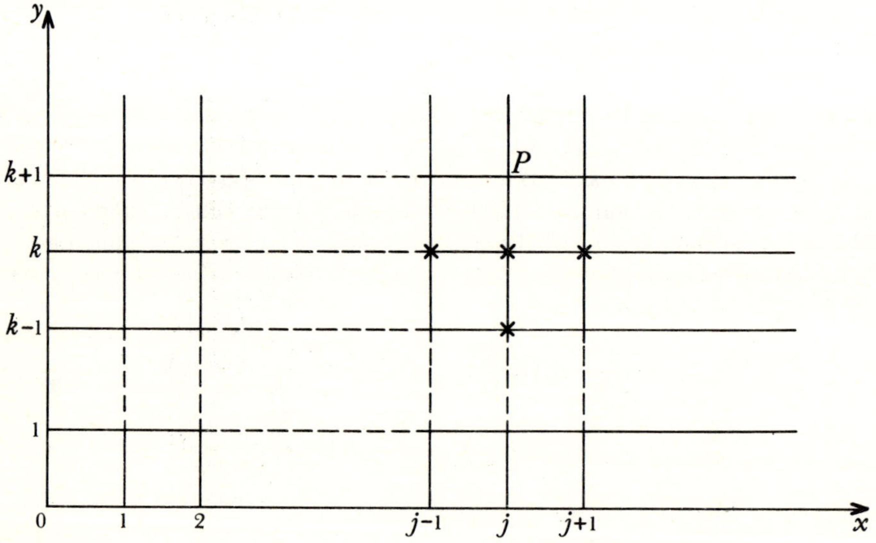

Fig. 12.6 Grid points used in the finite-difference solution of $\dfrac{\partial^2 u}{\partial x^2} - \dfrac{\partial^2 u}{\partial y^2} = 0$.

carried out initially for $k = 0$, using the given 'boundary' conditions. In this case equation (12.22) introduces the 'fictitious' value $u_{j,-1}$ and this value has to be eliminated. The first condition gives

$$u_{j,0} = \tfrac{1}{2}j(\Delta x)(1 - j\Delta x),$$

and the second gives

$$\left(\frac{\partial u}{\partial y}\right)_{j,0} = \frac{u_{j,1} - u_{j,-1}}{2(\Delta y)} = 0, \qquad \text{i.e., } u_{j,1} = u_{j,-1}.$$

Thus for the first step we have the special equation

$$u_{j,1} = (1 - \rho^2)\tfrac{1}{2}j(\Delta x)(1 - j\Delta x) + \tfrac{1}{2}\rho^2\Delta x[j - (j^2 + 1)\Delta x], \qquad [j = 1, 2, \ldots].$$

Alternatively $u_{j,1}$ can be obtained using the less accurate formula,

$$\left(\frac{\partial u}{\partial y}\right)_{j,0} = \frac{u_{j,1} - u_{j,0}}{(\Delta y)} = 0$$

i.e.,

$$u_{j,1} = \tfrac{1}{2}j(\Delta x)(1 - j\Delta x).$$

After calculating $u_{j,1}$ we use the formula (12.22) with $k \geqslant 1$, utilizing the known values of $u_{j,1}$ $(j = 0, 1, \ldots)$ and $u_{j,0}$ $(j = 1, 2, \ldots)$ to calculate $u_{j,2}$ $(j = 1, 2, \ldots)$ and hence $u_{j,k}$. The first few stages of the calculation are given in Table 12.3, where we have taken $\rho = 1$ which is the largest value of ρ permissible (see section 12.5 and exercise 7).

Table 12.3 Values of $u(x, y)$ satisfying (12.21) and given initial conditions

$x \backslash y$	0·0	0·1	0·2	0·3
0·0	0	0	0	0
0·1	0·045	0·040	0·030	0·020
0·2	0·080	0·075	0·060	0·040
0·3	0·105	0·100	0·085	0·060
0·4	0·120	0·115	0·100	0·075
0·5	0·125	0·120	0·105	0·080
0·6	0·120	0·115	0·100	0·075
0·7	0·105	0·100	0·085	0·060
0·8	0·080	0·075	0·060	0·040
0·9	0·045	0·040	0·030	0·020
1·0	0	0	0	0

With finite-difference methods it is necessary to show that the system is stable, i.e., any errors introduced (due to rounding) do not grow unboundedly so as to swamp the solution of the finite-difference equations. It must also be shown that the finite-difference solution converges, as the mesh size decreases, to the solution of the differential equation.

For hyperbolic partial differential equations the method of integrating along characteristics is usually the most convenient and accurate. Step-by-step methods can only be used for relatively simple equations and for this reason it is not proposed to discuss finite-difference methods any further for hyperbolic equations. Such methods are discussed in the books referred to at the end of the chapter and are similar to the methods discussed in section 12.5.

12.5 Parabolic partial differential equations

These are of the type

$$a\frac{\partial^2 u}{\partial x^2} + b\frac{\partial^2 u}{\partial x \partial y} + c\frac{\partial^2 u}{\partial y^2} = e,$$

where a, b, c, and e do not involve second order partial derivatives of u and $b^2 - 4ac = 0$. One example of this class of linear equations is

$$\frac{\partial^2 u}{\partial x^2} = \frac{\partial u}{\partial y}. \tag{12.23}$$

This is unsymmetric in y and thus the solution in the positive y-direction has different form to that in the negative y-direction. Further, since we have only one characteristic, we must resort to finite-difference methods. It is not difficult to set up a difference equation corresponding to the differential equation and no new concepts are involved. However, in any method we develop, it is extremely important that the difference equation we establish is stable and the solution of the difference equation converges to the differential equation. These questions we shall discuss for the parabolic system (12.23) taken with appropriate boundary and initial conditions.

We have, with the notation of the previous section,

$$\left(\frac{\partial^2 u}{\partial x^2}\right)_{jk} = \frac{u_{j+1,k} - 2u_{jk} + u_{j-1,k}}{(\Delta x)^2} + O[(\Delta x)^2].$$

Similarly $\partial u/\partial y$ can be represented by a difference formula, one of the simplest being the truncated forward-difference formula

$$\left(\frac{\partial u}{\partial y}\right)_{jk} = \frac{u_{j,k+1} - u_{jk}}{\Delta y} + O(\Delta y).$$

A finite-difference scheme for (12.23) is then

$$\frac{u_{j,k+1} - u_{jk}}{\Delta y} = \frac{u_{j+1,k} - 2u_{jk} + u_{j-1,k}}{(\Delta x)^2}. \tag{12.24}$$

A method using a formula such as (12.24) is called an *explicit* method, because $u_{j,k+1}$ is computed directly from known values of u at the kth y-level.

We now proceed to show that, subject to certain restrictions, this method is stable and that the computed values of u_{jk} obtained from (12.24) converge, as the mesh size decreases, to the true value of u given by the partial differential equation (12.23). In general a finite-difference scheme is stable for a restricted range of $\Delta y/(\Delta x)^2$ and by considering this restricted range the proof of convergence can be simplified.

There are two standard ways of investigating stability. In the first the errors are expressed as a finite Fourier series, while in the second the equations are expressed in matrix form and the eigenvalues of this matrix examined (in a similar manner to the methods of chapter 10). The Fourier series method is the easier of the two but is less rigorous since it neglects the boundary conditions. These methods will be illustrated by discussing various difference schemes (for a more detailed discussion see Smith [3]).

The Fourier method for the system (12.24) is as follows. Suppose that the true solution of the finite-difference system (12.24) is v_{jk} given by

$$v_{jk} = u_{jk} + \epsilon_{jk},$$

then we express the error ϵ_{jk} by a finite Fourier series of the form

$$\sum_k \mu^k \exp(i\beta_k x_j)$$

where $i = \sqrt{(-1)}$ and the number of terms is equal to the number of mesh points in the x-direction.

The finite-difference equations are linear and so independent solutions are additive. Thus we need only consider the effect of a single term of the form

$$\epsilon_{jk} = \mu^k e^{i\beta(j\Delta x)}.$$

This error will not increase with y provided that $|\mu|$ is less than or equal to 1, that is, the system will be stable for $|\mu| \leqslant 1$.

Since ϵ_{jk} satisfies the same difference equation as u and v,

$$\epsilon_{j,k+1} = \rho\epsilon_{j+1,k} + (1-2\rho)\epsilon_{jk} + \rho\epsilon_{j-1,k},$$

where $\rho = (\Delta y)/(\Delta x)^2$. Hence, by substitution,

$$\mu^{k+1} e^{i\beta(j\Delta x)} = [\rho e^{i\beta(\Delta x)} + (1-2\rho) + \rho e^{-i\beta(\Delta x)}]\mu^k e^{i\beta(j\Delta x)},$$

or

$$\mu - 1 = \rho(e^{i\beta(\Delta x)} - 2 + e^{-i\beta(\Delta x)}),$$

$$= 2\rho\,[\cos\beta(\Delta x) - 1] = -4\rho\sin^2(\tfrac{1}{2}\beta\Delta x),$$

and hence $|\mu| \leqslant 1$ provided that $0 \leqslant \rho \leqslant \frac{1}{2}$.

To decide convergence in this range of ρ we find an expression for the difference between the finite-difference solution and the true solution of the differential equation.

Now suppose w_{jk} is the true solution of the differential equation (12.23) at the mesh points and u_{jk} is the finite-difference solution of (12.24). The error ϵ_{jk} is given by $\epsilon_{jk} = w_{jk} - u_{jk}$. Substitution in (12.24) gives after simplification,

$$\epsilon_{j,k+1} = \rho\epsilon_{j+1,k} + (1-2\rho)\epsilon_{jk} + \rho\epsilon_{j-1,k} + w_{j,k+1} - w_{jk} - \rho(w_{j+1,k} - 2w_{jk} + w_{j-1,k}). \qquad (12.25)$$

Now Taylor's theorem gives, assuming first and second derivatives exist and are continuous,

$$w_{j,k+1} = w(x_j, y_k + \Delta y) = w_{jk} + (\Delta y)\left(\frac{\partial w}{\partial y}\right)_{j,\,k+\theta_1},$$

$$w_{j\pm1,k} = w(x_j \pm \Delta x, y_k) = w_{j,k} \pm (\Delta x)\left(\frac{\partial w}{\partial x}\right)_{j,\,k} + \frac{(\Delta x)^2}{2}\left(\frac{\partial^2 w}{\partial x^2}\right)_{j\pm\theta_r,\,k}$$

where $r = 2$ for $w_{j+1,k}$, $r = 3$ for $w_{j-1,k}$, and $0 < \theta_1, \theta_2, \theta_3 < 1$. Substitution in (12.25) gives

$$\epsilon_{j,k+1} = \rho\epsilon_{j+1,k} + (1-2\rho)\epsilon_{jk} + \rho\epsilon_{j-1,k} + \Delta y\left\{\left(\frac{\partial w}{\partial y}\right)_{j,\,k+\theta_1} - \left(\frac{\partial^2 w}{\partial x^2}\right)_{j+\theta_4,\,k}\right\}, \qquad (12.26)$$

where $-1 < \theta_4 < 1$.

Let E_k denote the modulus of the maximum error along the kth y-row and M the modulus of the maximum value of the expressions in curly brackets $\{\ldots\}$ in (12.26). If $0 < \rho \leqslant \frac{1}{2}$ we know that the finite-difference system is stable. Further in this range for ρ, all the coefficients of the ϵ terms in (12.26) are positive or zero. Thus

$$E_{k+1} \leqslant \rho E_k + (1-2\rho)E_k + \rho E_k + (\Delta y)M = E_k + (\Delta y)M,$$

and hence $E_{k+1} \leqslant E_0 + (k+1)(\Delta y)M$.

Initially the values of u and w are the same (this might not be the case if we are given derivative boundary conditions – see example at the end of this section), that is $E_0 = 0$. As $\Delta x \to 0$, $\Delta y = \rho(\Delta x)^2 \to 0$ and the error is $O[k(\Delta x)^2]$.

This proves that, for $0 \leqslant \rho \leqslant \frac{1}{2}$, u tends to w as Δx tends to zero provided that w is twice differentiable with respect to x and once differentiable with respect to y. It can be shown that if we take $\rho = \frac{1}{6}$ then the error $\epsilon_{j,k+1}$ is $O[(\Delta x)^4]$ (see exercise 13).

The stability criterion, $\rho \leqslant \frac{1}{2}$, imposes a severe restriction on the interval lengths, and forces the computer to take small steps in the y-direction which is a big drawback to using this type of method. For this reason it is usually preferable to use an *implicit* scheme, that is, one in which the unknown values of u_{jk} are not calculated directly. The simplest implicit scheme is obtained by replacing

$$(\partial^2 u/\partial x^2)_{j,k+1} \quad \text{by} \quad (u_{j+1,k+1} - 2u_{j,k+1} + u_{j-1,k+1})/(\Delta x)^2;$$

this gives

$$\frac{u_{j,k+1}-u_{jk}}{\Delta y}=\frac{u_{j+1,k+1}-2u_{j,k+1}+u_{j-1,k+1}}{(\Delta x)^2}$$

or

$$u_{jk}=-\rho u_{j+1,k+1}+(1+2\rho)u_{j,k+1}-\rho u_{j-1,k+1}, \qquad (12.27)$$

where

$$\rho=(\Delta y)/(\Delta x)^2.$$

This equation cannot be solved in a stepwise manner, as is the case with the explicit scheme, and is unsuitable for an infinite x-region. Problems in a finite range $a<x<b$ can be handled as a system of $[(b-a)/(\Delta x)]-1$ simultaneous linear equations in $[(b-a)/(\Delta x)]-1$ unknowns.

The proof of stability and convergences can be established by means of the usual Fourier or matrix methods and is similar to the analysis for the explicit scheme (see Fox [1], Forsythe and Wasow [4]).

Another, and more general, implicit scheme can be obtained by combining equations (12.24) and (12.27) by weighting as follows. We multiply (12.24) by $(1-\theta)$ and (12.27) by θ and add, to obtain

$$u_{j,k+1}-u_{jk}=[\theta\delta_j^2u_{j,k+1}+(1-\theta)\delta_j^2u_{jk}] \qquad (12.28)$$

where $0\leqslant\theta\leqslant 1$, and δ_j^2 represents the central difference operator (section 3.4) acting on the suffix j.

When $\theta=0$ we have the forward-difference (explicit) scheme; when $\theta=1$ we have the backward-difference (implicit) scheme; when $\theta=\frac{1}{2}$ we have the scheme devised by Crank and Nicolson, [5], and this form of equation, (12.28), is called the Crank–Nicolson equation.

As before, with $\epsilon_{jk}=\mu^k e^{i\beta(j\Delta x)}$, we have, for the scheme (12.28),

$$\epsilon_{j,k+1}-\epsilon_{jk}=\rho[\theta\delta_j^2\epsilon_{j,k+1}+(1-\theta)\delta_j^2\epsilon_{jk}]: \qquad (12.29)$$

Now

$$\begin{aligned}\delta_j^2\epsilon_{jk}&=\mu^k e^{i\beta(j\Delta x)}\,[e^{i\beta(\Delta x)}-2+e^{-i\beta(\Delta x)}],\\&=2\epsilon_{jk}[\cos\beta(\Delta x)-1],\\&=-4\epsilon_{jk}\sin^2[\tfrac{1}{2}\beta(\Delta x)].\end{aligned}$$

Equation (12.29) thus reduces to

$$\mu-1=\rho[-\mu\theta-(1-\theta)]\,.\,4\sin^2[\tfrac{1}{2}\beta(\Delta x)],$$

giving

$$\mu=\frac{1-z(1-\theta)}{1+z\theta}=1-\frac{z}{1+z\theta}, \qquad (12.30)$$

where

$$z=4\rho\sin^2[\tfrac{1}{2}\beta(\Delta x)]\geqslant 0.$$

For stability we require $|\mu| \leqslant 1$, that is $-1 \leqslant \mu \leqslant 1$. If $0 \leqslant \theta \leqslant 1$, (12.30) certainly gives $\mu \leqslant 1$, but this does not necessarily imply that $\mu \geqslant -1$. However, as z increases, μ tends asymptotically to $1-(1/\theta)$ and if $1-(1/\theta) \geqslant -1$, that is $\theta \geqslant \frac{1}{2}$, we always have a stable system. Further, since $\mu \geqslant -1$ implies $z \leqslant 2/(1-2\theta)$ we have stability for $\theta < \frac{1}{2}$, provided

$$4\rho \sin^2 [\tfrac{1}{2}\beta(\Delta x)] \leqslant 2/(1-2\theta).$$

If $\rho \leqslant 1/2(1-2\theta)$ this last inequality is always satisfied. We note in particular, that when $\theta = 0$, we are dealing with the explicit scheme and we obtain $\rho \leqslant \frac{1}{2}$, as before (Fig. 12.7).

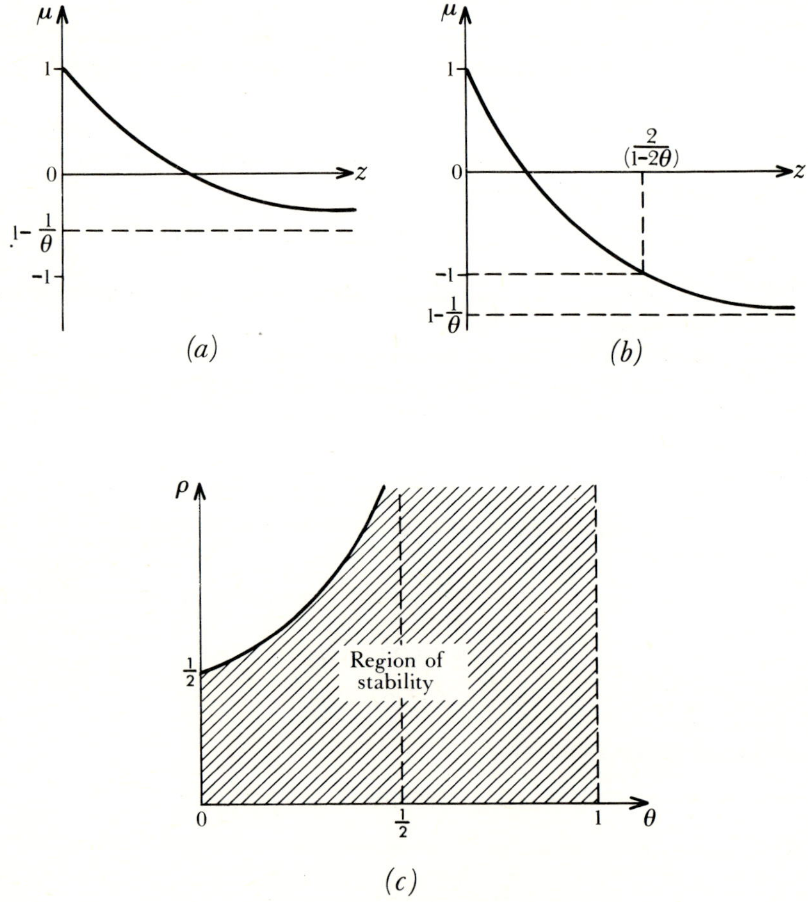

Fig. 12.7 Variation of (*a*) μ with $z, -1 < 1 - 1/\theta$,
(*b*) μ with $z, -1 > 1 - 1/\theta$,
(*c*) ρ with θ.

As an example consider the equation (12.23)

$$\frac{\partial^2 u}{\partial x^2} = \frac{\partial u}{\partial y},$$

subject to the conditions

$$\left.\begin{aligned} u &= 1 + x - \tfrac{1}{2}x^2 \quad &&\text{on} \quad y = 0, \ 0 \leqslant x \leqslant 1, \\ \frac{\partial u}{\partial x} &= u \quad &&\text{on} \quad x = 0, \ y \geqslant 0 \\ \frac{\partial u}{\partial x} &= 0 \quad &&\text{on} \quad x = 1, \ y \geqslant 0 \end{aligned}\right] \tag{12.31}$$

The Crank–Nicolson implicit scheme, with $\theta = \frac{1}{2}$ in (12.28), gives

$$\begin{aligned} -\rho u_{j-1,k+1} &+ (2 + 2\rho) u_{j,k+1} - \rho u_{j-1,k+1} \\ &= \rho u_{j-1,k} + (2 - 2\rho) u_{jk} + \rho u_{j+1,k} \end{aligned} \tag{12.32}$$

where $\rho = (\Delta y)/(\Delta x)^2$,

When $j = 0$ we introduce the fictitious terms $u_{-1,k}$ and $u_{-1,k+1}$ and eliminate these by means of the central difference approximation to the first boundary condition (cf. section 12.4), viz.,

$$u_{1k} - u_{-1,k} = 2(\Delta x) u_{0k} \, .$$

This gives

$$(2 + 2\rho + 2\rho\Delta x) u_{0,k+1} - 2\rho u_{1,k+1} = (2 - 2\rho - 2\rho\Delta x) u_{0k} + 2\rho u_{1k},$$

as the first equation to be solved.

Similarly, when $k = N$, where $N(\Delta x) = 1$, the Crank–Nicolson scheme introduces the fictitious terms $u_{N+1,k}$ and $u_{N+1,k+1}$. Once again the truncated central difference representation of the second boundary condition serves to eliminate these. The complete set of $(N + 1)$ equations that have to be solved in the unknowns $u_{0,k+1}, \ldots, u_{N,k+1}$ are, in matrix form,

$$A\mathbf{u}_{k+1} = (4I - A)\mathbf{u}_k, \tag{12.33}$$

where $\mathbf{u}'_k = (u_{0k}, u_{1k}, \ldots, u_{Nk})$, $(k = 0, 1, 2, \ldots)$, I is the $(N + 1) \times (N + 1)$ unit matrix, and A is the tridiagonal matrix.

$$A = \begin{bmatrix} 2 + 2\rho + 2\rho\Delta x & -2\rho & & & & \\ -\rho & 2 + 2\rho & -\rho & & & \\ & -\rho & 2 + 2\rho & -\rho & & \\ & & \ddots & \ddots & \ddots & \\ & & & -\rho & 2 + 2\rho & -\rho \\ & & & & -2\rho & 2 + 2\rho \end{bmatrix}$$

From (12.33), assuming det $A \neq 0$ (see exercise 12)

$$\mathbf{u}_{k+1} = (4A^{-1} - I)\mathbf{u}_k = M\mathbf{u}_k, \text{ say.}$$

We now look at the problem of stability, using the matrix method, which was mentioned earlier in the text. Suppose ϵ_k is the error in $\mathbf{u}_k$ ($k = 0, 1, \ldots$). Then we have from (10.10) (assuming exact arithmetic)

$$\epsilon_{k+1} = M\epsilon_k = \ldots = M^{k+1}\epsilon_0.$$

We also know from chapter 10 that the error vector ϵ_{k+1} will not grow unboundedly if the eigenvalues of M are less than or equal to unity in modulus. (Equality is allowed in this particular instance; why?) If λ is an eigenvalue of A, $1/\lambda$ is an eigenvalue of A^{-1}, and it follows that we require

$$|(4/\lambda) - 1| \leqslant 1$$

for stability, that is, $\lambda \geqslant 2$. This is always the case since Gerschgorin's theorem (Theorem 9·4, p. 163) states that the eigenvalues of the matrix A lie in the union of the circular discs

$$\left.\begin{array}{c} |\lambda - 2 - 2\rho - 2\rho\Delta x| \leqslant 2\rho \\ |\lambda - 2 - 2\rho| \leqslant 2\rho \end{array}\right]$$

which implies that $\lambda \geqslant 2$, for all the eigenvalues of A, i.e., the system (12.33) is stable for all ρ. This last result shows, incidentally, that we were quite justified in assuming det $A \neq 0$, earlier.

For simplicity we choose $\rho = 1$, $\Delta x = 0{\cdot}1$, and $\Delta y = 0{\cdot}1$ and at each stage of the computation we have eleven equations to solve. The solution for the first five y-levels is given in Table 12.4.

Table 12.4 Numerical solution of $\dfrac{\partial^2 u}{\partial x^2} = \dfrac{\partial u}{\partial y}$ subject to conditions (12.31)

x \ y	0·0	0·1	0·2	0·3	0·4
0·0	1·000	0·9905	0·9818	0·9733	0·9651
0·1	1·095	1·085	1·076	1·066	1·058
0·2	1·180	1·170	1·160	1·151	1·141
0·3	1·255	1·245	1·235	1·225	1·216
0·4	1·320	1·310	1·300	1·290	1·280
0·5	1·375	1·365	1·355	1·345	1·335
0·6	1·420	1·410	1·400	1·390	1·380
0·7	1·455	1·445	1·435	1·425	1·415
0·8	1·480	1·470	1·460	1·450	1·440
0·9	1·495	1·485	1·475	1·465	1·455
1·0	1·500	1·490	1·480	1·470	1·460

12.6 Elliptic partial differential equations

Elliptic equations in two dimensions arise frequently from equilibrium problems, for example, the heat-conduction equation

$$\frac{\partial^2 u}{\partial x^2} + \frac{\partial^2 u}{\partial y^2} = \frac{\partial u}{\partial t},$$

when $\partial u/\partial t \equiv 0$. The characteristics for this type of equation are complex and this influences the associated conditions required for a 'well-posed' problem. We shall assume that one condition is given at every point on the boundary, which is closed, and that a unique solution pertains.

The best known elliptic partial differential equations are Poisson's equation

$$\frac{\partial^2 u}{\partial x^2} + \frac{\partial^2 u}{\partial y^2} = f(x, y), \tag{12.34}$$

and Laplace's equation

$$\frac{\partial^2 u}{\partial x^2} + \frac{\partial^2 u}{\partial y^2} = 0. \tag{12.35}$$

We shall concern ourselves with the finite-difference solution of these equations. Most of the analysis given in the previous sections of this chapter can be applied here.

Consider, as an example, the solution of Laplace's equation (12.35) inside the L-shaped configuration shown in Fig. 12.8. Let the given boundary conditions be $u = y(1 - y)$ on $x = 0$, $u = x(1 - x)$ on $y = 0$, with $u = 0$ elsewhere on the boundary.

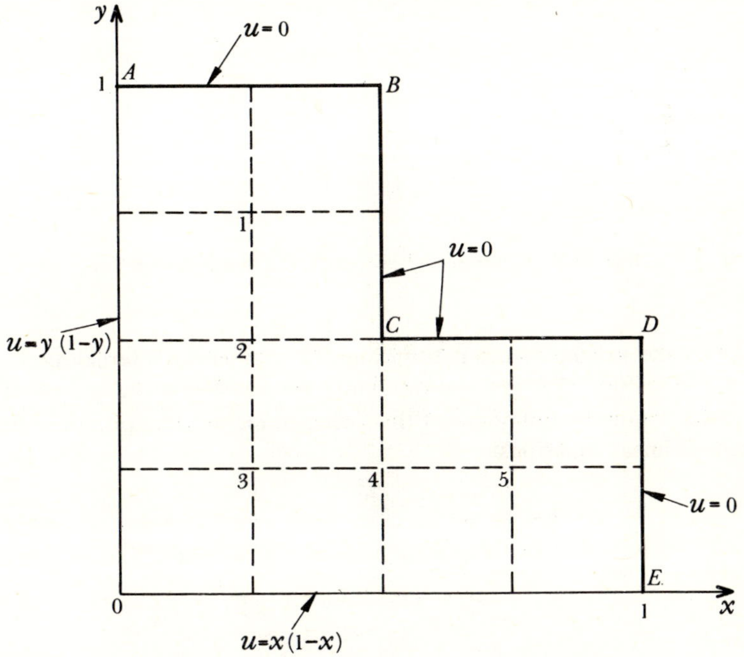

Fig. 12.8 Laplace's equation in an L-shaped region

For simplicity let us choose $\Delta x = \Delta y = 0{\cdot}25$. Replacing $(\partial^2 u/\partial x^2)_{jk}$ by $(u_{j+1,k} - 2u_{jk} + u_{j-1,k})/(\Delta x)^2$ and $(\partial^2 u/\partial y^2)_{jk}$ by a similar expression we have, for the general grid point u_{jk},

$$u_{j+1,k} + u_{j,k+1} - 4u_{jk} + u_{j,k-1} + u_{j-1,k} = 0. \tag{12.36}$$

Equation (12.36) is a five-point formula and, diagrammatically (Fig. 12.9), informs us that four times the value of u at P equals the sum of the values of u at the points marked x.

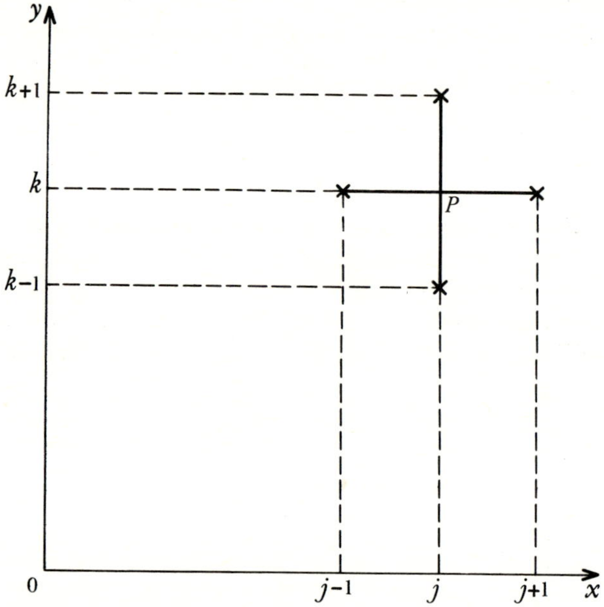

Fig. 12.9 Pictorial representation of a five-point formula

We now make a slight change in notation and number the internal mesh points 1, 2, 3, 4, 5 as in Fig. 12.8. Let u_i now denote the solution at the grid point i. Since u is known on the boundary of the L-shaped region, equation (12.36) yields the system of linear equations

$$\begin{aligned}
-4u_1 + u_2 \qquad\qquad\qquad\qquad &= -3/16,\\
u_1 - 4u_2 + u_3 \qquad\qquad\qquad &= -1/4,\\
u_2 - 4u_3 + u_4 \qquad\quad &= -3/8,\\
u_3 - 4u_4 + u_5 &= -1/4,\\
u_4 - 4u_5 &= -3/16.
\end{aligned}$$

These equations can be solved by the methods of chapters 8 and 10 (see section 10.5), the solution being

$$u_1 = u_5 = 0{\cdot}076\,92,$$
$$u_2 = u_4 = 0{\cdot}1202,$$
$$u_3 = 0{\cdot}1538.$$

The results $u_1 = u_5, u_2 = u_4$ could have been anticipated by noticing that the system is symmetrical about the line OC in Fig. 12.8. The additional information obtained from the symmetry properties reduces the number of equations and variables to three. In general, however, we do not have symmetry and the linear equations to be solved will not be in tridiagonal form. It can be shown that if all internal points are numbered systematically as in Fig. 12.10, we obtain a band matrix of the form

$$\begin{bmatrix} A_1 & B_1 & & & & \\ C_2 & A_2 & B_2 & & & \\ & C_3 & A_3 & B_3 & & \\ & & \ddots & \ddots & \ddots & \\ & & & C_{n-1} & A_{n-1} & B_{n-1} \\ & & & & C_n & A_n \end{bmatrix}$$

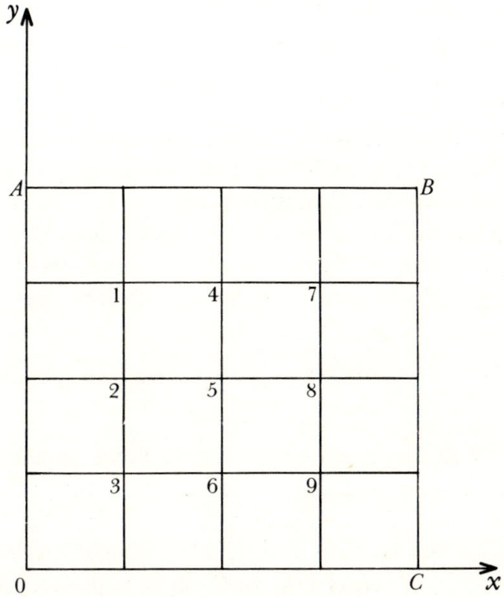

Fig. 12.10 Systematic numbering of internal grid points

where each of the matrices A_i, B_i, C_i are tridiagonal. For example, with a square boundary and internal points labelled as in Fig. 12.10, and specified boundary values, the five-point formula (12.36) yields the matrix equation

$$\begin{bmatrix} -4 & 1 & 0 & 1 & 0 & 0 & 0 & 0 & 0 \\ 1 & -4 & 1 & 0 & 1 & 0 & 0 & 0 & 0 \\ 0 & 1 & -4 & 0 & 0 & 1 & 0 & 0 & 0 \\ 1 & 0 & 0 & -4 & 1 & 0 & 1 & 0 & 0 \\ 0 & 1 & 0 & 1 & -4 & 1 & 0 & 1 & 0 \\ 0 & 0 & 1 & 0 & 1 & -4 & 0 & 0 & 1 \\ 0 & 0 & 0 & 1 & 0 & 0 & -4 & 1 & 0 \\ 0 & 0 & 0 & 0 & 1 & 0 & 1 & -4 & 1 \\ 0 & 0 & 0 & 0 & 0 & 1 & 0 & 1 & -4 \end{bmatrix} \begin{bmatrix} u_1 \\ u_2 \\ u_3 \\ u_4 \\ u_5 \\ u_6 \\ u_7 \\ u_8 \\ u_9 \end{bmatrix} = \begin{bmatrix} b_1 \\ b_2 \\ b_3 \\ b_4 \\ b_5 \\ b_6 \\ b_7 \\ b_8 \\ b_9 \end{bmatrix}$$

where the elements b_i depend on given values of u on the square boundary $OABC$. Clearly the matrix has the same form as the one given previously in terms of A_i, B_i, and C_i. Where we have a rectangular boundary it is worth noting that we can improve our computed results by recourse to the method of deferred correction (section 11.8), although this involves a considerable amount of extra work (see Fox [1]).

In practice, of course, we must use a smaller mesh size in order to get good results. With the problem of Laplace's equation over the L-shaped region which we encountered earlier a difficulty presents itself. At the corner C (Fig. 12.8) the boundary turns through an angle α ($> \pi$) and we can show that where this is the case, we have a disturbance at that corner (which is called a 're-entrant' corner). Milne [14] suggests the technique of subtracting out the singularity for solving this type of problem. Motz [15] and Woods [16] write Laplace's equation at a re-entrant corner in polar coordinates (r, θ) and, by the method of separation of variables, obtain solutions of the form $u = r^k \cos k\theta$, $u = r^k \sin k\theta$ for any k.

In our particular example, since $u = 0$ on the boundaries meeting at C we need only the terms in $\sin k\theta$ with positive k. We find, in fact, that

$$u(r, \theta) = \sum_{n=1}^{\infty} b_n r^{\frac{2}{3}n} \sin \tfrac{2}{3}n\theta .$$

With a finer mesh than that illustrated in Fig. 12.8, we use the standard finite-difference formula at most nodes but, in the neighbourhood of C, we take a few terms of the above series. The coefficients b_n can be found as linear combinations of u at points sufficiently far away from C as to be unaffected by the singularity. Hence the solution at grid points close to C can be expanded in terms of the solutions at the remote points. This theory and its application is extended to a more general elliptic problem by Fox and Sankar [17]. (Note that the general form of solution at a re-entrant corner is

$$u(r, \theta) = \sum_{n=0}^{\infty} r^{(n\pi/\alpha)} \left[a_n \cos\left(\frac{n\pi\theta}{\alpha}\right) + b_n \sin\left(\frac{n\pi\theta}{\alpha}\right)\right],$$

and hence $(\partial u/\partial r) \to \infty$ as $r \to 0$ if $\alpha > \pi$.)

In the previous examples we were fortunate that the boundaries were grid lines. With rectangular type boundaries this can usually be arranged by appropriate choice of Δx and Δy. If the boundary is not a grid line the formula (12.36) does not apply at points that are adjacent to the boundary, such as P in Fig. 12.11. In this case the

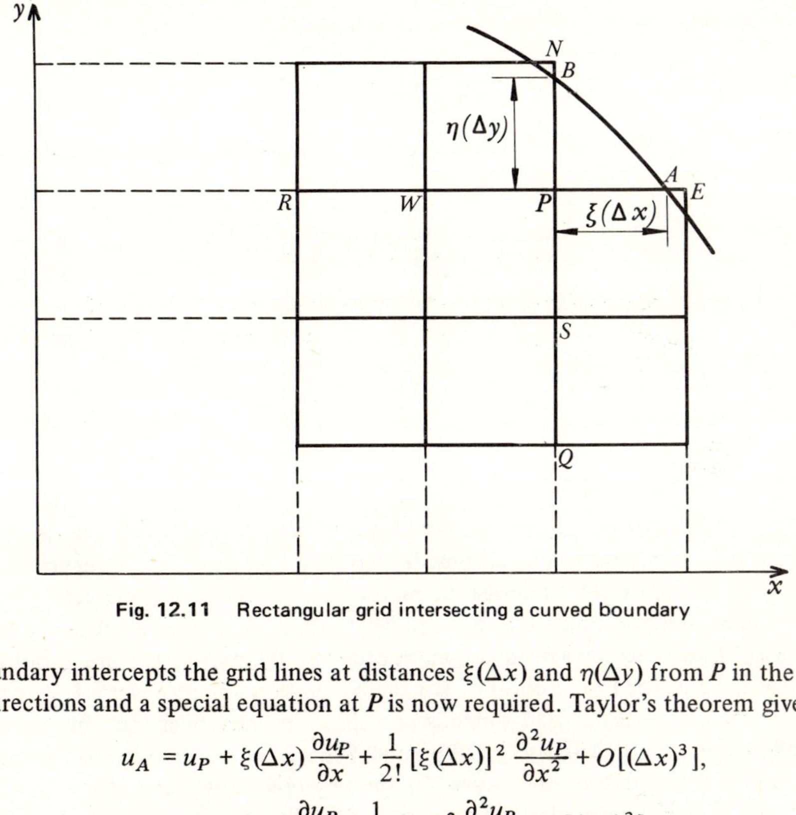

Fig. 12.11 Rectangular grid intersecting a curved boundary

boundary intercepts the grid lines at distances $\xi(\Delta x)$ and $\eta(\Delta y)$ from P in the x and y directions and a special equation at P is now required. Taylor's theorem gives

$$u_A = u_P + \xi(\Delta x)\frac{\partial u_P}{\partial x} + \frac{1}{2!}[\xi(\Delta x)]^2 \frac{\partial^2 u_P}{\partial x^2} + O[(\Delta x)^3],$$

$$u_W = u_P - (\Delta x)\frac{\partial u_P}{\partial x} + \frac{1}{2!}(\Delta x)^2 \frac{\partial^2 u_P}{\partial x^2} + O[(\Delta x)^3].$$

Elimination of $\partial u_P/\partial x$ gives

$$\frac{\partial^2 u_P}{\partial x^2} \doteqdot \frac{1}{(\Delta x)^2}\left\{\frac{2}{\xi(1+\xi)}u_A + \frac{2}{1+\xi}u_W - \frac{2}{\xi}u_P\right\}$$

with an error of $O(\Delta x)$, which is of lower accuracy than the usual formula (12.36).

A similar formula holds for $\partial^2 u_P/\partial y^2$, namely,

$$\frac{\partial^2 u_P}{\partial y^2} \mp \frac{1}{(\Delta y)^2}\left\{\frac{2}{\eta(1+\eta)}u_B + \frac{2}{1+\eta}u_S - \frac{2}{\eta}u_P\right\}$$

and we thus obtain the following formula for calculating u_P

$$\frac{2}{1+\eta}u_S + \frac{2\rho}{1+\xi}u_W - \left(\frac{2\rho}{\xi} + 2\eta\right)u_P = \frac{-2\rho u_A}{\xi(1+\xi)} - \frac{2u_B}{\eta(1+\eta)} \tag{12.37}$$

where $\rho = (\Delta y/\Delta x)^2$. Thus, in any problem, we can use formula (12.36) for points that are completely internal and (12.37) for points that are adjacent to the boundary.

An alternative (and simpler) approach is to use interpolation. With linear interpolation we immediately get the required extra equation

$$u_P \mp \frac{1}{1+\xi}u_A + \frac{\xi}{1+\xi}u_W .$$

However, this formula assumes variation of u with x only. As u is also a function of y is it better to use the weighted formula

$$u_P \mp \frac{1}{2}\left(\frac{u_A}{1+\xi} + \frac{\xi u_W}{1+\xi}\right) + \frac{1}{2}\left(\frac{u_B}{1+\eta} + \frac{\eta u_S}{1+\eta}\right) .$$

We can, of course, use quadratic interpolation expressing u_P in terms of the boundary value u_A and two mesh points u_W and u_R (or in terms of a weighted sum of u_A, u_W, u_R, u_B, u_S, and u_Q).

12.7 Summary

The purpose of this chapter has been to provide an introduction to the numerical solution of partial differential equations. In particular, we have been concerned with the problems arising out of numerical calculations, such as convergence and stability. There are two points that need to be stressed. First, the values of the step lengths used in the text have been chosen in order to obtain a simple presentation of the material. Thus, for the solution of Laplace's equation (12.35) over a L-shaped region, we took $\Delta x = \Delta y = 0{\cdot}25$ which is unrealistic in practice as the truncation error will be quite high. A fairly useful practical check to ensure that one has the correct degree of numerical accuracy is obtained by re-solving the problem with smaller mesh sizes (in particular half the original mesh size) and comparing the function values at the original mesh points. The second point concerns parabolic equations where we recommend implicit methods if we are concerned in any way (e.g., time considerations) about the small increments in the x-direction. However, if we have an infinite x-region then an explicit method must be used.

It should be noted that the finite-difference representation of the partial differential equations we have discussed is a system of linear simultaneous equations which can be represented in the matrix form $A\mathbf{x} = \mathbf{b}$, where A is a sparse matrix. The methods of chapter 10 can thus be used to good advantage in some problems, especially when storage difficulties present themselves. Otherwise the methods of chapter 8 should be used.

12.8 Notes

With finite-difference systems much work has been done to estimate the truncation error in a formula, and Fox [1] discusses some schemes which are designed to reduce this error. A finite-difference system is said to be compatible with the differential equation if the truncation error tends to zero as the mesh size (Δx and Δy) tend to zero. The results of Richtmyer [6] and John [7] indicate that stability and compatibility imply convergence and compatibility is easier to investigate than convergence.

Some authors (e.g., Froberg [8]) suggest that we solve the heat-conduction equation (12.23) by replacing $\partial^2 u/\partial x^2$ by the usual finite-difference approximation and we then consider the system of simultaneous ordinary differential equations of the type

$$\frac{\partial u_r}{\partial y} \doteqdot \frac{du_r}{dy} \doteqdot \frac{1}{(\Delta x)^2}(u_{r-1} - 2u_r + u_{r+1}), \qquad r = 1, 2, \ldots, n-1, \qquad (12.38)$$

where u_0 and u_n are obtained from the given conditions.

Elliott [9] represents the same heat-conduction equations by the system of simultaneous ordinary differential equations

$$\frac{u_{r+1} - u_r}{\Delta y} = \frac{1}{2}\left[\frac{d^2 u_r}{dx^2} + \frac{d^2 u_{r+1}}{dx^2}\right], \qquad (12.39)$$

and specifies boundary conditions at $x = \pm 1$ and solves the resulting two-point boundary value problem. He then expands the dependent variable directly in a series of Chebyshev polynomials $T_n(x)$. Equating coefficients of $T_n(x)$ on each side of the equation, for all n, gives a system of algebraic equations for the coefficients which can be solved numerically. Wragg [10] describes a variant of this method using the Lanczos τ-method but for a more complicated differential equation.

More difficult linear and non-linear equations are beyond the scope of this book although some of the methods given can be extended to treat these problems. We have only treated partial differential equations involving cartesian coordinates but many of the methods given are suitable for the solution of partial differential equations in polar coordinates (e.g., heat conduction along a cylinder – Smith [3], Albasiny [11]).

Apart from the numerical methods of solution, it is often useful to have some idea of the mathematical nature of the solution. This can, in some cases, be obtained by using the methods of separation of variables (exercise 24) and Laplace transforms (exercise 25). Details of these methods are given in Hildebrand [12] and Carslaw and Jaeger [13].

Finally, we remark that nothing whatever has been said in this chapter about an alternative numerical method of solution known as the finite-element method, about discontinuous boundary conditions and eigenvalue problems. The reader should consult reference [18] to learn more about the progress being made in these fields and for a general survey of the whole problem of the numerical solution of partial differential equations.

REFERENCES

[1] Fox, L., *Numerical Solution of Ordinary and Partial Differential Equations*, Pergamon Press, Oxford (1962).

[2] Hartree, D. R., *Numerical Analysis*, 2nd edn., Oxford University Press (1958).

[3] Smith, G. D., *Numerical Solution of Partial Differential Equations*, Oxford University Press (1965).

[4] Forsythe, G. E. and Wasow, W. R., *Finite-difference Methods for Partial Differential Equations*, J. Wiley and Sons, New York (1960).

[5] Crank, J. and Nicolson, P., 'A Practical Method for Numerical Evaluation of Solutions of Partial Differential Equations of the Heat Conduction Type', *Proc. Camb. Phil. Soc.*, **43** (1947) 50–67.

[6] Richtmyer, R. D., *Difference Methods for Initial-value Problems*, Interscience: J. Wiley and Sons, New York (1957).

[7] John, F., 'On Integration of Parabolic Equations by Difference Methods, I. Linear and Quasi-linear Equations for the Infinite Interval', *Commun. Pure & Appl. Math.*, **5** (1952) 155–211.

[8] Froberg, C. E., *Introduction to Numerical Analysis*, Addison-Wesley, Reading, Mass. (1965).

[9] Elliott, D., 'A Method for the Numerical Integration of the One-dimensional Heat Equation Using Chebyshev Series'. *Proc. Camb. Phil. Soc.*, **57** (1960) 823–32.

[10] Wragg, A., 'The Use of Lanczos τ-Methods in the Numerical Solution of a Stefan Problem', *Computer J.*, **9** (1966) 106–9.

[11] Albasiny, E. L., 'On the Numerical Solution of a Cylindrical Heat-conduction Problem', *Quart. J. Mech.*, **13** (1960) 374–84.

[12] Hildebrand, F. B., *Advanced Calculus for Applications*, Prentice-Hall, Inc., New Jersey (1962).

[13] Carslaw, H. S. and Jaeger, J. C., *Operational Methods in Applied Mathematics*, Oxford University Press, London (1941).

[14] Milne, W. E., *The Numerical Solution of Differential Equations*, J. Wiley and Sons, New York (1953).

[15] Motz, H., 'The Treatment of Singularities of Partial Differential Equations by Relaxation Methods', *Quart. Appl. Maths.*, **4** (1946) 371–77.
[16] Woods, L. C., 'The Relaxation Treatment of Singular Points in Poisson's Equation', *Quart. J. Mech.*, **6** (1953) 163–85.
[17] Fox, L. and Sankar, R., 'Boundary Singularities in Linear Elliptic Differential Equations', *J.I.M.A.*, **5** (1969) 340–50.
[18] *Proc. Roy. Soc., A*, **323** (1971).

Exercises

1. Classify the following second order differential equations into linear, quasi-linear or non-linear and hyperbolic, parabolic or elliptic type.

(*a*) $$\frac{\partial^2 u}{\partial x^2} + x\frac{\partial^2 u}{\partial y^2} = 0,$$

(*b*) $$\frac{\partial^2 u}{\partial x^2} - u\frac{\partial^2 u}{\partial x \partial y} + (1 + u^2)\frac{\partial^2 u}{\partial y^2} = 0,$$

(*c*) $$\frac{\partial^2 u}{\partial x^2} - e^y\frac{\partial}{\partial y}\left(e^y\frac{\partial u}{\partial y}\right) = x\frac{\partial u}{\partial x},$$

(*d*) $$\frac{\partial^2 u}{\partial x^2} + (1 - x^2 - y^2)\frac{\partial^2 u}{\partial y^2} = x^2 - y^2,$$

(*e*) $$\begin{cases} \dfrac{\partial u}{\partial t} + u\dfrac{\partial u}{\partial x} + 2v\dfrac{\partial v}{\partial x} = 0, \\[2ex] \dfrac{\partial v}{\partial t} + u\dfrac{\partial v}{\partial x} + \dfrac{1}{2}v\dfrac{\partial u}{\partial x} = 0. \end{cases}$$

2. The function $u(x, y)$ satisfies the differential equation

$$4x\frac{\partial u}{\partial y} - 2xy\frac{\partial u}{\partial x} + yu = xy$$

for all $x > 0$. On $y = 0$, u takes the value $2x^{\frac{1}{2}}$ for $2 \leqslant x \leqslant 5$. Find the value of u on $x = 1$ for $2 \leqslant y \leqslant 4$.

3. Use the method of characteristics to solve the partial differential equation (12.21) over a rectangular grid with mesh size $\Delta x = \Delta y = 0{\cdot}1$. Compare your results with those given in Table 12.3.

4. The differential equation

$$\frac{\partial^2 u}{\partial x^2} - u^2\frac{\partial^2 u}{\partial y^2} = 0,$$

is given with the initial conditions

$$5u - (5x)^2 = 1, \qquad \frac{\partial u}{\partial y} = 3x \qquad \text{on } y = 0, \qquad 0 \leqslant x \leqslant 1.$$

Use the method of characteristics to solve for u at characteristic grid points.

5. The equation of motion for steady isentropic flow of a compressible fluid in two dimensions is given by

$$(a^2 - p^2)\frac{\partial^2 u}{\partial x^2} - 2pq\frac{\partial^2 u}{\partial x \partial y} + (a^2 - q^2)\frac{\partial^2 u}{\partial y^2} = 0,$$

where u is the velocity potential, $p = \partial u/\partial x$ and $q = \partial u/\partial y$ are the Cartesian components of velocity, and a is the speed of sound in the fluid. Show that

$$\frac{\mathrm{d}y}{\mathrm{d}x} = -\left(\frac{a^2 - q^2}{a^2 - p^2}\right)\frac{\mathrm{d}q}{\mathrm{d}p},$$

and the differential equation is hyperbolic if the motion is supersonic.

6. The equation of motion for non-steady compressible fluid flow in one dimension is

$$\frac{\partial u}{\partial t} + u\frac{\partial u}{\partial x} + \frac{a^2}{\rho}\frac{\partial \rho}{\partial x} = 0,$$

and the equation of continuity is

$$\rho\frac{\partial u}{\partial x} + \frac{\partial \rho}{\partial t} + u\frac{\partial \rho}{\partial x} = 0,$$

where u denotes velocity potential and $a = a(\rho)$ is the velocity of sound in the fluid and ρ the density of the fluid. Determine the equation giving the characteristics and show that

$$\frac{\mathrm{d}u}{\mathrm{d}\rho} = \mp\frac{a}{\rho}.$$

7. Use the Fourier method (section 12.5) to show that the explicit difference scheme, given by equation (12.22), for solving the wave equation is stable provided $\rho \leqslant 1$.

8. Consider the wave equation $\dfrac{\partial^2 u}{\partial x^2} = \dfrac{\partial^2 u}{\partial y^2}$,

with

$$\left.\begin{aligned} u &= f(x) \\ \partial u/\partial y &= g(x) \end{aligned}\right\} \quad \text{on} \quad 0 \leqslant x \leqslant 1,\ y = 0,$$

and let w_{jk} be the exact solution at the grid point (j, k) and u_{jk} the solution of the explicit difference scheme (12.22), with $\rho \leqslant 1$. If $\epsilon_{jk} = w_{jk} - u_{jk}$ show that

$$\epsilon_{j,k+1} - \epsilon_{j,k-1} - \epsilon_{j+1,k} - \epsilon_{jk} \leqslant \tfrac{1}{6}(\Delta y)^4 E,$$

where

$$E = \max\left\{\left|\frac{\partial^4 w}{\partial x^4}\right|, \quad \left|\frac{\partial^4 w}{\partial y^4}\right|\right\}.$$

If

$$F = \max\left|\frac{\partial^2 w}{\partial y^2}\right|,$$

show that

$$|\epsilon_{jk}| \leqslant \tfrac{1}{12}k(k-1)(\Delta y)^4 E_{\max} + \tfrac{1}{12}(\Delta y)^2 F_{\max}.$$

9. Obtain an implicit scheme for solving the wave equation for a string tied at both ends by replacing $(\partial^2 u/\partial x^2)_{jk}$ by

$$\frac{1}{2}\frac{\partial^2}{\partial x^2}(u_{j,k+1} + u_{j,k-1}).$$

Show that the matrix form of the equations to be solved is

$$\mathbf{u}_{k+1} = A^{-1}B\mathbf{u}_k + A^{-1}C\mathbf{u}_{k-1},$$

where $\mathbf{u}_k$ is the vector of solutions at the level $y = y_k$ and find A, B, and C.

10. In exercise 9 write

$$\mathbf{v}_k = \begin{bmatrix} \mathbf{u}_k \\ \mathbf{u}_{k-1} \end{bmatrix}.$$

Hence show that the equations of the previous exercise reduce to

$$\mathbf{v}_{k+1} = P\mathbf{v}_k$$

where

$$P = \begin{bmatrix} A^{-1}B & A^{-1}C \\ I & O \end{bmatrix}.$$

11. If $\lambda_r, r = 1, 2, \ldots$ denote the eigenvalues of the matrix P of exercise 10 show that they satisfy the determinantal equation

$$|B - \lambda^{-1}C + \lambda A| = 0.$$

If $\mu_r = \lambda_r + 1/\lambda_r, r = 1, 2, \ldots, n-1$, show that

$$\mu_r = 2/[1 + 2\rho \cos^2(r\pi/2n)]$$

and this implies that the implicit scheme of exercise 9 is stable for all $\rho > 0$.

12. Let A_{N+1} denote the matrix which we called A in (12.33) and let B_{N+1} denote the same matrix but with $\Delta x = 0$ and $b_{12} = -\rho$. Show that

$$|A_{N+1}| = (2 + 2\rho + 2\rho\Delta x)|B_N| - 2\rho^2|B_{N-1}|,$$

and

$$|B_N| = (2 + 2\rho)|B_{N-1}| - \rho^2|B_{N-2}|, \qquad N > 2.$$

By solving this difference equation deduce that

$$|A_{N+1}| \neq 0.$$

13. If the explicit scheme (12.24), with $\rho = 1/6$, is used to solve the heat-conduction equation (12.23), show that the error in the finite-difference representation is $O(\Delta x)^4$.

14. Re-write the implicit difference equation (12.27) as

$$u_{j,k+1} = u_{jk} + \rho\delta_j^2 u_{j,k+1}.$$

If the numerical solution $u(x, y)$ is a maximum on the line $y = (k + 1)\Delta y$ when $x = J\Delta x$ show that

$$\max_j u_{j,k+1} \leqslant u_{Jk} \leqslant \max_j u_{jk}$$

and hence prove that the maximum value of the finite difference solution lies on the boundary. Deduce a similar result for the minimum value and hence show that the error in computation is bounded and therefore the implicit scheme is stable.

15. A suggested method (Richardson's method) for solving

$$\frac{\partial^2 u}{\partial x^2} = \frac{\partial u}{\partial y}$$

is to represent the partial differential equation by the difference formula

$$\frac{u_{j,k+1} - u_{j,k-1}}{2\Delta y} = \frac{u_{j-1,k} - 2u_{jk} + u_{j+1,k}}{(\Delta x)^2}.$$

Use the Fourier method to show that this difference scheme is always unstable.

16. Use the Crank–Nicolson method to solve

$$\frac{\partial^2 u}{\partial x^2} = \frac{\partial u}{\partial y} \qquad (0 \leqslant x \leqslant 1),$$

given that $u = \cos \frac{1}{2}\pi x$ when $y = 0$ and $\partial u/\partial x = 0$ on $x = 0$ and $x = 1$ when $y > 0$.

17. The radial flow of heat in a circular cylinder is given by

$$\frac{\partial u}{\partial t} = \frac{\partial^2 u}{\partial r^2} + \frac{1}{r}\frac{\partial u}{\partial r},$$

where r is the radius of the ring under consideration and t is the time. Assuming circular symmetry, that is $\partial u/\partial r = 0$ at $r = 0$, show that

$$\lim_{r \to 0} \frac{1}{r}\frac{\partial u}{\partial r} = \frac{\partial^2 u}{\partial r^2}.$$

18. Simulate the differential equation of exercise 17 subject to the conditions

$$\frac{\partial u}{\partial r} = 0 \quad \text{at} \quad r = 0, \qquad \frac{1}{u}\frac{\partial u}{\partial r} = -f(t) \quad \text{at} \quad r = 1$$

and $u = 1$ when $t = 0$ (all r), by explicit and implicit difference schemes. Hence find two sets of approximate solutions when $f(t) = \sin \frac{1}{2}\pi t$.

19. The function $u(x, y)$ satisfies Laplace's equation in two dimensions in the region $0 < x < 8, 0 < y < 5$. Along the edge $y = 0$, $u(x, y) = 80x - 10x^2$, and $u(x, y) = 0$ on the other three edges. Find approximate values for $u(x, y)$ at the nodes of a square mesh of unit side.

20. By using Taylor's theorem develop the second order, explicit, finite-difference approximation

$$u(x, t) = \tfrac{4}{3}(1-p)u(x, t-k) - \tfrac{1}{3}u(x, t-2k) + \tfrac{2}{3}p\{u(x-h, t-k) + u(x+h, t-k) - h^2 f(x, t)\}$$

for the solution of

$$\frac{\partial^2 u}{\partial x^2} - A\frac{\partial u}{\partial t} = f(x, t),$$

where A is a constant, $f(x, t)$ is a known function and $p = k/Ah^2$.

21. Obtain a numerical solution to Laplace's equation inside a semicircular region of unit radius and at the nodal points of a square mesh of side 0·2 given that $u = 0$ on the diameter $y = 0$ and $u = 100y$ on a semicircular arc centre (0, 0) and radius 1.

22. Laplace's equation is to be solved in a domain covered by a square mesh of size h. Derive the truncation errors for the five-point formula

(*a*) $$u_{00} = \tfrac{1}{4}(u_{-1,0} + u_{10} + u_{01} + u_{0,-1}),$$

(*b*) $$u_{00} = \tfrac{1}{4}(u_{-1,-1} + u_{1,-1} + u_{-1,1} + u_{11}).$$

Hence obtain a nine-point formula with error $O(h^6)$.

23. Solve the parabolic partial differential equation

$$\frac{\partial^2 u}{\partial x^2} = \frac{\partial u}{\partial y},$$

with

$$\begin{array}{llll} u = 0, & \text{when } y = 0 & (0 \leqslant x \leqslant 1) \\ u = 0, & \text{when } x = 0 & (y > 0) \\ u = y, & \text{when } x = 1 & (y > 0) \end{array}$$

by replacing $(\partial^2 u/\partial x^2)_{x=j\Delta x}$ by the finite difference approximation $(u_{j+1} - 2u_j + u_{j-1})/(\Delta x)^2$ and then solving the system of ordinary differential equations

$$\left(\frac{\partial u}{\partial y}\right)_{x=j\Delta x} = \frac{\mathrm{d}u_j}{\mathrm{d}y} = \frac{u_{j+1} - 2u_j + u_{j-1}}{(\Delta x)^2}, \qquad j = 1, 2, \ldots, n-1.$$

24. Assume that a solution exists of the form $u(x, y) = p(x)q(y)$ for

$$\frac{\partial^2 u}{\partial x^2} + \frac{\partial^2 u}{\partial y^2} = 0, \qquad 0 \leqslant x \leqslant \infty, \qquad 0 \leqslant y \leqslant 1,$$

with $u(x, 0) = u(x, 1) = u(\infty, y) = 0$. Show that

$$u(x, y) = \sum_{r=1}^{\infty} C_r \mathrm{e}^{-r\pi x} \sin r\pi y \qquad (C_r \text{ a constant}).$$

If $u(0, y) = U$, a constant, show that

$$u(x, y) = \frac{4U}{\pi}(\mathrm{e}^{-\pi x} \sin \pi y + \tfrac{1}{3}\mathrm{e}^{-3\pi x} \sin 3\pi y + \tfrac{1}{5}\mathrm{e}^{-5\pi x} \sin 5y + \ldots).$$

25. A uniform semi-finite transmission line has zero initial current and potential. At time $t = 0$, a constant e.m.f., E, is applied at the end $x = 0$. If R, L, C, G, I, and V denote respectively resistance, inductance, capacitance, leakage conductance, current, and potential then the equations satisfied at the point x of the line are

$$\left.\begin{aligned} L\frac{\partial I}{\partial t} + RI &= -\frac{\partial V}{\partial x} \\ C\frac{\partial V}{\partial t} + GV &= -\frac{\partial I}{\partial x} \end{aligned}\right] \quad t > 0$$

with $V = I = 0$ when $t = 0, x > 0$, $V = E$ at $x = 0, t > 0$. By multiplying throughout by e^{-st} $(s > 0)$ and integrating with respect to t from 0 to ∞, obtain the subsidiary equation

$$\frac{d^2\bar{V}}{dx^2} - q^2\bar{V} = 0,$$

where $\bar{V}$ denotes the Laplace transform of V and

$$q^2 = (Ls + R)(Cs + G).$$

Hence show that

$$\bar{V} = \frac{Ee^{-qx}}{s}, \qquad I = \sqrt{\left(\frac{Cs + G}{Ls + R}\right)}\,\frac{Ee^{-qx}}{s}.$$

By using contour integration obtain expressions for V and I.

13 The solution of simultaneous non-linear equations and optimization

13.1 Introduction

In this chapter we will be concerned with two problems. In the first we wish to determine the values of the n real variables $x_1, x_2, \ldots, x_n$ which simultaneously satisfy n given equations

$$f_i(\mathbf{x}) = f_i(x_1, x_2, \ldots, x_n) = 0, \qquad i = 1, 2, \ldots, n. \tag{13.1}$$

In the second, given a function $g(x_1, x_2, \ldots, x_n)$, we wish to determine the values of the n variables $x_1, x_2, \ldots, x_n$ which make the function g a maximum or a minimum. This is the problem of optimization. For this latter problem we will devote our attention to methods for the location of the minima of the function g. The maxima of g can be obtained by using the same methods but working with the function $-g$, since the maxima of the function g are the minima of the function $-g$.

Before proceeding to give methods for the solution of the above problems we note that, if no restrictions are imposed on the variables and our functions are differentiable, the two problems are equivalent. For, clearly, the solutions of (13.1) occur at the minima of the function

$$g(\mathbf{x}) = g(x_1, x_2, \ldots, x_n) = f_1^2 + f_2^2 + \ldots + f_n^2, \tag{13.2}$$

while, for a given function $g(\mathbf{x})$, maxima and minima can only occur where the n equations

$$\partial g/\partial x_i = 0, \qquad i = 1, 2, \ldots, n \tag{13.3}$$

are simultaneously satisfied. If we define D_i to be the determinant of the Hessian matrix G_i, where

$$G_i = \begin{bmatrix} \dfrac{\partial^2 g}{\partial x_1^2} & \dfrac{\partial^2 g}{\partial x_1 \partial x_2} & \cdots & \dfrac{\partial^2 g}{\partial x_1 \partial x_i} \\ \dfrac{\partial^2 g}{\partial x_2 \partial x_1} & \dfrac{\partial^2 g}{\partial x_2^2} & \cdots & \dfrac{\partial^2 g}{\partial x_2 \partial x_i} \\ \cdot\cdot & \cdot\cdot & \cdots & \cdot\cdot \\ \dfrac{\partial^2 g}{\partial x_i \partial x_1} & \dfrac{\partial^2 g}{\partial x_i \partial x_2} & \cdots & \dfrac{\partial^2 g}{\partial x_i^2} \end{bmatrix} \tag{13.4}$$

then a solution of (13.3) is a minimum if $D_i > 0$, $i = 1, 2, \ldots, n$, and it is a maximum if $D_i < 0$ for $i = 1, 3, 5, \ldots$ (all odd integers $\leqslant n$) and $D_i > 0$ for $i = 2, 4, 6, \ldots$ (all even integers $\leqslant n$). If these conditions on the D_i are not satisfied we assume that a solution of (13.3) is a *point of inflexion* or a *saddle point*, although this may not always be the case (exercise 1).

In chapter 2 we gave methods for finding the roots of a polynomial equation $f(x) = 0$, although some of the methods (e.g., the method of bisection and Newton's method) were not necessarily restricted to polynomial equations. Consequently, we will not dwell further on methods for the solution of the equation $f(x) = 0$, and we will turn out attention immediately to the minimization of a given function $g(x)$.

13.2 Minima of functions of one variable. Univariate and Fibonacci searches

The search methods we describe are very similar to the method of bisection (section 2.3) and the reader should also refer to this method. We consider first the basic method of univariate search. Suppose a minimum of the function $g(x)$ lies between $a = x_0$ and $b = x'_0$ ($x_0 < x'_0$). Let

$$y_0 = \tfrac{2}{3}x_0 + \tfrac{1}{3}x'_0, \qquad y'_0 = \tfrac{1}{3}x_0 + \tfrac{2}{3}x'_0,$$

i.e.,

$$x_0 < y_0 < y'_0 < x'_0.$$

We define a sequence of four points $\{x_k, y_k, y'_k, x'_k\}$ ($k = 1, 2, \ldots$) as follows. We set

$$\left.\begin{array}{ll}
\left.\begin{array}{ll} x_{k+1} = y_k, & x'_{k+1} = x'_k \\ y_{k+1} = y'_k, & y'_{k+1} = \tfrac{1}{2}(x'_k + y'_k) \end{array}\right] & \text{if } g(y_k) > g(y'_k), \\[2ex]
\left.\begin{array}{ll} x_{k+1} = x_k, & x'_{k+1} = y'_k \\ y_{k+1} = \tfrac{1}{2}(x_k + y_k), & y'_{k+1} = y_k \end{array}\right] & \text{if } g(y_k) < g(y'_k), \\[2ex]
\left.\begin{array}{ll} x_{k+1} = y_k, & x'_{k+1} = y'_k \\ y_{k+1} = \tfrac{2}{3}x_{k+1} + \tfrac{1}{3}x'_{k+1}, & y'_{k+1} = \tfrac{1}{3}x_{k+1} + \tfrac{2}{3}x'_{k+1} \end{array}\right] & \text{if } g(y_k) = g(y'_k).
\end{array}\right] \qquad (13.5)$$

Clearly, we have for each k,

$$x_k < y_k < y'_k < x'_k,$$

and the construction (13.5) ensures that the minimum of $g(x)$ always occurs between x_k and x'_k. Further, only one new function value need be computed at each stage of the calculation after the first, unless the exceptional stage $g(y_k) = g(y'_k)$ is reached.

As k increases x_k, y_k, y'_k, and x'_k tend to the same limiting value $\overline{x}$ (say) and the minimum value of g will be $g(\overline{x})$. We now apply the method to finding the minimum of

$$g(x) = x^4 - 2x^2 - 4x + 1,$$

given that it lies between $x = 1$ and $x = 2$. The results, obtained from a four decimal digit floating point computer are given in Table 13.1.

Table 13.1 Univariate search for minimum of $g(x) = x^4 - 2x^2 - 4x + 1$

k	x_k	y_k	y'_k	x'_k	$g(y_k)$	$g(y'_k)$
0	1·000	1·333	1·666	2·000	−4·728	−3·510
1	1·000	1·167	1·333	1·666	−4·537	−4·728
2	1·167	1·333	1·500	1·666	−4·728	−4·438
3	1·167	1·250	1·333	1·500	−4·683	−4·728
4	1·250	1·333	1·417	1·500	−4·728	−4·684
·	·	·	·	·	·	·
13	1·321	1·323	1·326	1·328	−4·730	−4·730
14	1·323	1·323	1·324	1·326	−4·730	−4·729

The minimum of $g(x)$ occurs at the value of x which is the real root of the polynomial equation $g'(x) = 4(x^3 - x - 1) = 0$. This has been determined by a number of methods in chapter 2 and the result obtained here is in agreement with that obtained by previous methods. Note that in iterations 13 and 14 the rounding error is significant so that the value of x marking the minimum of $g(x)$ is probably best given as $\frac{1}{2}(1{\cdot}323 + 1{\cdot}326) = 1{\cdot}325$ (Why? cf. section 9.9).

We now turn our attention to the more sophisticated method of univariate search known as Fibonacci search. This depends on the properties of the Fibonacci numbers which are defined by $F_0 = F_1 = 1$ and the recurrence relation

$$F_n = F_{n-1} + F_{n-2}, \qquad n \geqslant 2. \tag{13.6}$$

The Fibonacci numbers are thus 1, 1, 2, 3, 5, 8, 13, 21, 34, ... We can show by the methods given in section 11.3 that

$$F_n = \frac{1}{\sqrt{5}}\left\{\left(\frac{1+\sqrt{5}}{2}\right)^{n+1} - \left(\frac{1-\sqrt{5}}{2}\right)^{n+1}\right\}$$

$$\sim \frac{1}{\sqrt{5}}\left(\frac{1+\sqrt{5}}{2}\right)^{n+1} \quad \text{for large } n.$$

As with the method of univariate search we define a sequence of four points $\{x_k, y_k, y'_k, x'_k\}$ such that $x_0 = a, x'_0 = b$ and

$$\left.\begin{aligned} y_k &= \frac{F_{N-1-k}}{F_{N+1-k}}(x'_k - x_k) + x_k \\ y'_k &= \frac{F_{N-k}}{F_{N+1-k}}(x'_k - x_k) + x_k \end{aligned}\right] \qquad (13.7)$$

where k indicates the number of iterations performed (including the current one) and N is the total number of iterations necessary to produce the required accuracy. We take

$$\left.\begin{aligned} x_{k+1} &= y_k, \quad & x'_{k+1} &= x'_k \quad & \text{if} \quad & g(y_k) > g(y'_k) \\ x_{k+1} &= x_k, \quad & x'_{k+1} &= y'_k \quad & \text{if} \quad & g(y_k) \leqslant g(y'_k). \end{aligned}\right] \qquad (13.8)$$

Here again only one new function value is required at each stage of the calculation (exercise 2). At the $(k + 1)$th iteration the range of x is reduced to

$$\frac{F_{N-k}}{F_{N+1-k}}(x'_k - x_k),$$

and, at the Nth stage, we have

$$y'_N - x_N = \frac{1}{F_{N+1}}(x'_0 - x_0).$$

If accuracy ϵ is required then we want

$$\epsilon > \frac{1}{F_N}(x'_0 - x_0),$$

i.e., N is the subscript of the smallest Fibonacci number greater than $(x'_0 - x_0)/\epsilon$.

We now consider the Fibonacci method applied to locating the minimum of $g(x) = x^4 - 2x^2 - 4x + 1$ in $(1, 2)$, this time using an English Electric 4/50 computer and rounding results to four or five decimal places (Table 13.2). (These results should be compared with those in Table 13.1. See also Spang **[11]**, p. 348.) In order to obtain the minimum point correct to three decimal places we require $F_{N+1} > 1000$, i.e., $N = 15$, $F_{16} = 1597$. The minimum value occurs at $x = \frac{1}{2}(1{\cdot}3244 + 1{\cdot}3256) = 1{\cdot}325$ (approx.).

Table 13.2 The method of Fibonacci search

k	x_k	y_k	y'_k	x'_k	$g(y_k)$	$g(y'_k)$
0	1·0000	1·3820	1·6180	2·0000	−4·700 07	−3·854 10
1	1·0000	1·2361	1·3820	1·6180	−4·665 63	−4·700 07
2	1·2361	1·3820	1·4721	1·6180	−4·700 07	−4·526 23
3	1·2361	1·3262	1·3820	1·4721	−4·729 01	−4·700 07
4	1·2361	1·2918	1·3262	1·3820	−4·719 98	−4·729 01
.	.	.	.	.	.	.
12	1·3231	1·3244	1·3250	1·3262	−4·729 03	−4·729 03
13	1·3244	1·3250	1·3256	1·3262	−4·729 03	−4·729 02

13.3 Interpolation and gradient methods

Suppose the function $g(x)$ has a minimum between $x'_k = x_k - h_k$ and $x''_k = x_k + h_k$, for given x_k, h_k, and denote the function values at x_k, x'_k, and x''_k by g_{0k}, g_{1k}, g_{2k} respectively. It is convenient to think of the curve $y = g(x)$ as being parabolic in the region (x'_k, x''_k). Since three points are known on the 'parabola' its equation is readily determined and so is its minimum point from the 'interpolation' formula,

$$x_{k+1} = x_k - \frac{\frac{1}{2}h_k(g_{2k} - g_{1k})}{g_{2k} - 2g_{0k} + g_{1k}}. \qquad (13.9)$$

We now take $h_{k+1} < h_k$ and repeat the process until we find our minimum isolated in a very small region. For example, consider once again

$$g(x) = x^4 - 2x^2 - 4x + 1,$$

and take $x_0 = 1{\cdot}5$, $h_0 = 0{\cdot}5$, i.e., $x'_0 = 1$, $x''_0 = 2$, $g_{00} = -4{\cdot}438$, $g_{10} = -4$, $g_{20} = 1$. Formula (13.9) gives

$$x_1 = 1{\cdot}5 - (0{\cdot}25)(5)/5{\cdot}875 = 1{\cdot}288.$$

If we take $h_1 = 0{\cdot}05$, say (in general, $h_k = h_{k-1}/a^k$, $a = 2$ or 3 proves quite workable), and repeat the process, our next approximation is

$$x_2 = 1{\cdot}288 - (0{\cdot}025)(0{\cdot}041)/(-0{\cdot}061) = 1{\cdot}305, \text{ etc.}$$

With this particular method it is advisable to evaluate g_{0k}, g_{1k}, g_{2k} and the expressions $g_{2k} - 2g_{0k} + g_{1k}$, $g_{2k} - g_{1k}$ using double-length arithmetic. This is because g_{0k}, g_{1k}, and g_{2k} are nearly equal numbers for small h_k and significant digits would otherwise be lost in the arithmetical operations. In fact, double-length arithmetic at this stage gives $1{\cdot}325$ as the next approximation (to four significant digits). The above method is known as the method of quadratic interpolation

. . .

Now consider a function $g(x)$ whose value we know at x_k. Taylor's theorem gives the value of the function $g(x)$ at $x_{k+1} = x_k + h_k$ (for small h_k) viz.,

$$g(x_{k+1}) = g(x_k + h_k) = g(x_k) + h_k g'(x_k) + \tfrac{1}{2}h_k^2 g''(x_k + \theta h_k), \qquad 0 < \theta < 1.$$

If we take

$$h_k = -kg'(x_k), \qquad k \text{ a positive constant}, \qquad (13.10)$$

we see that

$$g(x_{k+1}) = g(x_k + h_k) = g(x_k) - k[g'(x_k)]^2 + \tfrac{1}{2}k^2[g'(x_k)]^2 g''(x_k + \theta h_k),$$

and we can, under certain conditions, choose k so that the term involving k^2 is small compared with $-k[g'(x_k)]^2$. In this case

$$g(x_{k+1}) \mp g(x_k) - k[g'(x_k)]^2 < g(x_k),$$

and $g(x_{k+1})$ is closer to the minimum value of $g(x)$ than is $g(x_k)$. Note that the step h_k is a multiple of the gradient of $g(x)$ at $x = x_k$. A method such as this, in which the step is a multiple of the gradient, is called a gradient method. Consider this method applied to finding the minimum of

$$g(x) = x^4 - 2x^2 - 4x + 1, \qquad g'(x) = 4(x^3 - x - 1),$$

starting from $x_0 = 1{\cdot}0$ and $k = 0{\cdot}1$. The stages in our numerical calculation are recorded in Table 13.3 (a four decimal digit floating point computer has been used to obtain results). The sequence of values of x eventually converges to the correct result $x = 1{\cdot}325$ although our results oscillate quite violently about it in the initial stages. This suggests a smaller value of k than $0{\cdot}1$ would have been preferable. Note also that there is no need to take the same fixed value of the constant k at each iteration. In fact Fletcher [16] has suggested taking k to be the first number in the sequence $1, 0{\cdot}1, 0{\cdot}1, \ldots$ satisfying

$$g(x_{k+1}) \leqslant g(x_k) + 0{\cdot}1(g'(x_k) + h_k).$$

Table 13.3 The gradient method

k	x_k	$g'(x_k)$	h_k
0	1·000	−4·000	0·4000
1	1·400	1·376	−0·1376
2	1·262	−1·008	0·1008
3	1·363	0·676	−0·0676
4	1·295	.	.

13.4 The solution of simultaneous non-linear equations in several variables. The method of Newton–Raphson

The method of Newton given in section 2.4 can be extended to obtain the solution of the simultaneous non-linear equations (13.1). This extension is termed the method of Newton–Raphson and is as follows. Let $\mathbf{x}_k$ be a good approximation to the true solution $\mathbf{x}$. Taylor's theorem for n variables gives for each function f_i,

$$0 = f_i(\mathbf{x}) = f_i(\mathbf{x}_k) + \sum_{j=1}^{n} (x_j - x_{jk})\left(\frac{\partial f_i}{\partial x_j}\right)_{\mathbf{x}=\mathbf{x}_k} + O\left[\sum_{j=1}^{n} (x_j - x_{jk})^2\right]. \qquad (13.11)$$

Hence, if second order terms can be neglected, we find $\mathbf{x}_{k+1}$ is a better approximation than $\mathbf{x}_k$, where $\mathbf{x}_{k+1}$ satisfies

$$\sum_{j=1}^{n} (x_{j,k+1} - x_{jk})\left(\frac{\partial f_i}{\partial x_j}\right)_{\mathbf{x}=\mathbf{x}_k} = -f_i(\mathbf{x}_k), \qquad i = 1, \ldots, n \qquad (13.12)$$

or, in matrix form,

$$A\delta\mathbf{x}_k = -\mathbf{f}_k, \tag{13.13}$$

where A is the matrix

$$\begin{bmatrix} \dfrac{\partial f_1}{\partial x_1}, & \dfrac{\partial f_1}{\partial x_2}, \ldots, & \dfrac{\partial f_1}{\partial x_n} \\ \dfrac{\partial f_2}{\partial x_1}, & \dfrac{\partial f_2}{\partial x_2}, \ldots, & \dfrac{\partial f_2}{\partial x_n} \\ \cdots & \cdots & \cdots \\ \dfrac{\partial f_n}{\partial x_1}, & \dfrac{\partial f_n}{\partial x_2}, \ldots, & \dfrac{\partial f_n}{\partial x_n} \end{bmatrix}$$

all derivatives are evaluated at $\mathbf{x} = \mathbf{x}_k$, and

$$\delta\mathbf{x}_k = \mathbf{x}_{k+1} - \mathbf{x}_k, \qquad \mathbf{f}_k = (f_{1k}, f_{2k}, \ldots, f_{nk})'.$$

The reader is warned that convergence is not guaranteed and that during the course of computation the matrix A is often almost singular.

We now apply the Newton–Raphson method to the solution of the simultaneous non-linear equations

$$\left.\begin{aligned} f_1(x, y) &= y - x^2 = 0 \\ f_2(x, y) &= x - xy + 1 = 0 \end{aligned}\right] \tag{13.14}$$

given $(x_0, y_0) = (1{\cdot}5, 1{\cdot}5)$ as a first approximation. We have from (13.14)

$$\partial f_1/\partial x = -2x, \qquad \partial f_1/\partial y = 1,$$
$$\partial f_2/\partial x = 1 - y, \qquad \partial f_2/\partial y = -x.$$

Better approximations are obtained by solving the simultaneous linear equations (13.13) with the latest available approximation (x_k, y_k). The results, with the exception of $(\partial f_1/\partial y)_k$ and $(\partial f_2/\partial y)_k$ (which have not been recorded for obvious reasons), are set out in Table 13.4. Note that equations (13.14) are equivalent to the polynomial equation $x^3 - x - 1 = 0$ encountered earlier. This explains the result 1·325 for x.

Table 13.4 The method of Newton–Raphson

k	x_k	y_k	$(\partial f_1/\partial x)_k$	$(\partial f_2/\partial x)_k$	f_{1k}	f_{2k}
0	1·500	1·500	−3·000	−0·5000	−0·7500	0·2500
1	1·325	1·755	−2·650	−0·7250	−0·0310	0·0390
2	1·325	1·755	..	..	..	..

13.5 Gradient methods for locating the minimum of functions of several variables

The gradient method given in section 13.3 for functions of one variable is easily extended to deal with functions of several variables. At the rth stage of the computation to find the minimum of $g(\mathbf{x})$ we take

$$h_{ir} = -k_r(\partial g/\partial x_i)_r, \tag{13.15}$$

for suitably small k_r. The Taylor expansion shows that, in certain cases

$$g(\mathbf{x}_{r+1}) < g(\mathbf{x}_r),$$

i.e., the point $\quad \mathbf{x}_{r+1} = \mathbf{x}_r + h_r \quad (h_r = (h_{1r}, h_{2r}, \ldots, h_{nr})),$

is nearer to the minimum than $\mathbf{x}_r$. This method is usually referred to as the method of steepest descent. Let us now locate the minimum of

$$g(x, y) = (y - x^2)^2 + (x - xy + 1)^2,$$

restricting ourselves to the range $1 < x < 2$, $1 < y < 2$. This is equivalent to the problem of solving the non-linear simultaneous equations (13.14) of the previous section. We note that

$$\frac{\partial g}{\partial x} = -4x(y - x^2) + 2(1 - y)(x - xy + 1),$$

$$\frac{\partial g}{\partial y} = 2(y - x^2) - 2x(x - xy + 1),$$

and for convenience we have chosen $k_r = 0{\cdot}1$ $(r = 0, 1, 2, \ldots)$ – see section 13.3 and reference [16] regarding choice of k_r. With $\mathbf{x}_0 = (1{\cdot}5, 1{\cdot}5)$ and a four decimal digit floating point computer the results in Table 13.5 were obtained.

Table 13.5 The steepest descent method for several variables

r	$\mathbf{x}_r$	$(\partial g/\partial x)_r$	$(\partial g/\partial y)_r$	$\mathbf{h}_r$
0	(1·500, 1·500)	4·250	−2·250	(−0·4250, 0·2550)
1	(1·075, 1·725)	−2·768	0·6630	(0·2768, −0·0663)
2	(1·352, 1·659)	0·7703	−0·6327	(−0·0770, 0·0633)
3	(1·275, 1·722)	−0·6037	−0·0095	(0·0604, 0·0010)
4	(1·335, 1·723)			

It will help us in our treatment of more sophisticated gradient methods to note that the basic iteration in the above method can be expressed in the form

$$\mathbf{x}_{r+1} = \mathbf{x}_r - Y_r^{-1}\mathbf{z}_r \tag{13.16}$$

where

$$\mathbf{z}_r = \text{grad } g(\mathbf{x}_r) = (\partial g/\partial x_1, \ldots, \partial g/\partial x_n)_r,$$

i.e., the components of each z are the direction ratios of the normal to the surface g = constant, and Y_r is a positive definite $n \times n$ weighting matrix which, in the present instance, is diagonal. The method can be represented pictorially by Fig. 13.1 (where the weighting constants k have been taken as 1 for convenience).

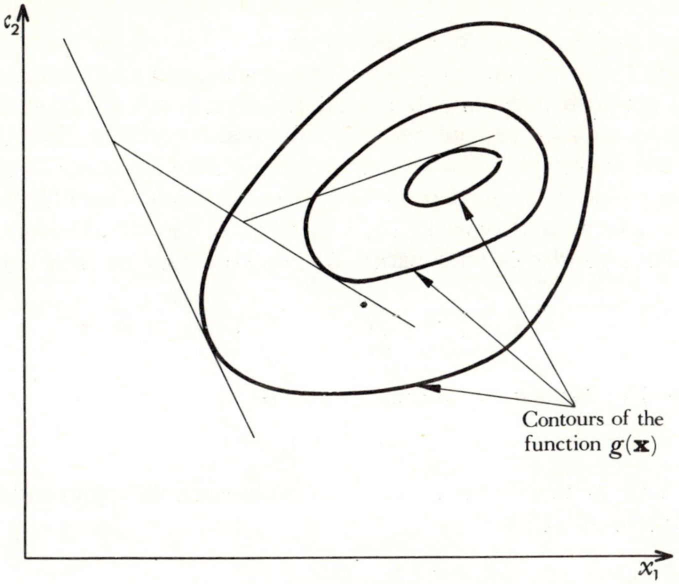

Fig. 13.1 Geometrical illustration of steepest descent

The basis for the derivation of the simple steepest descent method was the Taylor series expansion. This is also the basis for the method of optimization known as Newton's method. Let $\boldsymbol{\xi}$ be a minimum point of the function and $g(\boldsymbol{\xi})$ the corresponding minimum function value. Then the value of the function g at a point $\mathbf{x}$ near to $\boldsymbol{\xi}$ is approximately given by the truncated Taylor series

$$g(\mathbf{x}) = g(\boldsymbol{\xi}) + (\mathbf{x} - \boldsymbol{\xi})'\mathbf{z} + \tfrac{1}{2}(\mathbf{x} - \boldsymbol{\xi})'Y(\mathbf{x} - \boldsymbol{\xi})$$

where $\mathbf{z}$ = grad $g(\boldsymbol{\xi})$, and Y is the matrix G_n defined by (13.4), evaluated at $\mathbf{x} = \boldsymbol{\xi}$.

Since at a minimum $\dfrac{\partial g}{\partial x_1} = \dfrac{\partial g}{\partial x_2} = \ldots = \dfrac{\partial g}{\partial x_n} = 0$, it follows that $\mathbf{z} = \mathbf{o}$ and the truncated Taylor series can be written

$$g(\mathbf{x}) = g(\boldsymbol{\xi}) + \tfrac{1}{2}(\mathbf{x} - \boldsymbol{\xi})'Y(\mathbf{x} - \boldsymbol{\xi}). \tag{13.17}$$

From (13.17) it follows that

$$\text{grad } g(\mathbf{x}) = \text{grad } g(\boldsymbol{\xi}) + \text{grad } [\tfrac{1}{2}(\mathbf{x} - \boldsymbol{\xi})'Y(\mathbf{x} - \boldsymbol{\xi})],$$

or

$$\text{grad } g(\mathbf{x}) = Y(\mathbf{x} - \boldsymbol{\xi}),$$

and hence

$$\mathbf{x} - \boldsymbol{\xi} = Y^{-1} \operatorname{grad} g(\mathbf{x}). \tag{13.18}$$

Equation (13.18) gives rise in an obvious manner to the iterative formula

$$\mathbf{x}_{r+1} - \mathbf{x}_r = -Y_r^{-1} \mathbf{z}_r \tag{13.19}$$

for locating a minimum point of a function where Y_r denotes the matrix Y evaluated at $\mathbf{x} = \mathbf{x}_r$. If $g(\mathbf{x})$ is a quadratic function (possessing a minimum) then Y_r is constant at all points and positive definite. In general, however, $g(\mathbf{x})$ is not quadratic and Y_r is consequently not constant and not necessarily positive definite. This tends to make the method unstable and convergence to the minimum cannot be assured.

In a more sophisticated gradient method due to Davidon – see Fletcher and Powell [1] – a formula of type (13.16) is used but Y_r is modified at each stage so that it remains a positive definite matrix. We begin our iteration from a point $\mathbf{x}_0$ with

$$\mathbf{z}_0 = \left(\frac{\partial g}{\partial x_1}, \frac{\partial g}{\partial x_2}, \ldots, \frac{\partial g}{\partial x_n}\right)_{\mathbf{x}=\mathbf{x}_0},$$

and a matrix $Y_0 = I$ (the $n \times n$ unit matrix). We set

$$\mathbf{s}_i = -Y_i \mathbf{z}_i \qquad (i \geqslant 0) \tag{13.20}$$

and obtain α_i (>0) such that $g(\mathbf{x}_i + \alpha_i \mathbf{s}_i)$ is a minimum with respect to λ along $\mathbf{x}_i + \lambda \mathbf{s}_i$. With this α_i we put

$$\sigma_i = \alpha_i \mathbf{s}_i, \qquad \mathbf{x}_{i+1} = \mathbf{x}_i + \sigma_i,$$

and we evaluate $g(\mathbf{x}_{i+1})$ and $\mathbf{z}_{i+1}$. Note that $\mathbf{z}_{i+1}$ is orthogonal to σ_i, i.e., $\sigma'_i \,.\, \mathbf{z}_{i+1} = 0$ (Why?). We next form

$$\xi_i = \mathbf{z}_{i+1} - \mathbf{z}_i$$
$$Y_{i+1} = Y_i + A_i + B_i$$

where

$$A_i = \sigma_i \,.\, \sigma'_i / (\sigma'_i \,.\, \boldsymbol{\xi}_i)$$
$$B_i = -(Y_i \boldsymbol{\xi}_i) \,.\, (\boldsymbol{\xi}'_i Y_i) / \boldsymbol{\xi}'_i Y_i \xi_i)$$

and the procedure starting from (13.20) is repeated for the next value of i. After about five or six steps it is often desirable to reset Y_i to be I. The procedure is terminated when the absolute value of s's is less than a prescribed amount or each element of s is less in modulus than a given amount. If it is obvious that the minimum value is zero, which often happens when minimizing sums of squares, the procedure could be terminated when g is less in modulus than a prescribed amount. Fletcher and Powell show in their original paper that the process is stable (in the sense that the function to be minimized is decreased at each step) and quadratically convergent (i.e., if the original function $g(\mathbf{x})$ was a quadratic form in n variables the process converges in exactly n steps with the appropriate α_i). Another interesting feature of the method is that the first step is a steepest descent step while in later steps the matrix Y_r tends to G_n^{-1} where G_n is the Hessian matrix (13.4) (exercises 12–14).

It is suggested that α_i is estimated as follows. We let g_0 be the predicted value of the minimum. We take

$$u_i = \mathbf{z}'_i \mathbf{s}_i$$

$$\eta_i = \min\left[1, \frac{-2[g(\mathbf{x}_i) - g_0]}{u_i}\right],$$

$$t_i = \mathbf{x}_i + \eta_i \mathbf{s}_i.$$

We evaluate $\mathbf{z}$ at $\mathbf{x} = \mathbf{t}_i$ and $g(\mathbf{t}_i)$ and set

$$v_i = \mathbf{z}'(\mathbf{t}_i)\mathbf{s}_i$$

$$d_i = \{3[g_i - g(\mathbf{t}_i)]/\eta_i\} + u_i + v_i$$

$$w_i = (d_i^2 - u_i v_i)^{\frac{1}{2}}$$

Table 13.6 The method of Fletcher and Powell. Predicted value of $g = 0$

i	Y_i	$\mathbf{x}_i$	$g(\mathbf{x}_i)$	grad $g(\mathbf{x}_i)$	$\mathbf{s}_i$
0	$\begin{bmatrix} 1{\cdot}000 & 0{\cdot}000 \\ 0{\cdot}000 & 1{\cdot}000 \end{bmatrix}$	$\begin{bmatrix} 1{\cdot}0000 \\ 1{\cdot}0000 \end{bmatrix}$	1·0000	$\begin{bmatrix} 0{\cdot}000 \\ -2{\cdot}000 \end{bmatrix}$	$\begin{bmatrix} 0{\cdot}0000 \\ 2{\cdot}0000 \end{bmatrix}$
1	$\begin{bmatrix} 0{\cdot}3903 & 0{\cdot}4878 \\ 0{\cdot}4878 & 0{\cdot}8598 \end{bmatrix}$	$\begin{bmatrix} 1{\cdot}0000 \\ 1{\cdot}5000 \end{bmatrix}$	0·5000	$\begin{bmatrix} -2{\cdot}500 \\ 0{\cdot}000 \end{bmatrix}$	$\begin{bmatrix} 0{\cdot}9756 \\ 1{\cdot}2195 \end{bmatrix}$
2	$\begin{bmatrix} 0{\cdot}1335 & 0{\cdot}1019 \\ 0{\cdot}1019 & 0{\cdot}2848 \end{bmatrix}$	$\begin{bmatrix} 1{\cdot}2862 \\ 1{\cdot}8577 \end{bmatrix}$	0·0520	$\begin{bmatrix} -0{\cdot}8694 \\ 0{\cdot}6725 \end{bmatrix}$	

i	u_i	η_i	$\mathbf{t}_i$	$g(\mathbf{t}_i)$	grad $g(\mathbf{t}_i)$	v_i	d_i
0	−4·000	0·5000	$\begin{bmatrix} 1 \\ 2 \end{bmatrix}$	1·000	$\begin{bmatrix} -4 \\ 2 \end{bmatrix}$	4·0000	0·0000
1	−2·439	0·4100	$\begin{bmatrix} 1{\cdot}4 \\ 2 \end{bmatrix}$	0·1616	$\begin{bmatrix} 0{\cdot}576 \\ 1{\cdot}200 \end{bmatrix}$	2·0253	2·0624
2	.	.	.	.	.	.	.

i	w_i	α_i	ξ_i	A_i	B_i
0	4·0000	0·2500	$\begin{bmatrix} -2{\cdot}5000 \\ 2{\cdot}0000 \end{bmatrix}$	$\begin{bmatrix} 0{\cdot}0000 & 0{\cdot}0000 \\ 0{\cdot}0000 & 0{\cdot}2500 \end{bmatrix}$	$\begin{bmatrix} -0{\cdot}6097 & 0{\cdot}4878 \\ 0{\cdot}4878 & -0{\cdot}3902 \end{bmatrix}$
1	3·0321	0·2934	$\begin{bmatrix} 1{\cdot}6306 \\ 0{\cdot}6725 \end{bmatrix}$	$\begin{bmatrix} 0{\cdot}1158 & 0{\cdot}1448 \\ 0{\cdot}1448 & 0{\cdot}1810 \end{bmatrix}$	$\begin{bmatrix} 0{\cdot}3726 & 0{\cdot}5307 \\ 0{\cdot}5307 & 0{\cdot}7558 \end{bmatrix}$
2	.	.	.	.	.

and we finally obtain the following approximation for α_i,

$$\alpha_i = \eta_i \left[1 - \frac{v_i + w_i - d_i}{v_i - u_i + 2w_i} \right].$$

We now demonstrate the method for the function

$$g(x, y) = (y - x^2)^2 + (x - xy + 1)^2,$$

taking $\mathbf{x}_0 = (1, 1)$ as an initial approximation. The results are set out in Table 13.6.

13.6 Search methods for minimizing functions of several variables. Rosenbrock's method

The methods described earlier in section 13.2 for unimodal search do not readily extend themselves to a multimodal (or n-dimensional) search and for this reason their generalization will not be considered here. We give instead a method due to Rosenbrock [2] which is as follows. We start from the given function $g(x)$, an approximation $\mathbf{x}_0$ to the minimum, and a set of mutually orthogonal unit direction vectors $\boldsymbol{\xi}_{10}, \boldsymbol{\xi}_{20}, \ldots, \boldsymbol{\xi}_{n0}$ which, in the first instance, are taken to be the directions of the coordinate axes, viz., $(1, 0, 0, \ldots, 0)$, $(0, 1, 0, \ldots, 0)$, etc. At the $(k + 1)$th stage ($k = 0, 1, \ldots$) we take a step d_1 in the direction $\boldsymbol{\xi}_{1k}$ and evaluate the function g at the point $\mathbf{x}_k + d_1\boldsymbol{\xi}_{1k}$. If, for some given $\alpha > 1$,

$$\left.\begin{aligned} & g(\mathbf{x}_k + d_1\boldsymbol{\xi}_{1k}) \leqslant g(\mathbf{x}_k), \\ \text{and} \quad & g(\mathbf{x}_k + \alpha d_1\boldsymbol{\xi}_{1k}) > g(\mathbf{x}_k), \end{aligned}\right] \tag{13.21}$$

the step is regarded as being successful. If

$$g(\mathbf{x}_k + \alpha d_1\boldsymbol{\xi}_{1k}) \leqslant g(\mathbf{x}_k), \tag{13.22}$$

then we replace d_1 by αd_1 and again attempt to satisfy (13.21). On the other hand if initially

$$g(\mathbf{x}_k + d_1\boldsymbol{\xi}_{1k}) > g(\mathbf{x}_k), \tag{13.23}$$

we replace d_1 by $-\beta d_1$ $(0 < \beta < 1)$ and again attempt to satisfy (13.21). We must eventually be successful for, if (13.22) continually holds, we go outside the region under investigation and hence have to return to the boundary while if (13.23) holds continually d_1 is so small that, because of computer round-off, it has no effect on the value of g. We now proceed to take a step d_2 from the point $\mathbf{x}_k + d_1\boldsymbol{\xi}_{1k}$ in the direction $\boldsymbol{\xi}_{2k}$ carrying out the analogous tests to (13.21)–(13.23) to determine the value of d_2 which is 'successful'. Similarly we determine $d_3, \ldots, d_n$. We thus arrive, at the end of the $(k + 1)$th stage at an approximation $\mathbf{x}_{k+1}$ to the minimum point of g where

$$\mathbf{x}_{k+1} = \mathbf{x}_k + d_1\boldsymbol{\xi}_{1k} + d_2\boldsymbol{\xi}_{2k} + \ldots + d_n\boldsymbol{\xi}_{nk}, \tag{13.24}$$

and a new set of orthogonal direction vectors $\boldsymbol{\xi}_{1,k+1}, \boldsymbol{\xi}_{2,k+1}, \ldots, \boldsymbol{\xi}_{n,k+1}$ obtained from equations (13.25) and (13.26) (see Powell [3], concerning the 'stability' of the formulation),

$$\left.\begin{aligned} A_1 &= d_1\xi_{1k} + d_2\xi_{2k} + \ldots + d_n\xi_{nk} \\ A_2 &= \qquad\quad d_2\xi_{2k} + \ldots + d_n\xi_{nk} \\ &\ldots\ldots\ldots\ldots\ldots\ldots \\ A_n &= \qquad\qquad\qquad\qquad d_n\xi_{nk} \end{aligned}\right] \tag{13.25}$$

$$\left.\begin{aligned} & B_1 = A_1 \\ & \xi_{1,k+1} = B_1/|B_2| \\ & B_2A_2 - (A'_2 \cdot \xi_{1,k+1})\xi_{1,k+1} \\ & \xi_{2,k+1} = B_2/|B_2| \\ & \ldots\ldots\ldots\ldots\ldots\ldots \\ & B_n = A_n - \sum_{j=1}^{n-1} (A'_n \cdot \xi_{j,k+1})\xi_{j,k+1} \\ & \xi_{n,k+1} = B_n/|B_n| \end{aligned}\right] \tag{13.26}$$

This is essentially the Gram–Schmidt orthogonalization process. Note that A_1 is the vector joining $\mathbf{x}_k$ (the initial point of the $(k+1)$th stage) and $\mathbf{x}_{k+1}$ (the initial point of the next stage), while A_2 is the sum of the advances made in directions other than the first, etc. The effect of applying (13.25) and (13.26) is to ensure that $\xi_{1,k+1}$ lies in the direction of fastest advance to the minimum, $\xi_{2,k+1}$ is the best direction which can be found normal to $\xi_{1,k+1}$ and so on. The iteration procedure is terminated either after a finite number of stages or, alternatively, when $|\mathbf{x}_{k+1} - \mathbf{x}_k|$ is less than some prescribed amount.

13.7 Constraints

In the previous sections no restrictions (apart from isolating enquiries to a given region in space in order to reduce computation) were placed on the values of the variables $x_1, \ldots, x_n$ and the problem of minimization is, in such circumstances, termed one of unconstrained minimization. If, in addition, we must also satisfy requirements on the independent variables such as

$$h_i(\mathbf{x}) \geqslant 0, \qquad i = 1, \ldots, p \tag{13.27}$$

or

$$k_i(\mathbf{x}) = 0, \qquad i = 1, \ldots, m \tag{13.28}$$

or both these last two conditions simultaneously, then we have a constrained minimization problem. Note that (13.27) includes inequalities such as $l(\mathbf{x}) \leqslant 0$. Suppose conditions of type (13.28) are given. Then the constrained problem can be turned into an unconstrained one by forming a new function

$$\phi(x_1,\ldots,x_n,x_{n+1},\ldots,x_{n+m}) = g(x_1,\ldots,x_n) + \sum_{i=1}^{m} x_{n+i}\, k_i(x_1,\ldots,x_n), \qquad (13.29)$$

where $x_{n+1},\ldots,x_{n+m}$ are new independent variables termed Lagrange multipliers, and searching for the stationary values of ϕ. These occur where

$$\frac{\partial \phi}{\partial x_r} = 0, \qquad (r = 1,\ldots,n+m). \qquad (13.30)$$

The solution of the simultaneous non-linear equations (13.30) can be carried out by the methods of earlier sections to give the values of $x_1,\ldots,x_{n+m}$ which are stationary values of ϕ. We now have to examine the $(n+m) \times (n+m)$ Hessian matrix $\left(\frac{\partial^2 \phi}{\partial x_i \partial x_j}\right) i, j = 1,\ldots,n+m$ to decide which of the stationary values of ϕ make $g(\mathbf{x})$ a a minimum (or maximum). Suppose

$$\Delta_k = \det\left(\frac{\partial^2 \phi}{\partial x_i \partial x_j}\right) \qquad (i, j = k, k+1,\ldots,n+m) \qquad (13.31)$$

then sufficient conditions for a minimum of $g(\mathbf{x})$ are (see Gillespie [4])

(*a*) if m is even: $\Delta_1 > 0,\ \Delta_2 > 0,\ldots,\Delta_{n-m} > 0,$

(*b*) if m is odd: $\Delta_1 < 0,\ \Delta_2 < 0,\ldots,\Delta_{n-m} < 0.$

Sufficient conditions for a maximum of $g(\mathbf{x})$ are

(*a*) if n is even: $\Delta_1 > 0,\ \Delta_2 < 0,\ \Delta_3 > 0,\ldots,(-1)^{n-m}\Delta_{n-m} < 0,$

(*b*) if n is odd: $\Delta_1 < 0,\ \Delta_2 > 0,\ \Delta_3 < 0,\ldots,(-1)^{n-m}\Delta_{n-m} > 0.$

If $m = 0$ these conditions are equivalent to the conditions on $D_i (i = 1,\ldots,n)$ given in section 13.1. Note that the technique of Lagrange multipliers can be extended to deal with inequality constraints like (13.27) (see John [5]). Alternatively we can replace a constraint such as $-1 \leqslant x \leqslant 1$ by the equality constraint $x = \sin\alpha$ where α is a new variable.

The above method of Lagrange multipliers is of more theoretical than practical interest and has a number of disadvantages; one being that the original problem in n-dimensions has been converted to one in $(m+n)$-dimensions. A more practical technique is the Created Response Surface Technique which, in its basic form, is due to Carroll [6]. Suppose we wish to minimize $g(\mathbf{x})$ subject to (13.27). Then we create response surfaces $\phi(\mathbf{x}, \alpha_r)$ by forming

$$\phi(\mathbf{x}, \alpha_r) = g(\mathbf{x}) + \alpha_r \sum_{i=1}^{p} 1/h_i^2(\mathbf{x}), \qquad r = 1, 2, \ldots\ . \qquad (13.32)$$

By this construction we ensure that ϕ is very large near a constraint and a minimum is created inside the feasible region. We start with $\alpha_1 = 10$ (say) and a point $\mathbf{x}_0$ in the feasible domain and determine a point $\mathbf{x}_1$, in the feasible domain, which minimizes (13.32) for $r = 1$. Starting with $\mathbf{x}_1$ we choose $\alpha_2, 0 < \alpha_2 < \alpha_1$ and locate the minimum $\mathbf{x}_2$ of (13.32) for $r = 2$; and so on. We obtain a sequence of points $\mathbf{x}_r$ which, under fairly general conditions, tends to the true minimum of $g(\mathbf{x})$ as $\alpha_r \to 0$. The method can be extended to deal with equality constraints of the form (13.28) by considering the response surfaces

$$\phi(\mathbf{x}, \alpha_r) = g(\mathbf{x}) + \alpha_r \sum_{i=1}^{p} 1/h_i^2(\mathbf{x}) + (1/\alpha_r) \sum_{i=1}^{m} k_i^2(\mathbf{x}), \qquad r = 1, 2, \ldots .$$

13.8 Application of Rayleigh's quotient

In chapter 9 we saw that, for a positive definite matrix, A, the Rayleigh quotient $\mathbf{x}'A\mathbf{x}/\mathbf{x}'\mathbf{x}$ was always larger than the smallest eigenvalue. The Rayleigh quotient thus provides a convenient method for tackling the optimization problem of finding the smallest eigenvalue. For example, consider the determination of the normal modes of vibration of the double simple pendulum (Fig. 13.2). In the equilibrium position

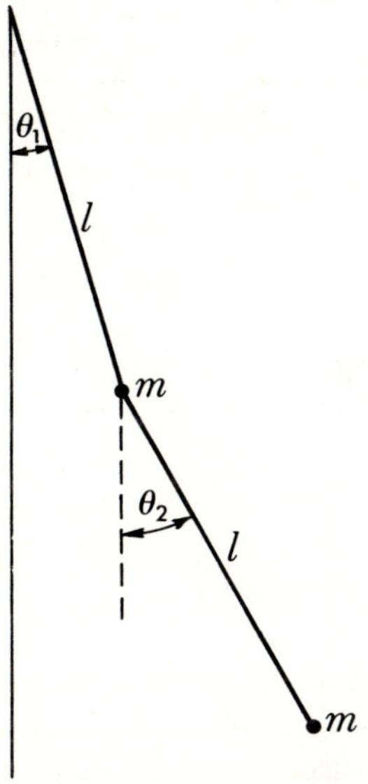

Fig. 13.2 The double simple pendulum

($\theta_1 = \theta_2 = 0$) we take the potential energy V to be zero. Denoting the kinetic energy by T, we have

$$V = mgl(1 - \cos\theta_1) + mgl\{(1 - \cos\theta_1) + (1 - \cos\theta_2)\}$$
$$= \tfrac{1}{2} mgl(2\theta_1^2 + \theta_2^2), \qquad \text{to second order terms,}$$

and

$$T = \tfrac{1}{2} ml^2(2\dot{\theta}_1^2 + 2\dot{\theta}_1\dot{\theta}_2 + \dot{\theta}_2^2) \qquad \text{to second order terms.}$$

If we put $\theta_1 = x_1 \sin wt, \theta_2 = x_2 \sin wt$, then

$$T_{\max} = \tfrac{1}{2} m l^2 (2x_1^2 + 2x_1 x_2 + x_2^2) w^2$$
$$V_{\max} = \tfrac{1}{2} mgl\,(2x_1^2 + x_2^2).$$

Putting $T_{\max} = V_{\max}$ and $\lambda = w^2$, we obtain the Rayleigh quotient

$$\lambda = \frac{g}{l} \frac{2x_1^2 + x_2^2}{2x_1^2 + 2x_1 x_2 + x_2^2}$$

and, if we set $\mu = \lambda l/g = w^2 l/g$,

$$\mu = \frac{2x_1^2 + x_2^2}{2x_1^2 + 2x_1 x_2 + x_2^2}\ . \tag{13.33}$$

The smallest μ, call it μ_1, gives the least w and the longest period. Now consider various trial values of x_1/x_2 in (13.33). We find (by using intelligent guesses)

$$x_1/x_2 = \frac{1}{2} \quad \text{gives} \quad \mu_1 = \frac{2+4}{2+4+4} = 0{\cdot}6,$$

$$x_1/x_2 = \frac{2}{3} \quad \text{gives} \quad \mu_1 = \frac{8+9}{8+12+9} = \frac{17}{29} \mp 0{\cdot}5862,$$

while $$x_1/x_2 = \frac{5}{7} \quad \text{gives} \quad \mu_1 = \frac{50+49}{50+70+49} = \frac{99}{169} \mp 0{\cdot}585\,798\,8.$$

The correct value of μ_1 is $2 - \sqrt{2} \mp 0{\cdot}585\,786$. This is always smaller than the value of μ given by (13.33) but, as one can see from our trial values, not by a very large amount.

13.9 Least squares minimization

Frequently, in scientific experiments, one is given a set of N values of the independent variable x and a corresponding set of N measured values of the dependent variable y. Further, one may have good reason to believe that a specified form of mathematical relationship connects y and x; e.g., we may suspect a polynomial relationship of the form

$$y = a_0 + a_1 x + a_2 x^2 + \ldots + a_m x^m \qquad (m < N). \tag{13.34}$$

If (13.34) is a plausible relationship then we would like to know what values should be assigned to $a_0, a_1, \ldots, a_m$ so that formula (13.34) is consistent, in some way, with the observed data. Let (X_k, Y_k) be the observed data points $(k = 1, 2, \ldots, n)$. Then, one criterion which has been proposed for consistency requires the sums of

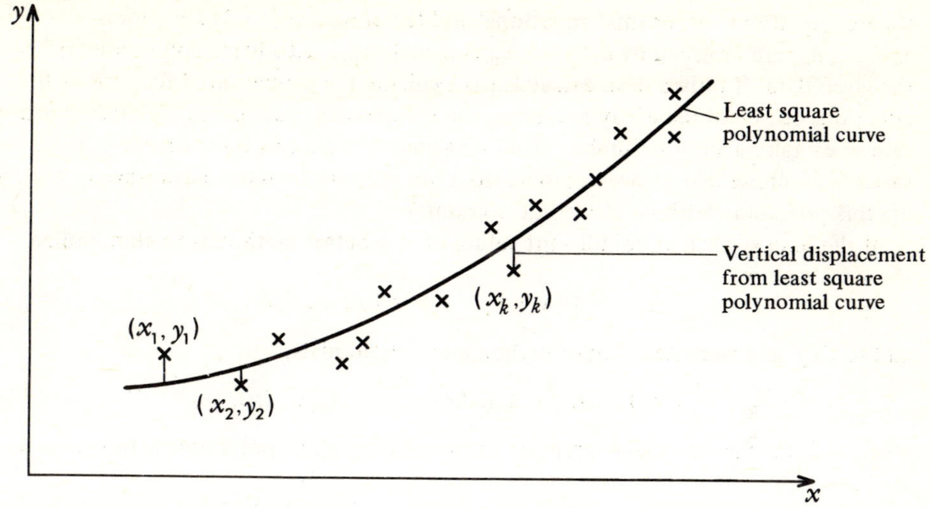

Fig. 13.3 The method of least squares

squares of the vertical displacements of the points (X_k, Y_k) from the curve (13.34) to be a minimum. This criterion is called the 'least squares criterion' and, in mathematical terms, requires

$$s = \sum_k (Y_k - a_0 - a_1 X_k - a_2 X_k^2 - \ldots - a_m X_k^m)^2, \qquad (13.35)$$

to be a minimum. This is the case when

$$\partial s/\partial a_i = 0, \qquad (i = 0, 1, \ldots, m),$$

i.e.,

$$\left.\begin{aligned} \sum_k Y_k &= \hat{a}_0\tau_0 + \hat{a}_1\tau_1 + \ldots + \hat{a}_m\tau_m \\ \sum_k Y_k X_k &= \hat{a}_0\tau_1 + \hat{a}_1\tau_2 + \ldots + \hat{a}_m\tau_{m+1} \\ &\ldots\ldots\ldots\ldots\ldots\ldots\ldots\ldots \\ \sum_k Y_k X_k^m &= \hat{a}_0\tau_m + \hat{a}_1\tau_{m+1} + \ldots + \hat{a}_m\tau_{2m} \end{aligned}\right] \qquad (13.36)$$

where $\tau_r = \sum_1^N X_k^r$, $(r = 0, 1, \ldots, 2m)$ and $\hat{a}_i$ denotes the value of a_i for which (13.34) is a minimum. Equations (13.35) are linear equations in $\hat{a}_0, \hat{a}_1, \ldots, \hat{a}_m$ and they can be solved by the methods of chapter 8. They are often called 'least squares equations' or 'normal equations' and the equation (13.34) in which $a_0, \ldots, a_m$ are replaced by $\hat{a}_0, \ldots, \hat{a}_m$ is called the 'least square polynomial fit' to the given data. If only a desk calculator is available then some simplification of the equations can be obtained by recording the independent variable at equally spaced values, by taking an odd number of observations and making a suitable change of variable (exercise 20). However, there does not seem to be much advantage in adopting this procedure with an electronic computer.

If the data is given at equidistant values of x, a better method is to standardize the x-scale so that

$$x_r = r - \tfrac{1}{2}(N+1) \qquad (r = 1, 2, \ldots, N)$$

and to fit y as a weighted sum of orthogonal polynomials,

$$y = A_0\phi_0(x) + A_1\phi_1(x) + \ldots + A_m\phi_m(x), \tag{13.37}$$

where $\phi_i(x)$ is the ith degree Gram or weighted Chebyshev polynomial. In particular, we have

$$\left.\begin{aligned} &\phi_0(x) = 1, \quad \phi_1(x) = x, \quad \phi_2(x) = x^2 - \tfrac{1}{12}(N^2 - 1), \\ &\phi_3(x) = x^3 - \tfrac{1}{20}(3N^2 - 7)x, \\ &\phi_4(x) = x^4 - \tfrac{1}{14}(3N^2 - 13)x^2 + \tfrac{3}{560}(N^2 - 1)(N^2 - 9), \end{aligned}\right] \tag{13.38}$$

and further polynomials can be obtained from the recurrence formula

$$\phi_{i+1} = \phi_1\phi_i - \frac{i^2(N^2 - i^2)}{4(4i^2 - 1)}\phi_{i-1}. \tag{13.39}$$

All these polynomials satisfy the orthogonality condition

$$\sum_{r=1}^{n} \phi_i(x_r)\phi_j(x_r) = 0, \qquad i \neq j. \tag{13.40}$$

The least squares estimate $\hat{A}_i$ for A_i is given by the formula

$$\hat{A}_i = \sum_{r=1}^{n} Y_r\phi_i(x_r) \Big/ \sum_{r=1}^{n} \{\phi_i(x_r)\}^2. \tag{13.41}$$

There is one difficulty concerning the estimation of the parameters $a_0, a_1, \ldots, a_m$ in (13.34), namely, that we do not in general know in advance the degree of the polynomial that we should fit. In this matter we can obtain some help by considering a statistical model of the form

$$y_k = a_0 + a_1 x_k + \ldots + a_m x_k{}^m + \epsilon_k, \qquad (k = 1, 2, \ldots, N), \tag{13.42}$$

instead of (13.34), where the 'error' random variables ϵ_k are assumed to be independent normal variates with zero mean and variance σ^2. Now the set of equations (13.42) can be written in the matrix form

$$\mathbf{y} = X\mathbf{a} + \boldsymbol{\epsilon}$$

where $\mathbf{y}$ is the $(N \times 1)$ matrix of known observation, $\boldsymbol{\epsilon}$ is the $(N \times 1)$ vector of error random variables, $\mathbf{a}$ is the $((m + 1) \times 1)$ vector of parameters, and X is the $(N \times (m + 1))$ matrix of known coefficients. The least square method in matrix form requires that we minimize the scalar sum of squares

$$S = \epsilon'\epsilon = (\mathbf{y} - X\mathbf{a})'(\mathbf{y} - X\mathbf{a}). \tag{13.43}$$

A necessary condition that (13.43) be minimized is that

$$\frac{\partial S}{\partial \mathbf{a}} = \mathbf{o},$$

that is

$$2X'(\mathbf{y} - X\mathbf{a}) = \mathbf{o}, \tag{13.44}$$

– which corresponds to (13.36) – and this gives for our least square estimator the vector

$$\hat{\mathbf{a}} = (X'X)^{-1}X'\mathbf{y}, \tag{13.45}$$

assuming $X'X$ is non-singular. Moreover, $\hat{\mathbf{a}}$ is an unbiased estimator of $\mathbf{a}$.

The importance of the orthogonal polynomial approach lies in the fact that $X'X$ becomes a diagonal matrix whose inverse is easily obtained. Further, for large m, we avoid inverting a matrix which can be ill-conditioned, since the matrix $X'X$ behaves very much like the Hilbert matrix for those large values (see Ralston [7]). Another advantage is the ease with which we can adopt a statistical test to 'confirm' that the degree we have selected for our polynomial is the correct one. This will be illustrated by the following example, where we assume a third degree polynomial provides a suitable fit to the data in Table 13.7. From the table we have

$$\hat{A}_0 = 157{\cdot}741/13 = 12{\cdot}1339,$$
$$\hat{A}_1 = 451{\cdot}558/182 = 2{\cdot}481\,088,$$
$$\hat{A}_2 = 443{\cdot}822/2002 = 0{\cdot}221\,689,$$
$$\hat{A}_3 = 162{\cdot}396/20\,592 = 0{\cdot}007\,886\,26.$$

Table 13.7 Data for least squares orthogonal polynomial fit

x_r	y_r	x'_r $(x_r - 7)$	$\phi_1(x'_r)$	$\phi_2(x'_r)$	$\phi_3(x'_r)$	*Accumulated totals*
1	1·604	−6	−6	22	−66	
2	2·167	−5	−5	11	0	
3	2·937	−4	−4	2	36	$\Sigma\phi_0^2 = 13$
4	3·961	−3	−3	−5	48	$\Sigma\phi_1^2 = 182$
5	5·286	−2	−2	−10	42	$\Sigma\phi_2^2 = 2002$
6	6·960	−1	−1	−13	24	$\Sigma\phi_3^2 = 205\,92$
7	9·030	0	0	−14	0	$\Sigma y_r\phi_0(x'_r) = 157{\cdot}741$
8	11·544	1	1	−13	−24	$\Sigma y_r\phi_1(x'_r) = 451{\cdot}558$
9	14·548	2	2	−10	−42	$\Sigma y_r\phi_2(x'_r) = 443{\cdot}822$
10	18·090	3	3	−5	−48	$\Sigma y_r\phi_3(x'_r) = 162{\cdot}396$
11	22·218	4	4	2	−36	$\Sigma y_r^2 = 3134{\cdot}0436$
12	26·978	5	5	11	0	
13	32·418	6	6	22	66	

Further, a residual estimate for σ^2 is obtained from

$$s^2 = \left(\sum_{r=1}^{N} y_r^2 - \sum_{i=0}^{m} \left\{ \hat{A}_i^2 \sum_{r=1}^{N} \phi_i^2(x_r) \right\} \right) \Big/ (N - m - 1).$$

If we wish to test the hypothesis $A_i = 0, i = 1, 2, \ldots, m$ assuming $A_i = 0, i > m$ then

$$\sum_{i=1}^{m} \hat{A}_i^2 \sum_r \phi_i^2(x'_r)$$

is distributed as the sum of m independent normal deviates each with variance σ^2. If we calculate the variance ratio

$$F = \left(\sum_{i=1}^{m} \hat{A}_i^2 \sum_r \phi_i^2 (x'_r) \right) / ms^2,$$

and compare our result with the percentage F-distribution with parameters $\upsilon_1 = m$, $\upsilon_2 = N - m - 1$ we can obtain some idea of the significance of the mth degree polynomial as a whole.

In the given example

$$\upsilon_1 = 3, \quad \upsilon_2 = 9, \quad s^2 \doteqdot 0{\cdot}000\,88, \quad \sum_1^m \hat{A}_i^2 \sum_r \phi_i^2 (x_r) \doteqdot 1220$$

and so

$$F \doteqdot 1220/3 \times 0{\cdot}000\,88 \doteqdot 4{\cdot}6 \times 10^5.$$

This is considerably larger than the upper 0·1% point of the F-distribution ($F_{3,9} = 13{\cdot}90$) so we can assert with some confidence that our third degree polynomial provides a satisfactory least squares fit to the data. The given data was in fact generated from the polynomial equation

$$y = 1{\cdot}2 + 0{\cdot}34x + 0{\cdot}056x^2 + 0{\cdot}007\,89x^3,$$

and the reader should compare this with the orthogonal polynomial obtained. For further statistical tests which can be applied see Pearson and Hartley [8].

Note also that the method can be extended to the case where the data is not given at equal intervals (exercise 30).

13.10 Summary and notes

The methods of this chapter fall into two main categories – search methods and gradient methods. Gradient methods are far superior to search methods as they utilize additional properties of the functions under discussion. Even when explicit formulae for derivatives are not available, we can use a technique due to Powell [9], or, if a considerable amount of computation is acceptable, we can use numerical approximations to derivatives (section 3.6). This extra computation in the calculation of the derivatives is still usually less than that required by search methods. Search methods can be very useful when the function to be minimized is given in the form of numerical data (as often happens in statistics, for instance) but here again we can resort to derivative methods by first finding a least square polynomial fit to the data. (See Davies **[10]**.) For the most recent work on gradient methods readers should consult Biggs **[13]**.

In all the methods given the results will be localized results. No methods have yet been devised where all solutions of a set of non-linear equations can be obtained or all minima of a given function recorded. Nevertheless, it is possible in certain special circumstances, e.g., in some physical problems, to infer that only one admissible solution can exist or that the minimum obtained is a global minimum. However, great care must be taken before making wild assertions as, in general, a special mathematical investigation of great complexity will have to be carried out to validate the assertion.

The field of optimization has received considerable attention in recent years and most useful surveys have been given by Spang **[11]** and, more recently, Powell **[15]**. The reader is also referred to the monograph of Box, Davies, and Swann **[12]**.

Finally, it is worth remarking that the approximation of data by a polynomial expression is part of the more general problem of linear regression, where an observation is represented by

$$y = a_0 + a_1X_1 + a_2X_2 + \ldots + a_rX_r + \epsilon$$

where $X_1, X_2, \ldots, X_r$ are independent variables, ϵ is a random normal deviate, and we wish to estimate the a_i. A full account of this is given in Kendall and Stuart **[14]** – the case considered in the text is the special one with $X_i = x^i$, $i = 1, 2, \ldots, r$.

REFERENCES

[1] Fletcher, R. and Powell, M. J. D., 'A Rapidly Convergent Descent Method for Minimization', *Computer J.*, **6** (1963) 163–68.

[2] Rosenbrock, H. H., 'An Automatic Method for Finding the Greatest or Least Value of a Function', *Computer J.*, **3** (1960) 175–84.

[3] Powell, M. J. D., 'On the Calculation of Orthogonal Vectors', *Computer J.*, **11** (1968) 302–4.

[4] Gillespie, R. P., *Partial Differentiation*, Oliver and Boyd, Edinburgh (1954).

[5] John, F., 'Extremum Problems with Inequalities as Subsidiary Conditions', *Courant Anniversary Volume* (1948), 187–204, Interscience Publ., New York.

[6] Carroll, C. W., 'The Created Response Surface Technique for Optimizing Non-linear Restraint Systems', *Operations Research*, **9** (1961).

[7] Ralston, A., *A First Course in Numerical Analysis*, McGraw-Hill Book Co., New York (1965).

[8] Pearson, E. S. and Hartley, H. O. (eds.), *Biometrika Tables for Statisticians*, Vol. 1, 3rd edn., Cambridge University Press (1966).
[9] Powell, M. J. D., 'An Efficient Method for Finding the Minimum of a Function of Several Variables without Calculating Derivatives', *Computer J.*, 7 (1964) 155–62.
[10] Davies, O. L. (ed.), *The Design and Analysis of Industrial Experiments*, 2nd edn., Oliver and Boyd, Edinburgh (1956).
[11] Spang, H. A., III, 'A review of Minimization Techniques for Non-linear Functions', *S.I.A.M. Review*, **4** (1962) 343–65.
[12] Box, M. J., Davies, D., and Swann, W. H., *Optimization Techniques*, I.C.I. Monograph No. 5, Oliver and Boyd Ltd., Edinburgh (1968).
[13] Biggs, M. C., 'Minimization Algorithms Making Use of Non-quadratic Properties of the Objective Function', *J. Inst. Math. Appl.*, **8** (1971), 315–27.
[14] Kendall, M. G. and Stuart, A., *The Advanced Theory of Statistics*, Vol. 2, Griffin, London (1961).
[15] Powell, M. J. D., 'A Survey of Numerical Methods for Unconstrained Optimization', *S.I.A.M. Review*, **21** (1970) 79–97.
[16] Fletcher, R., 'A New Approach to Variable Metric Algorithms', *Computer J.*, **13** (1970) 317–22.

Exercises

1. Show that the function $g(x) = (x-1)^3 \ln x$ has a minimum at $x = 1$ even though $d^2g/dx^2 = 0$.
2. In the method of Fibonacci search show that $y_{k+1} = y'_k$ if $g(y_k) > g(y'_k)$ and $y'_{k+1} = y_k$ if $g(y_k) \leqslant g(y'_k)$.
3. Find, by the methods of univariate or Fibonacci search, the minimum value of the function

$$g(x) = e^x - 2x + 1 - (x+2)\ln(x+2), \qquad (x > 0)$$

given that it lies in the region (0, 2). Check your result by a derivative method.
4. Find the solution of the simultaneous non-linear equations

$$x^2 + xy + y^2 = 3,$$
$$\sin x - y^2 = 0,$$

given $(x_0, y_0) = (1, 1)$ is a good initial approximation.
5. Find all four sets of solutions of the simultaneous non-linear equations

$$x^2 + y^2 = 1,$$
$$xy = 0{\cdot}4,$$

(*a*) by elimination of the variable x;
(*b*) by the methods given in this chapter.
6. Show that there are 2^n possible solutions of n independent simultaneous quadratic forms in n variables.

7. In a certain transistorized circuit the following equations hold

$$V_1 = r(1 + \beta^{-1})(i_1 + i_2) + K \ln\{(i_1 + I_s)/I_s\},$$
$$V_2 = r(1 + \beta^{-1})(i_1 + i_2) + K \ln\{(i_2 + I_s)/I_s\},$$
$$V_2 = R\beta' i_1 - R i_2 \beta^{-1},$$

where $K = 0{\cdot}025$ volt, $I_s = 5{\cdot}55 \times 10^{-14}$ amp, β and β' are the amplification factors of the transistors in the circuit, V_1 and V_2 represent input and output voltages, r and R are resistances (R variable), and i_1 and i_2 are the currents in various sections of the circuit. Determine the value of R necessary to make $V_1 = V_2 = 6{\cdot}25$ volt, given that $\beta = 100$, $\beta' = 20$, $r = 10^6$ ohm. What is the value of V_2 when $V_1 = 10$ volt and R has the value already found?

8. Find the maximum and minimum values of

$$\frac{x + y - 1}{x^2 + 2y^2 + 2}.$$

9. (Deist and Sefor, *Computer J.* (1967).) To solve the simultaneous equations

$$f_1(\mathbf{x}) = x_1^2 + x_2^2 + x_3^2 - 5 = 0,$$
$$f_2(\mathbf{x}) = x_1 + x_2 \qquad - 1 = 0,$$
$$f_3(\mathbf{x}) = x_1 \qquad + x_3 - 3 = 0,$$

we introduce the parameter a so that we have a new set of equations

$$f_1(\mathbf{x}, a) = x_1^2 + a(x_2^2 + x_3^2) - 5 = 0; \qquad f_2(\mathbf{x}, a) = f_2(\mathbf{x}); \qquad f_3(\mathbf{x}, a) = f_3(\mathbf{x}),$$

which for some convenient value of a (such as $a = 0$) are easy to solve. Starting with this solution we vary a in suitable steps to determine the solution of the original equations. Generalize the method.

10. A company which has a factory sited at (0, 0) and customers at $(x_1, y_1), \ldots, (x_N, y_N)$ intend to build an additional factory at (x, y) where (x, y) minimizes the cost, C, of direct distribution from the nearest factory, which is regarded as having the form,

$$C = K \sum_{i=1}^{N} \min\{(x_i^2 + y_i^2),\ (x - x_i)^2 + (y - y_i)^2\}, \qquad K \text{ a constant.}$$

Find (x, y) when the customers are located at (5, 4), (2, 3), (17, 4), (26, 10), (19, 18), (5, 25), (87, 5), (80, 40), (56, 35), (48, 60), (43, 51), (25, 65), (12, 70), (34, 28), (10, 26), (−10, 18), (−21, −7), (−40, 12), (12, −25), (−19, −37), (−29, −44).

11. Find the maximum value in the range $0 \leqslant z_i \leqslant 1$ $(i = 1, \ldots, n)$ of the function

$$g(\mathbf{z}) = \left(1 + \sum_1^n z_i\right)^2 - \left(1 + \sum_1^n z_i^2\right)\left(1 + \frac{n}{n+1}\left[1 + \sum_1^n z_i\right]\right),$$

for $n = 1, 2, 3$. Find also the minimum value of $g(\mathbf{z})$ if the z_i are subject to the conditions

$$z_i \leqslant z_{i+1} \leqslant 2z_i$$
$$z_i \leqslant \bar{z}_i$$

where $\bar{z}_i$ is the value of z_i for which $g(\mathbf{z})$ is a maximum in the first part of the question.

12. If $g(\mathbf{x})$ is a quadratic form having minimum value at $\mathbf{x}_0$ show that, for any point $\mathbf{x}$,

$$\mathbf{x}_0 - x = -G^{-1} \text{ grad } g(\mathbf{x}),$$

where G is the matrix of the quadratic form (the Hessian matrix).

13. In the method of Fletcher and Powell assuming $\mathbf{x}'Y_i\mathbf{x} \geqslant 0$ show that $\mathbf{x}'Y_{i+1}\mathbf{x} > 0$ (i.e., the procedure is stable).

14. In the method of Fletcher and Powell show that, if $g(\mathbf{x})$ is a quadratic form then $\boldsymbol{\sigma}_0, \boldsymbol{\sigma}_1, \ldots, \boldsymbol{\sigma}_k$ are linearly independent eigenvectors of $Y_{k+1}G$ with corresponding eigenvalue 1. Hence show that the sequence of matrices Y_i converges after exactly n steps to G^{-1}.

15. Find the minimum value of $x^2 + y^2$ subject to $x + y = 3$,
(*a*) by the method of Lagrange multipliers;
(*b*) by Carroll's method.

16. A rectangular closed box has sides of length x, y, z. Find the maximum volume of the box given that its surface area is 24 units.

17. If $\mathbf{a} = (a_1, a_2)$, $\mathbf{b} = (b_1, b_2)$ are two points on $x^4 + y^4 = 1$, the lattice Λ generated by $\mathbf{a}, \mathbf{b}$ is said to be admissible if no point of the form

$$l\mathbf{a} + m\mathbf{b} = (la_1 + mb_1,\ la_2 + mb_2)$$

satisfies $x^4 + y^4 < 1$, where l and m are integers and $(l, m) \neq (0, 0)$. Find the values of $\mathbf{a}$ and $\mathbf{b}$ which make

$$|a_1b_2 - a_2b_1| = |\det \Lambda|$$

a minimum (see Bigg, M. D. *Proc. Camb. Phil. Soc.*, **59**, (1963), 523–30).

18. Find the minimum value of the function

$$100(x_2 - x_1^2)^2 + (1 - x_1)^2$$

by the method of Rosenbrock. Take $\alpha = 3, \beta = 0{\cdot}5$ and $\alpha = 5, \beta = 0{\cdot}5$ and compare your results, starting in both cases from $x_1 = -1{\cdot}2, x_2 = 1$. Find the minimum of the same function by the method of Fletcher and Powell (using the same starting values).

19. Given a set of data $(X_1, Y_1), \ldots, (X_N, Y_N)$, show that the estimates $\hat{a}_0, \hat{a}_1$ of the coefficients in the least squares linear approximation $y = a_0 + a_1x$ are given by

$$\hat{a}_0 = \frac{(\sum X_i^2)(\sum Y_i) - (\sum X_iY_i)(\sum X_i)}{N\sum X_i^2 - (\sum X_i)^2}, \qquad \hat{a}_1 = \frac{N\sum X_iY_i - (\sum X_i)(\sum Y_i)}{N\sum X_i^2 - (\sum X_i)^2}.$$

20. By first making the change of variable $X = X' + 3$ find the least squares linear approximation to the data

X	0	1	2	3	4	5	6
Y	2·4	2·1	3·2	5·6	9·3	14·6	21·9

Find also, by the orthogonal polynomial method, the least square linear and parabolic approximation to the data. Use a statistical test to determine the suitability of these approximations.

21. Show, by considering data at $x_i = \frac{1}{4} + \frac{3}{4}\frac{i}{n}$ $(i = 0, 1, \ldots, n)$ $(n \to \infty)$ that the least square line which best approximates the curve $y = x^{\frac{1}{2}}$ in the range $(\frac{1}{4}, 1)$ is

$$y = \frac{88}{135}x + \frac{10}{27}.$$

(Hint: Use the result of exercise 19.)

22. Show, by considering data at $x_i = \frac{1}{2} + \frac{1}{2}\frac{i}{n}$ $(i = 0, 1, \ldots, n)$ $(n \to \infty)$ that the least square line which best approximates the curve $y = 1/x$ in the range $(\frac{1}{2}, 1)$ is

$$y = (48 - 72 \ln 2)x + 56 \ln 2 - 36.$$

23. Develop a method for finding a rational polynomial approximation of the form

$$y = P(x)/Q(x),$$

to given data $(X_1, Y_1), \ldots, (X_N, Y_N)$. Assuming that the following data is suitably represented by an expression of the form $y = (a + bx)/(c + dx)$ estimate a, b, c, d.

x	0·0	0·1	0·2	0·3	0·4	0·5	0·6
y	0·4043	0·4667	0·5349	0·6098	0·6923	0·7838	0·8857

x	0·7	0·8	0·9	1·0
y	1·0000	1·1290	1·2759	1·4444

24. (Linear regression problem.) The table shows the weights X_1 (to the nearest pound), the heights X_2 (to the nearest inch), and ages X_3 (to the nearest year), of 12 boys. Assuming the existence of a relationship of the form

$$X_1 = a_1 + a_2 X_2 + a_3 X_3$$

find estimates for a_1, a_2, a_3. Hence estimate the weight of a boy 10 years old and 55 inches tall.

X_1	64	71	53	67	55	58	77	57	56	51	76	68
X_2	57	59	49	62	51	50	55	48	52	42	61	57
X_3	8	10	6	11	8	7	10	9	10	6	12	9

25. Let Z_k be the value of a dependent variable corresponding to values X_k, Y_k of the independent variables $(k = 1, 2, \ldots, N)$. If it is assumed that Z satisfies a relationship of the form

$$Z = a_0 + a_1 X + a_2 Y + a_3 X^2 + a_4 XY + a_5 Y^2$$

find least squares estimates for $a_0, a_1, \ldots, a_5$.

26. By finding $f'(x)$ at the tabular points and using inverse interpolation to find where $f'(x) = 0$, find the value of x for which $f(x)$ is a maximum and deduce this maximum value.

x	0·0	0·1	0·2	0·3	0·4	0·5
$f(x)$	1·000 00	1·051 14	1·089 12	1·114 24	1·127 06	1·128 38

x	0·6	0·7	0·8	0·9	1·0
$f(x)$	1·119 18	1·100 55	1·073 67	1·039 75	1·000 00

Find the maximum value of $f(x)$ by the alternative method of fitting a least square polynomial to the data and finding the maximum value of the polynomial in the given range.

27. (Prony's method for fitting sums of exponentials.) Let

$$y = a_1 e^{\alpha_1 x} + a_2 e^{\alpha_2 x} + \ldots + a_m e^{\alpha_m x}$$

be the equation representing data (x_k, y_k), $k = 0, 1, \ldots, n$, where $x_k = x_0 + kh$. Let $b_k = a_k \exp \alpha_k x_0$, $z_k = \exp \alpha_k h$. Then

$$y_i = b_1 z_1^i + b_2 z_1^i + \ldots b_m z_m^i.$$

We now determine, by the method of least squares, quantities $s_1, s_2, \ldots, s_m$ such that the linear equations

$$y_i s_m + y_{i+1} s_{m-1} + \ldots + y_{i+m-1} s_1 + y_{i+m} = 0, \qquad i = 0, \ldots, n-m$$

are satisfied. If this is the case then, ideally, $z_1, \ldots, z_m$ are the roots of the polynomial equation

$$z^m + s_1 z^{m-1} + s_2 z^{m-2} + \ldots + s_m = 0.$$

Once the z_k are known the α_k are readily determined and the a_k can be deduced by another application of the method of least squares. Use this method to analyse the data

x	0	2	4	6	8	10	12	14	16	18	20
$f(x)$	3000	2695	1995	1332	845	531	341	230	163	121	92

(Buckingham, R. A., *Numerical Methods*, Pitman (1957) – see also Lanczos, C, *Applied Analysis*, Pitman (1957) for general comments about fitting by exponentials.)

28. Repeat exercise 18 by Rosenbrock's method using formula (13.9) to estimate the total distance d_1 travelled in the direction ξ_1.

29. Given $\phi_0(x) = 1$, $\phi_1(x) = x$, establish the results (13.38) for higher order Gram–Chebyshev polynomials by noting that the $\phi_i(x)$ satisfy

$$\sum_{\substack{r=1 \\ i \neq j}}^{N} \phi_i(x_r)\phi_j(x_r) = 0,$$

where the data points x_r satisfy $x_r = r - \frac{1}{2}(N+1)$.

30. Given $\psi_0(x) = 1$, $\psi_1(x) = x$, construct a sequence of orthogonal polynomials $\psi_2(x)$, $\psi_3(x)$, . . . which satisfy the orthogonality conditions

$$\sum_{\substack{r=1 \\ i \neq j}}^{N} \phi_i(x_r)\phi_j(x_r) = 0$$

where $x_r = x'_r - \bar{x}'$, x'_r $(r = 1, \ldots, N)$ are the given data points (not necessarily equally spaced) and

$$\bar{x}' = \frac{1}{N}\sum_r x'_r.$$

14 Monte Carlo methods

14.1 Introduction. Hit-or-miss Monte Carlo

Monte Carlo techniques have their origins in the Second World War. Problems in neutron diffusion had proved intractable by conventional methods and a probabilistic approach was developed to solve them. It was later found that this probabilistic approach could be used to solve deterministic problems and, in particular, it is useful in evaluating integrals in several dimensions, where deterministic methods require a large number of function evaluations.

For simplicity we shall restrict Monte Carlo applications to the evaluation of a one-dimensional integral in the standard form

$$I = \int_0^1 f(x)\,\mathrm{d}x, \qquad f(0) = 0, \qquad f(1) = 1, \tag{14.1}$$

and for comparison purposes we will use the example

$$I = \int_0^1 \frac{x^2 \mathrm{e}^x\,\mathrm{d}x}{1 - x + x\mathrm{e}^x}. \tag{14.2}$$

Extension of the methods to higher dimensions is fairly straightforward. Throughout this chapter we will assume that we have a source of (pseudo) random numbers available and in section 14.6 we append some notes on the generation and testing of pseudo-random numbers.

We now describe the crudest Monte Carlo technique – the 'hit-or-miss' technique. We imagine the square region $0 \leqslant x, y \leqslant 1$ in Fig. 14.1 to be a large dartboard. Darts are thrown at random at the dartboard, that is, in such a way that all points of the target area are equally likely to be struck. A 'hit' is registered if a dart lands in the shaded area and we have

$$Pr\,(\text{'hit'}) = \text{probability of a hit} = (\text{shaded area})/(\text{target area})$$

i.e.,

$$I = \text{shaded area} = Pr\,(\text{'hit'}) = p, \text{ say.} \tag{14.3}$$

A simple experiment is carried out to estimate p, the probability of a hit. n darts are thrown at random at the target. If r land in the shaded area then an estimate $\hat{p}$ of p is given by

$$\hat{p} = r/n. \tag{14.4}$$

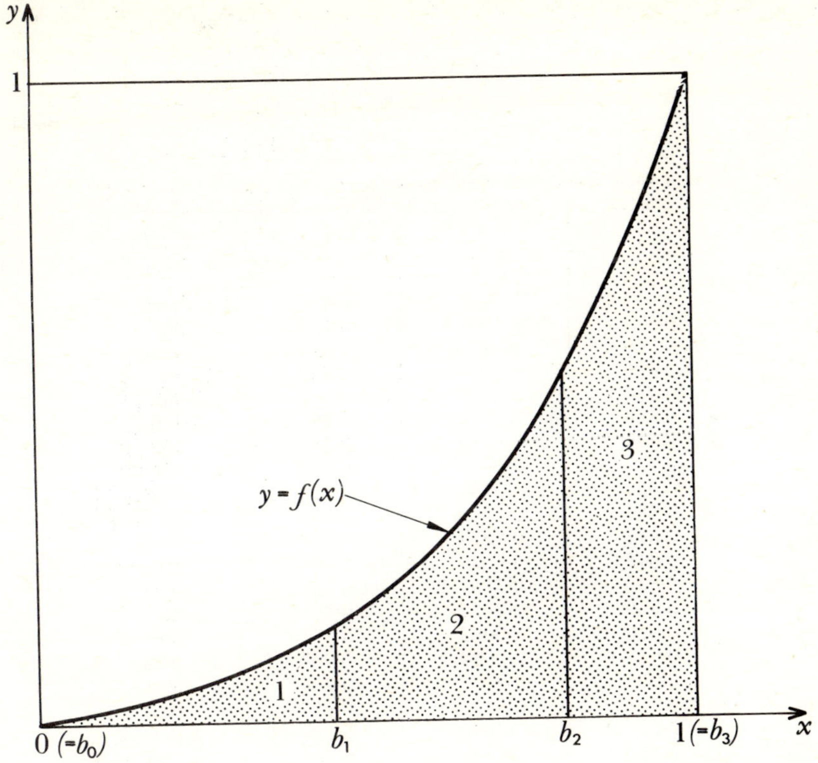

Fig. 14.1 Area enclosed by standard integral. (Divided into three strata)

This estimate is distributed about the true value p with standard error given by

$$s = \sqrt{\{p(1-p)/n\}}$$

estimated by
$$\hat{s} = \sqrt{\{\hat{p}(1-\hat{p})/n\}} \tag{14.5}$$

The dart throwing experiment can be turned into a mathematical calculation to evaluate the integral in (14.1) the steps being:

1. Select two numbers u_1, u_2 at random from the interval $(0, 1)$.
2. Take the first of these as the x-coordinate and determine the boundary between the shaded and unshaded areas by computing

$$y_{\text{boundary}} = f(x) = f(u_1).$$

3. u_2 is taken as the y-coordinate of the 'dart' and we have a hit if $u_2 < f(u_1)$.

Table 14.1 displays this calculation for the example (14.2) with $n = 25$. We obtain

$$I \doteqdot r/n = 0{\cdot}36$$

and

$$\hat{s} = 0{\cdot}096.$$

Table 14.1. Sample calculation for hit-or-miss/simple Monte Carlo

u_1 x	$f(u_1)$ y_{boundary}	u_2 y	*Hit/Miss*
0·457	0·2608	0·418	miss
0·573	0·4035	0·113	hit
0·309	0·1170	0·994	miss
0·496	0·3064	0·243	hit
0·284	0·0980	0·880	miss
0·648	0·5046	0·996	miss
0·770	0·6764	0·406	hit
0·929	0·9018	0·026	hit
0·059	0·0037	0·488	miss
0·968	0·9560	0·697	hit
0·877	0·8287	0·214	hit
0·048	0·0024	0·164	miss
0·520	0·3357	0·727	miss
0·859	0·8032	0·708	hit
0·016	0·0003	0·374	miss
0·057	0·0034	0·984	miss
0·096	0·0100	0·256	miss
0·513	0·3271	0·712	miss
0·233	0·0646	0·197	miss
0·350	0·1516	0·467	miss
0·459	0·2631	0·121	hit
0·487	0·2956	0·091	hit
0·279	0·0944	0·141	miss
0·150	0·0255	0·339	miss
0·153	0·0266	0·510	miss

14.2 Improving accuracy. Simple and stratified Monte Carlo

The estimate $\hat{s}$ for the standard error is a pointer to the accuracy of the computed value of I. Clearly, by reducing $\hat{s}$, we get a better value of I. We can do this by increasing n, but a hundredfold increase in n only results in a tenfold decrease in $\hat{s}$ so this is not very helpful. An alternative method is to try and reduce p. This can be done by approximating the given function $f(x)$ by a function $g(x)$, which can be integrated analytically, writing

$$I = \int_0^1 \{f(x) - g(x)\}\, dx + \int_0^1 g(x)\, dx = I_1 + \int_0^1 g(x)\, dx, \tag{14.6}$$

and then using the hit-and-miss technique to evaluate I_1. Thus, for the example (14.2), we can put $g(x) = e^{-1}x^2 e^x$ and concentrate on evaluating

$$I_1 = \int_0^1 \frac{x^2 e^x \, dx}{1 - x + xe^x} - e^{-1}\int_0^1 x^2 e^x \, dx$$

$$= \int_0^1 x^2 e^x \left(\frac{1}{1 - x + xe^x} - \frac{1}{e}\right) dx$$

by the hit-or-miss technique.

However, we can achieve more by noting that each $y_i = f(u_i)$ is an independent estimate of the required integral I and consequently the mean $\bar{f}$,

$$\bar{f} = \frac{1}{n}\sum_{i=1}^{n} f(u_i), \tag{14.7}$$

is an unbiased estimator of I with estimated variance s^2 where

$$s^2 = \frac{1}{n-1}\sum_{i=1}^{n} \{f(u_i) - \bar{f}\}^2 . \tag{14.8}$$

Applying this to the data in Table 14.1 we obtain the estimate

$$I = \bar{f} = 0{\cdot}2984$$

with $s = 0{\cdot}3100$ and standard error $0{\cdot}062$.

Of course a further reduction in standard error can be obtained by applying this method (known as simple Monte Carlo) in conjunction with the previous idea contained in (14.6) – the method then being called the method of control variates.

A further variance reduction can be achieved by stratification. We divide the x-axis between (0, 1) into l separate parts, or strata (see Fig. 14.1), and we estimate I_j,

$$I_j = \int_{b_{j-1}}^{b_j} f(x) \, dx$$

by taking n_j random numbers u_k from the interval (0, 1), forming $v_{jk} = b_{j-1} + (b_j - b_{j-1})u_k$ and $f(v_{jk})$, and evaluating

$$\bar{f} = \frac{1}{n_j}\sum_{k=1}^{n_j} f(v_{jk}). \tag{14.9}$$

An estimate for I is then $\bar{f}_{\text{stratified}}$ where

$$\bar{f}_{\text{stratified}} = \sum_{j=1}^{l} (b_j - b_{j-1})\bar{f}_j . \tag{14.10}$$

The variance of the estimate is given by

$$s^2_{\text{stratified}} = \sum_{j=1}^{l} (b_j - b_{j-1})^2 \sum_{k=1}^{n} \frac{(f(v_{jk}) - \bar{f}_j)^2}{n_j - 1} . \qquad (14.11)$$

In practice the number of samples in each stratum is taken to be the same and the width of each stratum is chosen to be equal, although it would be preferable to arrange the boundaries so that the difference between the maximum and minimum values of f in each stratum is equal. l is left for the individual to choose but to a large extent it depends on how much extra labour is acceptable. To estimate this we define the power of a method A to be

$$(\text{Power})_A = 1/\{(\text{computational labour})_A \times (\text{variance})_A\} \qquad (14.12)$$

so that, in comparing two methods A and B (or the same method with different sample sizes) we have

$$\frac{(\text{Power})_A}{(\text{Power})_B} = \frac{(\text{computational labour})_B}{(\text{computational labour})_A} \times \frac{(\text{variance})_B}{(\text{variance})_A}$$

or

$$\text{Efficiency ratio } A:B = \text{labour ratio } B:A \times \text{variance ratio } B:A. \qquad (14.13)$$

Thus, if A represents Simple Monte Carlo and B represents hit-or-miss Monte Carlo, we have using the data in Table 14.1

$$\begin{aligned}\text{Efficiency ratio} &= \text{labour ratio} \times (0{\cdot}096/0{\cdot}062)^2 \\ &= \text{labour ratio} \times 2{\cdot}4.\end{aligned}$$

The computational labour in both methods is practically the same so that Simple Monte Carlo is about 2·4 times more powerful a method than Hit-or-Miss.

14.3 Importance sampling

In the method of importance sampling we require some information about the behaviour of $f(x)$. The difference between this and previous methods is that sampling now takes place from a distribution which is chosen to mimic the integrand as closely as possible and not from the interval (0, 1). We now have

$$I = \int_0^1 f(x)\,\mathrm{d}x = \int_0^1 \left(\frac{f(x)}{p(x)}\right) p(x)\,\mathrm{d}x$$

$$= \int_0^1 F(x)\,\mathrm{d}\,P(x), \qquad (14.14)$$

where
$$F(x) = f(x)/p(x)$$
$$P(x) = \int_0^x p(t)\,\mathrm{d}t$$

and $p(t)$ is a positive function of t such that

$$\int_0^1 p(t)\,\mathrm{d}t = 1,$$

that is, $p(t)$ is a probability density function and $P(t)$ is the corresponding cumulative function. If $p(x)$ mimics $f(x)$ closely then $F(x)$ will vary only slightly in $(0, 1)$ and the variance will be reduced. In practice, it is very difficult to find a function $p(x)$ which satisfies the requirements that it can be integrated analytically, that it mimics $f(x)$ and that the generation of sample values from $P(x)$ is fairly easy. For monotonic functions $f(x)$ some simple possibilities are (u_i a random number in $(0, 1)$)

$$p_1(x) = 2x, \quad P_1(x) = x^2, \quad u_{1i} = \sqrt{u_i}$$
$$p_2(x) = 3x^2, \quad P_2(x) = x^3, \quad u_{2i} = \sqrt[3]{u_i}$$
$$p_3(x) = 4x^3, \quad P_3(x) = x^4, \quad u_{3i} = \sqrt[4]{u_i}.$$

Since the standard example (14.2) contains a factor x^2 it is to be expected that $p_2(x)$ would be the function that would mimic $f(x)$ best. This is borne out by Table 14.2 where a sample calculation is given for each of the three above alternatives.

Another probability density function which turns out to be useful, particularly if the function has a maximum near the middle of the interval $(0, 1)$, is

$$p(x) = \begin{cases} 4x, & 0 < x \leqslant 0{\cdot}5 \\ 4(1-x), & 0{\cdot}5 < x \leqslant 1 \end{cases} \qquad \text{(triangular distribution)}$$

which gives

$$P(x) = \begin{cases} 2x^2, & 0 < x \leqslant 0{\cdot}5 \\ 4x - 2x^2 - 1, & 0{\cdot}5 < x \leqslant 1. \end{cases}$$

Samples from this can be generated by two simple methods *either* by taking

$$x_i = \begin{cases} \tfrac{1}{2}\sqrt{u_i}, & 0 < u_i \leqslant 0{\cdot}5 \\ 1 - \tfrac{1}{2}\sqrt{(1-u_i)}, & 0{\cdot}5 < u_i \leqslant 1 \end{cases}$$

or by taking

$$x_i = \tfrac{1}{2}(u_i + u_j),$$

where all the u's are sampled from the interval $(0, 1)$.

It is worth mentioning that importance sampling can be quite effective when used in conjunction with the method of control variates. We write our integrand in the form (14.6) and we then search for a probability density function which mimics $f(x)-g(x)$.

Table 14.2 Sample calculation for Importance Sampling with $p(x) = 2x$, $3x^2$, and $4x^3$

u	$u_1 = \sqrt{u}$	$f_1 = \frac{f(u_1)}{2u_1}$	$u_2 = \sqrt[3]{u}$	$f_2 = \frac{f(u_2)}{3u_2^2}$	$u_3 = \sqrt[4]{u}$	$f_3 = \frac{f(u_3)}{4u_3^3}$
0·457	0·676	0·4020	0·770	0·3803	0·822	0·3377
0·573	0·757	0·4346	0·831	0·3686	0·870	0·3108
0·309	0·556	0·3429	0·676	0·3964	0·746	0·3871
0·496	0·704	0·4140	0·792	0·3762	0·839	0·3278
0·284	0·533	0·3302	0·657	0·3992	0·730	0·3982
0·648	0·805	0·4511	0·865	0·3615	0·897	0·2967
0·770	0·877	0·4726	0·917	0·3509	0·937	0·2776
0·929	0·964	0·4929	0·976	0·3384	0·982	0·2576
0·059	0·243	0·1451	0·389	0·4151	0·493	0·6320
0·968	0·984	0·4970	0·989	0·3356	0·991	0·2533
0·877	0·936	0·4871	0·957	0·3424	0·968	0·2636
0·048	0·219	0·1294	0·363	0·4135	0·468	0·6667
0·520	0·721	0·4209	0·804	0·3738	0·849	0·3222
0·859	0·927	0·4848	0·951	0·3438	0·963	0·2658
0·016	0·126	0·0706	0·252	0·4000	0·356	0·8709
0·057	0·239	0·1424	0·385	0·4149	0·489	0·6377
0·096	0·309	0·1898	0·458	0·4162	0·557	0·5540
0·513	0·716	0·4189	0·801	0·3745	0·846	0·3238
0·233	0·483	0·3010	0·615	0·4049	0·695	0·4248
0·350	0·592	0·3618	0·705	0·3919	0·769	0·3710
0·459	0·677	0·4027	0·771	0·3800	0·823	0·3372
0·487	0·698	0·4114	0·787	0·3771	0·835	0·3300
0·279	0·582	0·3275	0·653	0·3998	0·727	0·4005
0·150	0·387	0·2411	0·513	0·4131	0·622	0·4869
0·153	0·391	0·2436	0·535	0·4129	0·625	0·4840
Mean	0·3446		0·3832		0·4087	
Variance	0·016 580		0·000 701		0·024 291	
s	0·1288		0·0265		0·1559	
$s/\sqrt{n}$	0·0258		0·0053		0·0312	

14.4 Antithetic variates

This method depends on the well known statistical result that

$$\text{Variance}\,\tfrac{1}{2}(X+Y)=\tfrac{1}{4}\,\text{variance}\,X+\tfrac{1}{4}\,\text{variance}\,Y+\tfrac{1}{2}\,\text{covariance}\,(X,Y) \quad (14.15)$$

where
$$\text{covariance}\,(X,Y)=\frac{1}{n}\sum_{i=1}^{n}(X_i-\bar{X})(Y_i-\bar{Y}).$$

We choose X and Y to be estimates for I which are negatively correlated (i.e., the covariance is negative). It then follows that $\frac{1}{2}(X + Y)$ is also an estimate for I and, from (14.15), we see that $\frac{1}{2}(X + Y)$ has a smaller variance than either X or Y.

If the function $f(x)$ is monotonic in (0, 1) (Fig. 14.2 (*a*)) then we can choose X and Y to be

$$X = \frac{1}{n}\sum_{1}^{n} f(u_i), \qquad Y = \frac{1}{n}\sum_{1}^{n} f(1 - u_i)$$

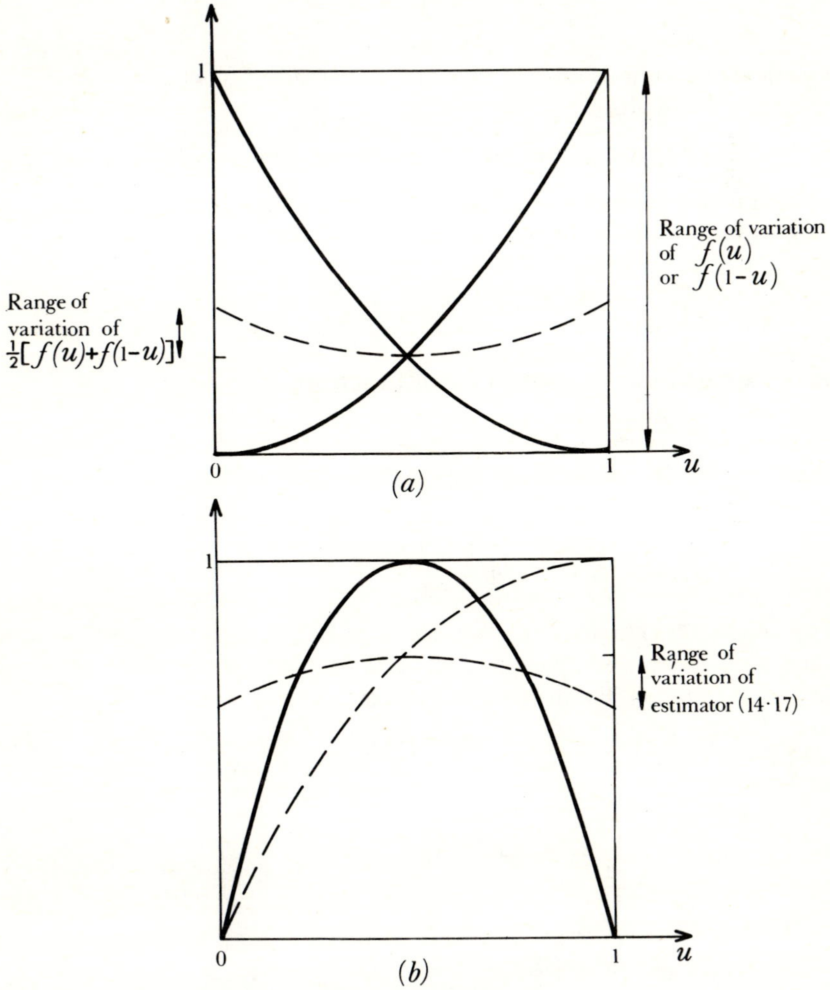

Fig. 14.2 Variation with antithetic variates
(*a*) monotonic function
(*b*) function with a maximum.

where u_i are numbers taken at random in (0, 1). $f(u_i)$ and $f(1-u_i)$ are known as antithetic variates since one is increasing when the other is decreasing. We then have

$$I_{\text{antithetic}} = \frac{1}{2n} \sum_{i=1}^{n} [f(u_i) + f(1-u_i)]. \tag{14.16}$$

If the function $f(x)$ possesses a maximum at the middle of the range (Fig. 14.2 (*b*)) then the above method can be applied to the two parts $(0, \frac{1}{2})$, $(\frac{1}{2}, 1)$. We have

$$I_{\substack{\text{antithetic}\\ \text{(stratified)}}} = \frac{1}{4n} \sum_{i=1}^{n} [f(\tfrac{1}{2}u_i) + f(\tfrac{1}{2} - \tfrac{1}{2}u_i) + f(\tfrac{1}{2} + \tfrac{1}{2}u_i) + f(1 - \tfrac{1}{2}u_i)]. \tag{14.17}$$

The method of antithetic variates applied to the standard function (14.2) with 25 samples gave the results

$$I_{\text{antithetic}} = 0{\cdot}3810, \qquad \text{standard error} = 0{\cdot}0127.$$

The power of this method in relation to hit-or-miss is thus 28·6. By increasing the number of strata a further increase in efficiency can be obtained.

14.5 Application of orthonormal functions

Variance reductions can also be brought about by a technique due to Ermakov and Zolotukhin [1] which uses orthonormal functions. We write

$$I = \int_0^1 f(x)\,dx = \int_0^1 [f(x) - f^*(x)]\,dx + \int_0^1 f^*(x)\,dx \tag{14.18}$$

and we choose $f^*(x)$ to make the first part of (14.18) integrable analytically while $\int_0^1 f^*(x)\,dx$ is determined by importance sampling from $\{\phi(x)\}^2$ where $\phi(x)$ is an orthonormal function. For instance, if we take

$$f^*(x) = f(x) - (1-x)f(0) - xf(1), \qquad \phi(x) = \sqrt{[6x(1-x)]} \tag{14.19}$$

then

$$\left.\begin{aligned} \int_0^1 [f(x) - f^*(x)]\,dx &= \tfrac{1}{2}[f(0) + f(1)] \\ \int_0^1 f^*(x)\,dx &\doteqdot \frac{1}{n} \sum_{i=1}^{n} \frac{f^*(v_i)}{6v_i(1-v_i)} \end{aligned}\right] \tag{14.20}$$

where each v_i is selected by taking any three numbers u_1, u_2, u_3 at random in (0, 1), ordering them so that $u_1 \leqslant u_2 \leqslant u_3$ and choosing v_i to be u_2 (see exercise 6).

We can go a step further and take

$$\left.\begin{aligned} f^*(x) &= f(x) - (1-x)(1-2x)f(0) + x(1-2x)f(1) - 4x(1-x)f(\tfrac{1}{2}) \\ \phi(x) &= (1-2x)\surd[30x(1-x)] \end{aligned}\right] \quad (14.21)$$

and we have

$$\left.\begin{aligned} \int_0^1 [f(x) - f^*(x)]\,\mathrm{d}x &= \tfrac{1}{6}[f(0) + 4f(\tfrac{1}{2}) + f(1)] \\ \int_0^1 f^*(x)\,\mathrm{d}x &\doteqdot \frac{1}{n}\sum_{i=1}^{n} \frac{f^*(w_i)}{30w_i(1-w_i)(1-2w_i)^2} \end{aligned}\right] \quad (14.22)$$

where w_i is selected by taking five numbers $u_1, \ldots, u_5$ at random in the interval (0, 1) and arranging them so that

$$|u_1 - \tfrac{1}{2}| \leqslant |u_2 - \tfrac{1}{2}| \leqslant \ldots \leqslant |u_5 - \tfrac{1}{2}|$$

and then choosing w_i to be u_4 with probability $\frac{3}{4}$ or u_3 with probability $\frac{1}{4}$.

Finally, we mention that a further variance reduction can be achieved by combining the above techniques with that of antithetic variates. Thus with $f^*(x)$ and $\phi(x)$ as in (14.19) we have

$$\left.\begin{aligned} \int_0^1 [f(x) - f^*(x)]\,\mathrm{d}x &= \tfrac{1}{2}[f(0) + f(1)] \\ \int_0^1 f^*(x)\,\mathrm{d}x &\doteqdot \frac{1}{n}\sum_{i=1}^{n} \frac{f(v_i) + f^*(1-v_i)}{12v_i(1-v_i)} \end{aligned}\right] \quad (14.23)$$

It was found applying the method (14.23) to the standard integral, that the power of this method was $O(10^5)$ times that of the hit-or-miss method.

Hammersley and Handscomb [2] should be consulted for applications of Monte Carlo methods to other deterministic problems, and for further details about the methods we have given for integration. It is worth stressing that Monte Carlo methods should be used sparingly and that deterministic methods (if they exist) are always to be preferred.

14.6 Appendix. Generation and testing of psuedo-random numbers

In a large scale Monte Carlo calculation as many as 10^6 random numbers may be required. In practice, no numbers are random, and the best that we can do is to generate a series of numbers which satisfy certain statistical tests. The numbers are then called pseudo-random numbers (see Taussky and Todd [3]).

Most random number generators are congruential difference equations. The most widely used are probably multiplicative congruential generators of the form

$$u_{i+1} \equiv \alpha u_i + \beta \qquad (\text{mod } m) \tag{14.24}$$

where α, β, m, and u_0 are suitably chosen. One suggestion due to Greenberger [4] is to take $\beta = 0$, $\alpha = 2^{\frac{1}{2}n+1}$, $m = 2^n$ which has advantages on a binary machine as no actual multiplications are required to evaluate the u_i in the sequence. Some other generators are given in the exercises including an additive congruential generator. It should be noted that the numbers obtained by the above generators have to be normalized to give a number in the interval (0, 1).

The statistical tests which are usually performed on the generated numbers can be found in [5].

REFERENCES

[1] Ermakov, S. M. and Zolotukhin, V. G., 'Polynomial Approximations and the Monte Carlo Method', published in translation, *Theor. Prob. and Appl.*, **5** (1960) 428–31.

[2] Hammersley, J. M. and Handscomb, D. C., *Monte Carlo Methods*, Methuen, London (1964).

[3] Taussky, O. and Todd, J., 'The Generation and Testing of Pseudo-random Numbers', *Symposium on Monte Carlo Methods*, J. Wiley & Sons, New York (1956).

[4] Greenberger, M., 'Notes on a new Pseudo-random Number Generator', *J.A.C.M.*, 8 (1965) 177–79.

[5] Miller, J. C. P. and Prentice, M. J., 'Additive Congruential Pseudo-random Number Generators', *Computer J.*, **11** (1968) 341–46.

Exercises

1. Compare the effect of the two different antithetic variate transforms given in the text by evaluating

$$I = \int_0^1 \sin \pi x \, dx$$

using the same 25 samples in each case.

2. If formula (14.36) is applied to the function

$$f(x) = \sum_0^n a_i x^i \qquad (n \leqslant 5),$$

prove that the expected variance of the estimate is zero.

3. One method for the generation of random numbers is the multiplicative congruence method. One example (Downham and Roberts, *Computer J.* (1967)) is provided by the formulae

$$x_{i+1} \equiv 8192x_i \pmod{67\,099\,547},$$
$$z_i = x_i/67\,099\,547.$$

Starting with $x_0 = 400$ generate 10 000 numbers z_i $(0 \leqslant z_i < 1)$. Apply some simple tests to indicate the pseudo-random behaviour of the numbers generated.

4. An alternative method for generation of random numbers is the additive congruence method. An example is provided by the formulae

$$x_{i+3} \equiv x_{i+1} + x_i \pmod{2003},$$
$$z_{4i} = x_{4i}/2003. \qquad \text{(see [5])}$$

Starting with $x_0 = 3$, $x_1 = 0$, $x_2 = 2$ generate 10 000 numbers z_{4i} $(0 \leqslant z_{4i} < 1)$ and apply the same tests used in exercise 3.

5. Given a four digit integer, square it and then extract the middle 4 digits to obtain a new four digit integer. Repeat the above process to obtain a sequence of, say, 100 numbers. Decide whether this is a good method for generating pseudo-random numbers.

6. If u_1, u_2, u_3 are numbers selected at random in the interval (0, 1) and ordered such that $u_1 \leqslant u_2 \leqslant u_3$ show that u_2 has probability density function $6x(1-x)$.

Miscellaneous exercises

1. Let $x_1, x_2, \ldots, x_5$ be the roots of a quintic polynomial equation and suppose the relative error in x_i is ϵ_i. What is the maximum relative error in

$$P = \prod_{i<j}^{5} (x_i - x_j)?$$

Calculate the (real) roots $x_1, \ldots, x_5$ of the polynomial equation

$$x^5 + 2x^4 - 5x^3 - 2x^2 + 4x - 1 = 0.$$

Evaluate
$$P = \prod_{i<j}^{5} (x_i - x_j),$$

and use the above analysis to obtain the maximum relative error in your value of P.

2. Show that the equation

$$z^m = e^{0\cdot3(z-1)},$$

has exactly m roots. Let $\eta^m = 1$ and w_i be a mth root of unity. Then the roots of the above equation can be found by solving the equation

$$z = w_i e^{0\cdot3(z-1)/m}.$$

Find the roots by a numerical method when $m = 3$. Show that the roots can also be obtained by considering

$$z = \eta e^{0\cdot3(z-1)/m}$$

and obtaining an expansion for z in terms of η.

3. In a particular section of a coal seam the depths of the seam at the points (x, y) of a rectangular grid are given in the table below

x \ y	0	1	2	3	4	5
0	252	184	172	216	316	472
1	222	134	102	126	206	342
2	228	120	68	62	108	192
3	270	142	70	54	100	184
4	348	200	108	72	92	168
5	462	294	182	126	126	182

Estimate the minimum depth of the seam and the volume of coal in the region investigated. Find also the depth of the seam at $x = 2\cdot5$, $y = 3\cdot5$.

4. Evaluate the integral (Riley, N., *Proc. Roy. Soc.*, *A* **(260)**, (1961))

$$\int_0^\infty \left\{ e^y \int_y^\infty \frac{e^{-t}}{t}\,dt + e^{-y}\left(\ln y + \gamma + \int_0^y \frac{e^t - 1}{t}\,dt\right)\right\} dy$$

where γ is Euler's constant.

5. Evaluate the functions ϕ_1, ϕ_2 defined by

$$\pi\phi_1 = x\int_0^\infty \ln\left[\frac{\sinh \pi u}{\sinh \pi u + k\pi u}\right] \frac{x^2+y^2+u^2}{(x^2+y^2+u^2)^2 - 4y^2u^2}\,du$$

$$\pi\phi_2 = y\int_0^\infty \ln\left[\frac{\sinh \pi u}{\sinh \pi u + k\pi u}\right] \frac{u^2-x^2-y^2}{(x^2+y^2+u^2)^2 - 4y^2u^2}\,du$$

for $x = 1(1)10, y = 0, k = \pm(2/\pi)$. If $z = x + iy$ evaluate

$$(z:k)! = (\psi_1 + i\psi_2)\exp(\phi_1 + i\phi_2)$$

where $\psi_1 = \mathcal{R}[z!]$, $\psi_2 = \mathcal{I}[z!]$. Repeat the calculations for z being the first 2 roots of

$$\sin \pi z + 2z = 0.$$

(See Doran, H. E. and Buchwald, N. T., *J. Inst. Math. & its Applications*, **5** (1969).)

6. Show that the value of α for which

$$\int_0^{\frac{3}{2}\pi} u^{-\alpha}\cos u\,du = 0$$

is approximately 0·308 44. (See Church, R. F., *Math. Comp.*, **19** (1965), p.501 and Luke, Y. L., *et al.*, loc. cit.)

7. Obtain a table of abscissae and weights for a Gaussian quadrature formula to integrate

$$\int_0^1 -\ln x\, f(x)\,dx.$$

8. The Riemann zeta function is defined by

$$\zeta(s) = \sum_{n=1}^{\infty} \frac{1}{n^s}.$$

If $s = \frac{1}{2} + i\tau$ ($\tau > 0$) show that $\zeta(s)$ is zero for some value of τ between $\tau = 14{\cdot}1$ and $\tau = 14{\cdot}2$ and estimate this value. Show that the next zero on the line $\mathcal{R}s = \frac{1}{2}$ is at $\tau = 21{\cdot}022\,039\,63$.

9. Consider

$$\int_{x_0}^{x_n} f(x)\,dx,$$

where $f(x)$ is a continuous function which is monotonic (increasing or decreasing) in the intervals $[x_0, x_1], [x_1, x_2], \ldots, [x_{n-1}, x_n]$. Let m_r be the minimum value of $f(x)$ in $[x_{r-1}, x_r]$ and M_r the maximum value. Show

$$\sum_{r=1}^{n} m_r(x_r - x_{r-1}) \leqslant \int_{x_0}^{x_1} f(x)\,dx \leqslant \sum_{r=1}^{n} M_r(x_r - x_{r-1}).$$

Use this result to find an interval estimate for

$$\int_1^2 \frac{dx}{x^2}.$$

10. Denote by $g(X)$ the interval estimates for the function g in a range of values $x \in X$. Show, from Taylor's theorem that an interval estimate of $f(x_0 + t)$ is given by

$$f(x_0 + t) = f(x_0) + tf'(x_0) + \ldots + \frac{t^{k-1}}{(k-1)!} f^{(k-1)}(x_0) + \frac{t^k}{k!} f^{(k)}(X),$$

where $X = [x_0, x_0 + t]$. Hence obtain a formula for evaluating

$$\int_{x_0}^{x_0+h} f(x)\,dx$$

and use this result with $k = 2$, $h = 0{\cdot}1$ to evaluate

$$\int_1^2 dx/x^2.$$

11. Obtain an interval version of the Gauss–Legendre k point quadrature formula.

12. Suggest a method for evaluating the Fredholm integral equation of the second kind

$$\int_a^b k(x, y)f(y)\,dy = \lambda f(x) + g(x).$$

Hence evaluate $f(x)$ approximately when $f(x)$ satisfies the equation

$$\int_0^1 (x + y)f(y)\,dy = f(x) - x.$$

13. Given
$$J_0(x) = \sum_{n=0}^{\infty} \frac{(-1)^n}{(n!)^2} \left(\frac{x}{2}\right)^{2n}$$

and the property that $J_0(x)$ is oscillatory, estimate

$$\int_0^\infty J_0(x)\,dx.$$

If $J_1(x)$ is defined as in chapter 7, exercise 21, also evaluate

$$\int_0^{\infty} (J_1(x)/x)\,dx \qquad \text{and} \qquad \int_0^{\infty} (J_1(x)/x)^2\,dx$$

14. Find the inverse of the matrix $I + S$ where I is the unit matrix and S is the partitioned matrix

$$S = \begin{bmatrix} 0 & -\mathbf{x}' \\ \mathbf{x} & 0 \end{bmatrix} \qquad \mathbf{x} = (x_1, \ldots, x_{n-1})'.$$

15. Given an interval square matrix, A, obtain a method for finding an interval estimate of the inverse. Also give a method for determining interval estimates of the solution of a set of equations $A\mathbf{x} = \mathbf{b}$ where $\mathbf{b}$ is an interval vector.

16. Find the inverse of the 4 x 4 matrix

$$A = \begin{bmatrix} 2 & 1 & 0 & 0 \\ 1 & 2 & 1 & 0 \\ 0 & 1 & 2 & 1 \\ 0 & 0 & 1 & 2 \end{bmatrix}.$$

If $B = (b_{ij})$, $b_{41} = 1$, $b_{ij} = 0$ elsewhere and X, Y, Z are the matrices

$$X = \begin{bmatrix} A & B & 0 & 0 \\ B & A & B & 0 \\ 0 & B & A & B \\ 0 & 0 & B & A \end{bmatrix}, \qquad Y = \begin{bmatrix} A & 0 & 0 & 0 \\ 0 & A & 0 & 0 \\ 0 & 0 & A & 0 \\ 0 & 0 & 0 & A \end{bmatrix}, \qquad Z = X - Y$$

show that

$$X^{-1} = Y^{-1}(I + Y^{-1}Z)^{-1}$$

and the matrix $Y^{-1}Z$ has eigenvalues less than one. Hence, by expansion of $(I + Y^{-1}Z)^{-1}$, evaluate X^{-1} and estimate the error if the expansion is terminated at $(-Y^{-1}Z)^n$.

17. Let $p(x)$ be a polynomial of degree n such that $p(x) = 0$ does not have multiple roots. Let $p_1(x) = p'(x)$ and let q_1 be the quotient when $p(x)$ is divided by $p_1(x)$. We form the sequence

$$p(x) = q_1 p_1(x) - p_2(x)$$
$$p_1(x) = q_2 p_2(x) - p_3(x)$$
$$\cdot \quad \cdot \quad \cdot \quad \cdot$$
$$p_{m-2}(x) = q_{m-1} p_{m-1}(x) - p_m \qquad (p_m \text{ a constant}).$$

Show that the sequence $p(x), p_1(x), \ldots, p_m$ is a Stürm sequence. What happens if $p(x)$ has a multiple root?

18. Let $x_0 = a, x_n = b$. A cubic spline function $u_r(x)$ is defined in $[x_r, x_{r+1}]$ by the formula

$$u_r(x) = u_{r-1}(x) + d_r(x - x_r)^3 \qquad (r \geqslant 1)$$
$$u_0(x) = a + b(x - x_0) + c(x - x_0)^2 + d_0(x - x_0)^3.$$

Show that the first and second derivatives of $u_r(x)$ are continuous across the nodes $x_1, \ldots, x_r$. If $x_0 = 0, x_1 = 0{\cdot}5, x_2 = 1{\cdot}0$ find a cubic spline approximation for e^{-x} in $[0, 1]$.

19. If u satisfies the differential equation

$$u'' + u + 1 = 0, \qquad u = 0 \text{ at } x = 0 \text{ and } 1,$$

estimate u at $x = 0{\cdot}5$ by representing the function as a cubic spline function. Compare your answer with the exact answer at $x = 0{\cdot}5$. (See Bickley, W. G., *The Computer Journal*, **11** (1968), 206–8.)

20. Let Y_1 be the exact solution of the differential equation

$$y' = f(x, y), \qquad y(x_0) = y_0,$$

at $x = x_1 = x_0 + h$ and let y_1^+, y_1^- be approximations to Y_1 given by the formulae

$$k_1 = hf(x_0, y_0),$$
$$k_2^+ = k_1, \qquad k_2^- = hf(x_1, y_0 + k_1),$$
$$y_1^+ = y_0 + k_2^+, \qquad y_1^- = y_0 + k_2^-.$$

Show that

$$Y_1 - y_1^+ = \tfrac{1}{2}h^2(f_x + ff_y)_0 + O(h^3),$$
$$Y_1 - y_1^- = -\tfrac{1}{2}h^2(f_x + ff_y)_0 + O(h^3).$$

Develop similar two-sided formulae of third order and give uses for this type of method. (Gorbunov, A. D. and Shakhov, Yu. A., *USSR Journal of Computational Math. and Math. Phys.* (1963) 316–35, published in translation by Pergamon Press, Oxford.)

21. Show that the solution of the differential equation

$$y' = -10^8(y - \sin x) + \cos x, \quad y(0) = 1$$

is $y = \sin x + \exp(-10^8 x)$. This is an example of a 'stiff' differential equation, the term $\exp(-10^8 x)$ decaying very rapidly, so that for $x > 10^{-7}, y \approx \sin x$. Suggest a method for the numerical integration of such an equation.

22. Obtain an interval adaptation of the Taylor series method for the solution of differential equations.

23. Assume that the remainder when

$$y = x + a_2 x^2 + a_3 x^3 + a_4 x^4 + a_5 x^5,$$

is substituted in

$$r(x) = xy'' + xy' + y,$$

has the form

$$r(x) = k_5 T_5(x).$$

Estimate a_2, a_3, a_4, a_5 to find an approximation to the solution of the differential equation

$$xy'' + xy' + y = 0, \qquad y(0) = 0, \qquad y'(0) = 1 \text{ in } (0, 1).$$

What is the maximum error in the approximation? What improvement could be obtained if $r(x)$ is taken to have the form

$$r(x) = k_5 T_5(x) + k_6 T_6(x)?$$

24. Show that the formula

$$y_{n+1} - y_n = hf_n + \frac{h^2}{2} f'_n + \frac{h^3}{6} \frac{4(f''_n)^2}{4f''_n - hf'''_n}$$

for integrating $y' = f(x, y)$ has a truncation error of

$$\tfrac{1}{24} h^5 [-\tfrac{1}{5} y^{(5)} + \tfrac{1}{4} (y^{(4)})^2 / y^{(2)}].$$

(Lambert, J. D. and Shaw, B., *Math. Comp.*, **19** (1965), 456–62.)

25. Over a square region the partial differential equation

$$\frac{\partial^2 u}{\partial x^2} + \frac{\partial^2 u}{\partial y^2} = \lambda u$$

holds. By dividing the region up into a 6 x 6 mesh, obtain a finite difference approximation to the partial differential equation and solve the resulting algebraic eigenvalue problem.

26. An article has a weight x which is normally distributed with mean ω and standard deviation 1. The expected profit, E, in pounds sterling of selling 10 000 articles is given by the formula

$$E = 7000 - (500\omega + 0{\cdot}05 N\omega) - 0{\cdot}3N,$$

where N is the expected number of articles of weight less than 8. For what value of ω is E a maximum and what is the maximum value of E?

27. Given that x_0, y_0 is an approximation to the solution of the simultaneous non-linear equations

$$f(x, y) = 0, \qquad g(x, y) = 0,$$

and $x, x_0 \in X$, $y, y_0 \in Y$ where X and Y are intervals containing the solution, show that the formulae

$$-f(x,y) = (x-x_0)\frac{\partial f}{\partial x}(X,Y) + (y-y_0)\frac{\partial f}{\partial y}(X,Y)$$

$$-g(x,y) = (x-x_0)\frac{\partial g}{\partial x}(X,Y) + (y-y_0)\frac{\partial g}{\partial y}(X,Y)$$

provide an interval estimate for the exact solution.

28. A periodic function $f(x)$, of period 2π, is evaluated at $x = 0^{\circ}, 30^{\circ}, \ldots, 330^{\circ}$. Its values in the order above are respectively 2·0000, 4·5981, 6·9641, 7·0000, 3·9641, −0·5981, −4·0000, −4·5981, −2·9641, −1·0000, 0·0359, 0·5981. Given that $f(x)$ has the form

$$f(x) = \tfrac{1}{2}a_0 + a_1 \cos x + a_2 \cos 2x + b_1 \sin x + b_2 \sin 2x$$

calculate the values of a_0, a_1, a_2, b_1, and b_2.

29. Show that the coefficients in the Fourier–Chebyshev series (p. 75) can be evaluated from the formula

$$a_n = \frac{1}{\pi i}\int_C \frac{f(z)\,dz}{\sqrt{(z^2-1)}[z+\sqrt{(z^2-1)}]^n}, \; n \geqslant 0$$

where C is any contour enclosing the real axis between −1 and +1 on (and within) which $f(z)$ is regular. Hence find the coefficients in the Fourier–Chebyshev series for

$$e^{ax}\,(a>0), \qquad J_1(kx)/x, \qquad \sin(kx)/x, \qquad \operatorname{erf}(kx).$$

Hint: Use Laplace transforms – see Elliott, D. and Szekeres, G., *Math. Comp.*, **19** (1965) 25–32.

Answers to selected Exercises

Exercises on chapter 1

1. Let I denote the integer part of the number and F the fractional part of the number in the scale N and let I' and F' have the corresponding meaning in the scale N'. Then the digits comprising I and F satisfy

$$N'I_{k+1} = I_k + a'_k, \qquad I_0 = I;$$
$$F_{k+1} = N'F_k + a'_{-k}, \qquad F_1 = F.$$

2. $136{\cdot}7 = 10\,001\,000{\cdot}101\,100\ldots = 210{\cdot}54\ldots.$

4. From exercise 2, $136{\cdot}7 = 0{\cdot}100\,010\,001\,1 \times 2^8$. The representational error is less than 2^{-2} in modulus.

6. Use
$$\cosh x - \sinh x = 1/(\cosh x + \sinh x).$$

7. (a) 1·606, (b) 1·606. (a) differs from the true value by an amount less than or equal to $4{\cdot}3435 \times 10^{-3}$ while (b) differs from the true amount by $\leqslant 3{\cdot}395 \times 10^{-4}$. Thus (b) is more accurate.

8. Show that $\displaystyle\sum_{2^N}^{2^{N+1}-1} \frac{1}{r} > \frac{1}{2}$. By the comparison test $\displaystyle\sum_1^\infty \frac{1}{r}$ diverges. If N_0 is the integer for which $1/N_0$ makes no additional contribution to the computed sum, then the sum is approximately $\ln N_0$.

9. $[-0{\cdot}2880, 0{\cdot}2160]$.

12. $x_2 \in [0{\cdot}970\,99, 1{\cdot}031\,26]$, $x_1 \in [0{\cdot}9527, 1{\cdot}0371]$. From the second formula $x_1 \in [0{\cdot}949\,96, 1{\cdot}041\,86]$.

13. $[\frac{10}{11}, \frac{4}{3}]$. The exact range for y is $\frac{12}{11} \leqslant y \leqslant \frac{10}{9}$. The reason for the differing estimates is failure of the distributive law for intervals (section 1.8).

15. For $N = 2$,

$$fl\sum_1^n x_r = \sum_1^n x_r(1+\epsilon_r), \qquad (1-\tfrac{3}{2}2^{-t})^{n+1-r} \leqslant 1+\epsilon_r \leqslant (1+\tfrac{3}{2}2^{-t})^{n+1-r}$$

$$fl\prod_1^n x_r = \prod_1^n x_r(1+E), \qquad (1-2^{-t})^{n-1} \leqslant 1+E \leqslant (1+2^{-t})^{n-1}$$

$$fl\sum_1^n x_r y_r = \sum_1^n x_r y_r(1+\eta_r), \qquad (1-\tfrac{3}{2}2^{-t})^{n-r+2} \leqslant 1+\eta_r \leqslant (1+\tfrac{3}{2}2^{-t})^{n-r+2}.$$

18. If α and β are the roots, $\alpha < \beta$, then we have convergence to β for all x_1. The iterative formula $x_{n+1} = 1/(18 - x_n)$ can be used to calculate α. Iteration formulae formed in this way sometimes fail to converge, e.g., $x_{n+1} = x_n^3 - 1$ for finding the real root of $x^3 - x - 1 = 0$. However, $x_{n+1} = (x_n + 1)^{\frac{1}{3}}$ converges to the correct answer. The criterion for the convergence of $x_{n+1} = f(x_n)$ is $|f'(x_n)| < 1$.

19. $J_0(1) = 146\,181\,170\tau$ where $\tau = 1/191\,037\,130$, i.e., $J_0(1) = 0{\cdot}765\,197\,69$. 'Exact' results can be found from

$$J_n(1) = \sum_{r=0}^{\infty} \frac{(-1)^r(\frac{1}{2})^{n+2r}}{r!(n+r)!}.$$

22. Estimate the value of the worst ϵ such that $x_n = a^{\frac{1}{2}}(1+\epsilon)$, and $fl\frac{1}{2}(x_n + a/x_n)$ does not improve the estimate for $a^{\frac{1}{2}}$.

23. Use the result of exercise 22.

25. Taylor's theorem gives

$$e^x = 1 + x + \frac{x^2}{2!} + \ldots + \frac{x^{k-1}}{(k-1)!} + \frac{x^k}{k!}e^{\theta} \qquad (-\infty \leqslant \theta \leqslant 0).$$

Hence result.

Exercises on chapter 2

2. See Wilkinson [1].
5. $g(0) = \ln 2$.
6. If x_n is near the root ξ which has multiplicity r then approximately

$$f(x_n) = \frac{(x_n - \xi)^r}{r!} f^{(r)}(\xi) + O(x_n - \xi)^{r+1},$$

$$f'(x_n) = \frac{(x_n - \xi)^{r-1}}{(r-1)!} f^{(r)}(\xi) + O(x_n - \xi)^r,$$

and

$$\frac{rf(x_n)}{f'(x_n)} = (x_n - \xi) + O(x_n - \xi)^2,$$

which leads to the result. The double root is at $x = 1$.

7. If ξ is the root then $\xi \in X_n$ and

$$\xi = x_n - f(x_n)/f'[\xi + \theta(x_n - \xi)], \qquad 0 < \theta < 1.$$

Since $\xi + \theta(x_n - \xi) \in X_n$ the stated result follows.

8. $X_0 = [0, 1]$, $X_1 = \frac{1}{2} - (e^{\frac{1}{2}} - \frac{3}{2})/(e^{[0,1]} - 3)$. Now use $e^{\frac{1}{2}} = 1 + \frac{1}{2} + \frac{1}{8} + \frac{1}{24}[1, 2]$ to find $X_1 = [\frac{7}{12}, 1]$, etc.

9. Use $f(x) = 0 = f(a) + (x-a)f'(a) + \frac{1}{2}(x-a)^2 f''[a + \theta(x-a)]$ $0 < \theta < 1$.

12. Roots are 6·326, 0·4573, 0·012 58, 0·002 453. The deflated polynomial is $x^3 - 0{\cdot}472x^2 + 0{\cdot}009\,128x + 0{\cdot}014\,05$. The product of the roots is −0·014 05 whereas it should be 0·000 014 11. The roots of the deflated polynomial are approximately −0·1437 and 0·3079 ± 0·0533i.

21. Show $z_{k+1} - \xi = 0(z_k - \xi)^3$ where ξ is the root.

24. If there are no zeros inside $|z| = \rho$, then

$$\frac{1}{2\pi}\int_0^{2\pi} \ln |F(\rho e^{i\theta})|\, d\theta = \ln |F(0)|.$$

By increasing ρ we can find the first value of ρ for which this formula is inconsistent and hence we can determine the smallest root (s). Further increasing ρ determines the larger roots. The advantage of the method is the determination of the zeros in increasing order. A disadvantage is the difficulty in computing the integral.

Exercises on chapter 3

3. $f(5) = 38{\cdot}7$, $f(6) = 57{\cdot}6$.
6. $f(0{\cdot}35) = 3{\cdot}35$, $f(9{\cdot}4) = 151{\cdot}78$.
8. $f(4{\cdot}3) = 27{\cdot}85$, $f(5{\cdot}8) = 53{\cdot}5$.
9. $\sin^{-1}(0{\cdot}35) = 20{\cdot}487$.
10. 0·456 55.
12. $|\delta_0^4 + \delta_1^4|$ must be less than 625.
16. Gauss's backward difference formula is

$$f_p = f_1 + (p-1)\delta_{\frac{1}{2}} + \tfrac{1}{2}p(p-1)\delta_1^2 + \frac{p(p-1)(p-2)}{3!}\delta_{\frac{1}{2}}^3 + \frac{p(p-1)(p-2)(p+1)}{4!}\delta_1^4 .$$

18. (*a*) 0·002 66, (*b*) 0·000 08. Linear extrapolation is not better in this case. One should not assume that questions are always correctly posed. A set of data for which linear extrapolation does give a better result is

x	0·2	0·4	0·6	0·8	1·0
$f(x)$	0·555 56	0·238 10	0·121 95	0·072 46	0·047 62

Linear extrapolation produces $f(0) = 1{\cdot}111\,12$ and the Lagrange polynomial gives 1·301 62. This data was generated from $f(x) = 1/(1+20x^2)$ which is a function exhibiting the Runge phenomenon – see Handscomb (ed.), *Methods of Numerical Approximation*, Pergamon Press (1966).

24. 3·1278.
25. $\mathrm{Si}(0) \doteqdot 0{\cdot}000\,07$.
26. $\mathrm{Si}(0{\cdot}42) = 0{\cdot}4159$, $\mathrm{Si}'(0{\cdot}42) = 0{\cdot}9709$.

Exercises on chapter 5

1. 344 55 ft.
2. −0·100 34.
3. (*a*) 1·0472, (*b*) 1·6857.
4. $(-1)^n n!/(n+1)^{n+1}$; 0·7834.

5. $$\int_0^1 f_p \, dp = (\mu - \tfrac{1}{12}\mu\delta^2 - \tfrac{11}{720}\mu\delta^4 - \ldots) f_{\frac{1}{2}}$$

7. −8/3.
9. 0·635 26.
10. 0·2943.
13. $-\frac{1}{45} f^{iv}(\xi)$, $x_0 < \xi < x_4$; $\frac{95}{144} f^{iv}(\xi)$, $x_1 < \xi < x_3$.

16. $$\frac{2^{4n+1}[(2n)!]^3}{(4n+1)[(4n)!]^2} f^{(2n)}(\eta), \qquad 0 < \eta < 1.$$

17. x_k are the roots of $x^{-\frac{1}{2}} P_{2n+1}(x^{\frac{1}{2}}) = 0$; $\alpha_k = 2\alpha'_k x_k$ where α'_k is the weight corresponding to $x_k^{\frac{1}{2}}$ in the Gauss–Legendre formula of order $2n+1$;

$$R = \frac{2^{4n+3}[(2n+1)!]^4}{(4n+3)[(4n+2)!]^2(2n)!} f^{(2n)}(\eta), \qquad 0 < \eta < 1.$$

18. $x_k = \cos^2 [(2k-1)\pi/(4n+2)]$, $\alpha_k = 2\pi x_k/(2n+1)$,

$$R = \pi f^{(2n)}(\eta)/2^{4n+1}(2n)!, \qquad 0 < \eta < 1.$$

19. $\frac{1}{2}\pi - 1 = 0{\cdot}570\,796$. Romberg integration can be used. For the case of Gauss–Legendre quadrature with $n = 1$, we have the familiar application of Romberg integration in conjunction with the mid-point rule.
20. The integral provides an alternative definition for the Bessel function $J_0(x)$. You can check your answers by evaluating $J_0(x)$, $0(\frac{1}{4}\pi)\frac{7}{4}\pi$. See chapter 7, exercises 21, 22.
22. $I = I_M(N) + Ah^{\frac{3}{2}} + Bh^{\frac{5}{2}} + Ch^{\frac{7}{2}} + \ldots$. $I_M(N)$ can be replaced by $I_T(N)$ or $I_S(N)$ in this formula.
23. 0·595 80, −0·239 81.
24. −0·4401.
25. The exact answer is $\frac{1}{2}\pi$.
26. The exact value of the first integral is $-\frac{1}{2}\pi \ln 2$. The value of the second integral is −0·3292.
27. $\lim_{n\to\infty} f(1/n) = \frac{1}{2}\pi$. An alternative method for evaluating $f(0) = \int_0^\infty \frac{\sin x}{x} \, dx$ is to apply Euler's method for the summation of alternating series (section 6.2, see also chapter 7, exercise 23).

28. Show that $I(\lambda)$ satisfies the differential equation

$$I - I' = \tfrac{1}{2}\sqrt{(\pi/\lambda)}, \qquad I(0) = \tfrac{1}{2}\pi.$$

Hence deduce that

$$I(\beta^2) = \sqrt{\pi}\, e^{\beta^2} \int_\beta^\infty e^{-\beta^2}\, d\beta, \qquad (\lambda = \beta^2),$$

and use the methods of sections 7.1 or 7.4 to evaluate $I(\beta^2)$. Alternatively, use Gauss–Hermite quadrature.

29. $A = 4/3$, $B = C = D = E = 2/3$.

30. $1 + \dfrac{1}{2{\cdot}2!} + \dfrac{1}{3{\cdot}3!} + \dfrac{1}{4{\cdot}4!} + \ldots$.

31. 0·6748.

32. See section 7.6. The integral is the function $f(1)$ defined by equation (7.23).

Exercises on chapter 6

1. Consider the first 2^k terms of the series $\sum v_r$ and the first 2^k terms of $\sum(-1)^{r-1}u_r$. Show that the difference in magnitude between these two quantities is less than an arbitrary amount $\epsilon(k)$ which $\to 0$ as $k \to \infty$. Hint: You may use the fact that $\sum v_r$ is convergent implies $v_r < c/r^\delta$ $(\delta > 1)$. 1·6449, 1·2021, 2·6124.

4. $(1+x)^{-1} \sum_{r=0}^{\infty} \left(-\dfrac{x}{1+x}\Delta\right)^r u_0$, 0·781 65.

7. 1·4936.

8. 0·8225, 0·9243, 0·6049.

9. 2·6124, 2·7444, 0·1317.

15. Hint: Show that $\sum_{k=1}^{\infty} 1/(k+a)(k+b) = (\psi(b) - \psi(a))/(b-a)$.

where ψ denotes the digamma function (chapter 7, exercise 19). Evaluate these expressions to check your numerical results.

Exercises on chapter 7

1. $1 + \tfrac{1}{2}x + \tfrac{3}{8}x^2 + \ldots + \dfrac{(2n)!}{2^n(n!)^2}x^n + \ldots$, 1·010 152 5, 1·414 214.

2. $2\left(x+\frac{x^3}{3}+\frac{x^5}{5}+\ldots+\frac{x^{2k+1}}{2k+1}+\ldots\right)$,

$$\ln(n/m)=2\left[\left(\frac{n-m}{n+m}\right)+\frac{1}{3}\left(\frac{n-m}{n+m}\right)^3+\frac{1}{5}\left(\frac{n-m}{n+m}\right)^3+\ldots\right],\quad 1{\cdot}098\,612.$$

5. The values for $t = 0$, $0{\cdot}5$ and $1{\cdot}0$ are respectively -1, $-0{\cdot}887\,922$, $-0{\cdot}796\,767$.

6. $e^x = e^{x_0}(1+h+\frac{1}{2}h^2+\frac{1}{6}h^3)$ $-0{\cdot}05 < h \leqslant 0{\cdot}05$, $25{\cdot}7903$. The correct value is given by

$$(e^{x_0}+\epsilon_1)(1+h+\tfrac{1}{2}h^2+\tfrac{1}{6}h^3+\epsilon_2)$$

where ϵ_1 is the representation error in e^{x_0} and ϵ_2 is the error in calculating e^h from the given formula.

7. $C(t) = t - \frac{t^5}{5\times 2!}+\frac{t^9}{9\times 4!}-\ldots+\frac{(-1)^n t^{4n+1}}{(4n+1)(2n)!}+\ldots$,

$$S(t)=\tfrac{1}{3}t^3-\frac{t^7}{7\times 3!}+\frac{t^{11}}{11\times 5!}-\ldots+\frac{(-1)^n t^{4n+3}}{(4n+3)(2n+1)!}+\ldots.$$

8.

t	1·0	1·1	1·2	1·3	1·4	1·5
$I(t)$	0·529 743	0·477 075	0·432 528	0·394 416	0·361 491	0·332 806

t	1·6	1·7	1·8	1·9	2·0
$I(t)$	0·307 627	0·285 380	0·265 609	0·247 945	0·232 088

9. For $n = 10$, $a_0 = 27\,720\tau$, $a_2 = -4756\tau$, $a_4 = 816\tau$, $a_6 = -140\tau$, $a_8 = 24\tau$, $a_{10} = -4\tau$, $\tau = 1/19\,601$; $a_{2r+1} = 0$.

10. $\sqrt{2}(\frac{1}{2}T_0+\lambda T_2+\lambda^2 T_4+\lambda^3 T_6+\ldots)$, $\lambda = -3+2\sqrt{2}$. Use exercise 8, chapter 4 to obtain the result for $\tan^{-1}x$.

15. $\int_t^\infty \cos u^2\,du \sim \frac{1}{2t}\left(-\sin t^2+\frac{\cos t^2}{2t^2}+\frac{3\sin t^2}{8t^4}-\frac{15\cos t^2}{8t^6}-\ldots\right)$,

$$\int_t^\infty \sin u^2\,du \sim \frac{1}{2t}\left(\cos t^2+\frac{\sin t^2}{2t^2}-\frac{3\cos t^2}{4t^4}-\frac{15\sin t^2}{8t^6}+\ldots\right).$$

16. $x-\frac{x^3}{3\cdot 3!}+\frac{x^5}{5\cdot 5!}-\frac{x^7}{7\cdot 7!}+\frac{x^9}{9\cdot 9!}-\ldots$.

$$\frac{\cos x}{x}+\frac{\sin x}{x^2}-\frac{2!\cos x}{x^3}-\frac{3!\sin x}{x^4}+\ldots.$$

17. Use the result

$$\int_Z^X (e^{-u}/u)\,du=\int_Z^X\frac{du}{u}-\int_Z^X\frac{1-e^{-u}}{u}\,du,$$

and find the limit as $X\to\infty$.

18. $\mathrm{ei}(5) \sim 0{\cdot}001\,147$; $\mathrm{ei}(5) = 0{\cdot}001\,148\,3$.

19. Note $\frac{1}{2}u(\coth\frac{1}{2}u-1) = u/(e^u-1)$.

20. $\psi(10) = 2{\cdot}351\,752\,589\,067$.
21. See the tables of Bessel functions in Abramowitz and Stegun (Bibliography).
22. 2·4048, 5·5201, 8·6537, 11·7915.
23. One method of evaluating $I(t)$ is to consider it as composed of the two integrals

$$\int_0^1 + \int_1^\infty .$$

The first may be integrated satisfactorily by Fox's method of extended Romberg integration (section 5.7) and the second by considering subdivisions $(1, \pi/t)$, $(\pi/t, 2\pi/t)$, $(2\pi/t, 3\pi/t)$, ... These integrals, call them $I_0, I_1, I_2, \ldots$, can be evaluated by Gaussian quadrature (compare evaluations by fourth and fifth order formulae). Next, apply the Euler transformation to the alternating series beginning with, say, I_7. Results obtained in this manner agree to six decimal places with those of exercise 8. (Cohen, A. M. Paper 36 Computational Physics Conference, H.M.S.O., (1969).)

Exercises on chapter 8

4. $\alpha = (A - BD^{-1}C)^{-1}$, $\quad \gamma = -D^{-1}C\alpha$,
$\delta = (D - CA^{-1}B)^{-1}$, $\quad \beta = -A^{-1}B\delta$.

$$\begin{bmatrix} \frac{3}{4} & -\frac{3}{8} & \frac{1}{4} & -\frac{1}{8} \\ -\frac{1}{2} & \frac{3}{4} & -\frac{1}{2} & \frac{1}{4} \\ \frac{1}{4} & -\frac{1}{8} & \frac{3}{4} & -\frac{3}{8} \\ -\frac{1}{2} & \frac{1}{4} & -\frac{1}{2} & \frac{3}{4} \end{bmatrix} \begin{bmatrix} -\frac{3}{5} & \frac{2}{5} & \frac{2}{5} & 1 & -\frac{2}{5} \\ \frac{2}{5} & -\frac{3}{5} & \frac{2}{5} & -2 & \frac{3}{5} \\ \frac{2}{5} & \frac{2}{5} & -\frac{3}{5} & 1 & -\frac{2}{5} \\ 0 & 0 & 0 & 1 & 0 \\ 0 & 0 & 0 & -2 & 1 \end{bmatrix}$$

17. $\begin{bmatrix} \frac{7}{2} & -\frac{3}{2} \\ -2 & 1 \end{bmatrix} \begin{bmatrix} 1 & 0 & 0 \\ 0 & 2 & 0 \\ 0 & 0 & 10 \end{bmatrix} \begin{bmatrix} 1 & 0 & 0 \\ -\frac{1}{2} & 1 & 0 \\ -\frac{1}{4} & -\frac{1}{2} & 1 \end{bmatrix}$

The column condition numbers are 55, 10, 3·5.
18. With row norms, $K = 83{\cdot}8$, rel $x \leqslant 0{\cdot}0144$. With column norms, $K = 110$, rel $x \leqslant 0{\cdot}0472$.
22. E can be computed very accurately, using double length arithmetic to evaluate the inner products. If the accuracy of the computed inverse is sufficiently high then we have approximately that

$$\|A^{-1} - X\|_\infty \leqslant \|E\|_\infty \, \|X\|_\infty$$

and this provides a good estimate of the accuracy of the computed inverse.
23. $\|E_{n+1}\| \leqslant \|E_n\|^2$. It follows that $\|E_0\| \leqslant 1$ is sufficient to ensure convergence.

24. $0{\cdot}04 - 0{\cdot}0004\epsilon$.

25. The error in each element of $A\mathbf{x}$ is just the error in computing an inner product. See section 1.10.

Exercises on chapter 9

1. The determinant is the Vandermonde determinant

$$\begin{vmatrix} 1 & 1 & 1 & \ldots & 1 \\ \lambda_1 & \lambda_2 & \lambda_3 & \ldots & \lambda_n \\ \lambda_1^2 & \lambda_2^2 & \lambda_3^2 & \ldots & \lambda_n^2 \\ \cdot & \cdot & \cdot & \cdot & \cdot \\ \lambda_1^{n-1} & \lambda_2^{n-1} & \lambda_3^{n-1} & \ldots & \lambda_n^{n-1} \end{vmatrix} = \prod_{i<j} (\lambda_i - \lambda_j).$$

3. Modify the argument of Theorem 9-2.

4. If B is non-singular then

$$\det(BA - \lambda I) = \det[B(AB - \lambda I)B^{-1}] = \det(AB - \lambda I),$$

which implies that the eigenvalues of AB and BA are identical. See [**1**], page 54, for a proof for general A, B. If $\mathbf{x}$ is the eigenvector of AB corresponding to the eigenvalue λ, $B\mathbf{x}$ is the eigenvector of BA corresponding to λ, provided that $B\mathbf{x} \neq \mathbf{o}$.

5.

$$P = \begin{bmatrix} 1 & -1 & -2 \\ -1 & 3 & 2 \\ -2 & 2 & 6 \end{bmatrix}.$$

6. If A has a complete system of independent eigenvectors then $X^{-1}AX = \Lambda$. Hence show that

$$X^{-1}(A^n + s_1A^{n-1} + \ldots + s_{n-1}A + s_nI)X = 0$$

and deduce the Cayley–Hamilton result. See [**10**] for a general proof.

8. If μ_i are the eigenvalues of B,

$$|\det B|^2 = |\prod \mu_i|^2 \leqslant \left(\frac{\sum|\mu_i|^2}{n}\right)^n \leqslant \left(\frac{\sum\sum|b_{ij}|^2}{n}\right)^n \leqslant 1$$

by Schur's theorem. Now use $\det A = \det B . \det D$.

9. If $a = c = d = 1$, $b = 5$ we have a better result than Gerschgorin's. The result does not generalize in an obvious manner. One would expect something like

$$\max|\lambda| \leqslant \sum_i |a_{ii}| + \frac{1}{n}\sum_{i,j}|a_{ij}| \quad \text{to hold,}$$

but a counter-example is provided by

$$A = \begin{bmatrix} 1 & 0 & 10 \\ 0 & 1 & 0 \\ 10 & 0 & 1 \end{bmatrix}.$$

14. $\lambda_1 = 8{\cdot}2695,\ \lambda_2 = 4{\cdot}1153,\ \lambda_3, \lambda_4 = -0{\cdot}1925 \pm 0{\cdot}8349i$

$\mathbf{x}_1 = (0{\cdot}7591,\ 1{\cdot}0000,\ 0{\cdot}9378,\ 0{\cdot}5457)'$
$\mathbf{x}_2 = (1{\cdot}0000,\ 0{\cdot}2788,\ -0{\cdot}4222,\ -0{\cdot}5072)'$
$\mathbf{x}_3, \mathbf{x}_4 = (1,\ -0{\cdot}7981 \pm 0{\cdot}2087i,\ 0{\cdot}0944 \mp 0{\cdot}3332i,\ 0{\cdot}1433 \pm 0{\cdot}1813i)'.$

17. If $\begin{bmatrix} \mathbf{u}_1 \\ \mathbf{v}_1 \end{bmatrix}$ is one eigenvector corresponding to λ_1 then $\begin{bmatrix} -\mathbf{v}_1 \\ \mathbf{u}_1 \end{bmatrix}$ is the eigenvector corresponding to the repeated λ_1.

18. Eigenvalues are $2 + \sqrt{2}$, $2 - \sqrt{2}$. Eigenvectors are

$$\begin{bmatrix} \sqrt{2} \\ 1 \end{bmatrix} + i\begin{bmatrix} 0 \\ -1 \end{bmatrix} \quad \text{and} \quad \begin{bmatrix} -\sqrt{2} \\ 1 \end{bmatrix} + i\begin{bmatrix} 0 \\ -1 \end{bmatrix}$$

20. $\lambda_1 = 0{\cdot}5069,\ \lambda_2 = 0{\cdot}2278,\ \lambda_3 = -0{\cdot}3465.$

$\mathbf{x}_1 = (0{\cdot}4836 - 0{\cdot}5068i,\ 0{\cdot}1905 - 0{\cdot}6195i,\ -0{\cdot}1398 - 0{\cdot}2639i)'$
$\mathbf{x}_2 = (0{\cdot}3833 - 0{\cdot}3400i,\ -0{\cdot}0616 + 0{\cdot}1536i,\ 0{\cdot}3311 + 0{\cdot}7749i)'$
$\mathbf{x}_3 = (0{\cdot}3144 + 0{\cdot}3849i,\ -0{\cdot}6779 - 0{\cdot}3047i,\ 0{\cdot}4184 - 0{\cdot}1596i)'.$

22. $\lambda_1 = 2, \quad \mathbf{x}_1 = (1, 0{\cdot}875, 0{\cdot}625)'$
$\lambda_2 = -2, \quad \mathbf{x}_2 = (0, 1, -1)'.$

24. Eigenvector can be found by iterating with a new vector, say $y = (1, 0, 0, 0)'$. Show A has its two largest eigenvalues equal to 3·4142 and eigenvectors $(0{\cdot}4142, 1, 1, 0{\cdot}4142)'$, $(1, 0{\cdot}7071, 0, -0{\cdot}7071)'$. Compare with exercise 18.

25. $3{\cdot}3453 \pm 1{\cdot}3188i$,
$\mathbf{x}_1, \mathbf{x}_2 = (0{\cdot}8435 \pm 1{\cdot}0366i,\ 1, 0{\cdot}6583 \mp 0{\cdot}7544i)'.$

26. (*a*) $\lambda_1 = 30{\cdot}2887,\ \lambda_2 = 3{\cdot}8581,\ \lambda_3 = 0{\cdot}8431,\ \lambda_4 = 0{\cdot}0102$

$\mathbf{x}_1 = (0{\cdot}5209,\ 0{\cdot}5520,\ 0{\cdot}5286,\ 0{\cdot}3803)'$
$\mathbf{x}_2 = (0{\cdot}6254,\ 0{\cdot}2716,\ -0{\cdot}6148,\ -0{\cdot}3963)'$
$\mathbf{x}_3 = (0{\cdot}5676,\ -0{\cdot}7603,\ 0{\cdot}3017,\ -0{\cdot}0933)'$
$\mathbf{x}_4 = (0{\cdot}1237,\ -0{\cdot}2086,\ -0{\cdot}5016,\ 0{\cdot}8304)'.$

(*b*) $\lambda_1 = 15,\ \lambda_2 = 5,\ \lambda_3 = 5,\ \lambda_4 = -1$

$\mathbf{x}_1 = (1, 1, 1, 1)', \qquad \mathbf{x}_2 = (1, 0, 0, -1)',$
$\mathbf{x}_2 = (0, 1, -1, 0)', \qquad \mathbf{x}_4 = (1, -1, -1, 1)'.$

(*c*) $\lambda_1 = 10,\ \lambda_2 = 5,\ \lambda_3 = 2,\ \lambda_4 = 1$

$\mathbf{x}_1 = (1, 1, 0{\cdot}5, 0{\cdot}5)', \qquad \mathbf{x}_2 = (-0{\cdot}5, -0{\cdot}5, 1, 1)'$
$\mathbf{x}_3 = (0, 0, 1, -1)', \qquad \mathbf{x}_4 = (1, -1, 0, 0)'.$

(*d*) $\lambda_1 = 12,\ \lambda_2 = 1 + 5i,\ \lambda_3 = 1 - 5i,\ \lambda_4 = 2$

$$\mathbf{x}_1 = (1, -1, 1, 1)', \qquad \mathbf{x}_2 = (i, 1, 1, -i)',$$
$$\mathbf{x}_3 = (-i, 1, 1, i)', \qquad \mathbf{x}_4 = (1, 1, -1, 1)'.$$

(*e*) $\lambda_1 = -8{\cdot}028\,66,\ \lambda_2 = 7{\cdot}932\,98,\ \lambda_3 = 5{\cdot}668\,86,\ \lambda_4 = -1{\cdot}573\,19$

$$\mathbf{x}_1 = (0{\cdot}26346,\ 0{\cdot}65904,\ -0{\cdot}19963,\ -0{\cdot}67557)'$$
$$\mathbf{x}_2 = (0{\cdot}56014,\ 0{\cdot}21163,\ 0{\cdot}77671,\ 0{\cdot}19538)'$$
$$\mathbf{x}_3 = (0{\cdot}37870,\ 0{\cdot}36242,\ -0{\cdot}53793,\ 0{\cdot}66020)'$$
$$\mathbf{x}_4 = (0{\cdot}68805,\ -0{\cdot}62412,\ -0{\cdot}25980,\ -0{\cdot}26375)'.$$

28. See [1] pp. 584 and 606.

29. $C = (c_{ij})$ where $c_{ij} = (-1)^{j-1}\binom{i-1}{j-1}, j \leqslant i, c_{ij} = 0, j > i$. The eigenvalues are 332·84, 15·553, 2·0435 and their reciprocals.

30. It is best to confine θ to the region $|\theta| \leqslant \frac{1}{4}\pi$. Iterations should continue until all off-diagonal elements are insignificant.

33. Use Theorem 9-1.

Exercises on chapter 11

1. $y_n = (1{\cdot}1)^n y_0 (1 + E)^{n-3},\ n \geqslant 3,\ |E| \leqslant \frac{1}{2}10^{-3}$. Note that it is not good to make h too small. While it is true that with exact arithmetic

$$y_n = (1 + h)y_{n-1} = y_0(1 + h)^n \to y_0 e^{nh}, \qquad \text{as } h \to 0,$$

– the correct solution – $h = 10^{-4}$ (for example) has a disastrous effect on this computer and gives $y_n = 1$ (all n).

3. $y_n = n + 1$. If $\mu_1, \mu_2, \ldots, \mu_{n-k}$ are the distinct roots and μ is the root repeated k times then the general solution is given by

$$y_m = \sum_{r=1}^{n-k} A_r \mu_r^{\,m} + \mu^m \sum_{r=0}^{k-1} (B_r m^r),$$

where the A's and B's are constants.

4. $I_k(\phi) = \sin k\phi / \sin \phi$.

5. It is better to solve as a boundary value problem since the results obtained in this way are smooth at the end points 0·5 and 2.

6.

x	0·1	0·2	0·3	0·4	0·5
y	0·9002	0·8011	0·7033	0·6069	0·5121

x	0·6	0·7	0·8	0·9	1·0
y	0·4185	0·3259	0·2342	0·1428	0·0517

8. See exercise 6.

9. The exact solution is $y^2 = -2x + 1$.

10. $a = 1,\ b = 1/3,\ c = 4/3,\ d = 1/3$.

14. Let $y_{n+1}^{[m]}$ be the value of y_{n+1} after the mth application of the corrector. Then

$$\left|y_{n+1}^{[m+1]} - y_{n+1}^{[m]}\right| \leqslant K\left|y_{n+1}^{[m]} - y_{n+1}^{[m-1]}\right|,$$

where $K = \frac{1}{12}hL|(5-a)|$ and L is the constant from the Lipschitz condition. We must have $K < 1$ for convergence

i.e., $$hL < 12/|(5-a)|.$$

For $-1 \leqslant a \leqslant 1$ convergence is clearly possible by appropriate choice of h.

16. $$k_4 = hf_0 + h^2 D_4 f_0 + h^3\left(\tfrac{1}{2}D_4^2 f + f_y \sum_{j=2}^{3} \beta_{4j} D_j f\right)_0$$

$$+ h^4\,[\tfrac{1}{6}D_4^3 f + \tfrac{1}{2} f_y \sum_{j=2}^{3} \beta_{4j} D_j^2 f + \beta_{32}\beta_{43}(f_y)^2 D_2 f + D_4 f_y \sum_{j=2}^{3} \beta_{4j} D_j f\,]_0$$

$$+ h^5\,[\tfrac{1}{24}D_4^4 f + \tfrac{1}{2}\beta_{32}\beta_{43}(f_y)^2 D_2^2 f + \beta_{42}\beta_{43} f_{yy} D_2 f D_3 f + \tfrac{1}{6} f_y \sum_{j=2}^{3} \beta_{4j} D_j^3 f$$

$$+ \tfrac{1}{2} D_4 f_y \sum_{j=2}^{3} \beta_{4j} D_j^2 f + \tfrac{1}{2} f_{yy} \sum_{j=2}^{3} \beta_{4j}^2 (D_j f)^2 + \tfrac{1}{2} D_4^2 f_y \sum_{j=2}^{3} \beta_{4j} D_j f$$

$$+ \beta_{32}\beta_{43} f_y D_2 f \sum_{j=3}^{4} D_j f_y\,]_0 + O(h^6).$$

$$k_5 = hf_0 + h^2 D_5 f_0 + h^3\left(\tfrac{1}{2}D_5^2 f + f_y \sum_{j=2}^{4} \beta_{5j} D_j f\right)_0$$

$$+ h^4\,[\tfrac{1}{6}D_5^3 f + \tfrac{1}{2} f_y \sum_{j=2}^{4} \beta_{5j} D_j^2 f + D_5 f_y \sum_{j=2}^{4} \beta_{5j} D_j f + f_y^2 \sum_{i=j+1}^{4} \sum_{j=2}^{3} \beta_{5i}\beta_{ij} D_j f\,]_0$$

$$+ h^5[\tfrac{1}{24}D_5^4 f + \tfrac{1}{6} f_y \sum_{j=2}^{4} \beta_{5j} D_j^3 f + \tfrac{1}{2} D_5^2 f_y \sum_{j=2}^{4} \beta_{5j} D_j f + \tfrac{1}{2} f_{yy} \sum_{j=2}^{4} \beta_{5j}^2 (D_j f)^2$$

$$+ \tfrac{1}{2} D_5 f_y \sum_{j=2}^{4} \beta_{5j} D_j^2 f + f_y D_5 f_y \sum_{i=j+1}^{4} \sum_{j=2}^{3} \beta_{5i}\beta_{ij} D_j f + \beta_{54}\beta_{43}\beta_{32} f_y^3 D_2 f$$

$$+ f_{yy} \sum_{i=j+1}^{4} \sum_{j=2}^{3} \beta_{5i}\beta_{ij} D_i f D_j f + \tfrac{1}{2} f_y^2 \sum_{\substack{i=3 \\ (i \neq j)}}^{4} \sum_{j=2}^{3} \beta_{5i}\beta_{ij} D_j^2 f$$

$$+ f_y \sum_{\substack{i=3 \\ (i \neq j)}}^{4} \sum_{j=2}^{3} \beta_{5i}\beta_{ij} D_i f_y D_j f\,]_0 + O(h^6).$$

17(*a*) $$w_1 = 1 + \frac{2 - 3(\alpha_2 + \alpha_3)}{6\alpha_2\alpha_3}, \qquad w_2 = \frac{3\alpha_2 - 2}{6\alpha_2(\alpha_3 - \alpha_2)},$$

$$w_3 = \frac{2-3\alpha_2}{6\alpha_3(\alpha_3-\alpha_2)}, \qquad \beta_{21} = \alpha_2,$$

$$\beta_{31} = \frac{3\alpha_2\alpha_3(1-\alpha_2)-\alpha_3^2}{\alpha_2(2-3\alpha_2)}, \qquad \beta_{32} = \frac{\alpha_3(\alpha_3-\alpha_2)}{\alpha_2(2-3\alpha_2)}.$$

(*b*) $\alpha_2 = \alpha_3 = \frac{2}{3}$, $w_1 = \frac{3}{4}$, $\beta_{21} = \frac{2}{3}$, $w_2 = \frac{3}{4} - w_3$, $\beta_{32} = \frac{1}{4}w_3$, $\beta_{31} = \frac{2}{3} - (\frac{1}{4}w_3)$.

18. $\alpha_4 = 1$, $\beta_{21} = \alpha_2$,

$$w_4 = \frac{\frac{1}{2}\alpha_2\alpha_3 - \frac{1}{3}\alpha_2 - \frac{1}{3}\alpha_3 + \frac{1}{4}}{(\alpha_2-1)(\alpha_3-1)}$$

$$w_3 = \frac{\frac{1}{6}\alpha_2 - \frac{1}{12}}{\alpha_3(1-\alpha_3)(\alpha_2-\alpha_3)}, \qquad w_2 = \frac{\frac{1}{6}\alpha_3 - \frac{1}{12}}{\alpha_2(\alpha_3-\alpha_2)(1-\alpha_2)}$$

$$\beta_{43} = \frac{\frac{1}{6}\alpha_2 - \frac{1}{12}}{\alpha_3 w_4(\alpha_2-\alpha_3)}, \qquad \beta_{32} = \frac{\frac{1}{24}}{\omega_3\alpha_2(1-\alpha_3)}$$

$$\beta_{42} = \frac{\frac{1}{24}(\alpha_2 - 4\alpha_3^2 + 5\alpha_3 - 2)}{w_4\alpha_2(\alpha_3-1)(\alpha_2-\alpha_3)}, \qquad \beta_{31} = \alpha_3 - \beta_{32},\ \beta_{41} = 1 - \beta_{42} - \beta_{43}.$$

When α_2 and α_3 are equal, $\alpha_2 = \alpha_3 = \frac{1}{2}$, $\alpha_4 = 1$, $w_4 = \frac{1}{6}$, $w_2 = \frac{2}{3} - w_3$, $\beta_{32} = 1/6\,w_3$, $\beta_{43} = 3w_3$, $\beta_{42} = 1 - 3w_3$.

19. The standard Runge–Kutta formulae for sovling the pair of simultaneous equations

$$y' = f(x, y, y' = z), \qquad y(0) = y_0,$$
$$z' = g(x, y, z), \qquad z(0) = y'(0)$$

are

$$k_1 = hf(x_0, y_0, z_0), \qquad k'_1 = hg(x_0, y_0, z_0),$$
$$k_2 = hf(x_0 + \tfrac{1}{2}h,\ y_0 + \tfrac{1}{2}k_1,\ z_0 + \tfrac{1}{2}k'_1), \qquad k'_2 = hg(x_0 + \tfrac{1}{2}h,\ y_0 + \tfrac{1}{2}k_1,\ z_0 + \tfrac{1}{2}k'_1),$$
$$k_3 = hf(x_0 + \tfrac{1}{2}h,\ y_0 + \tfrac{1}{2}k_2,\ z_0 + \tfrac{1}{2}k'_2), \qquad k'_3 = hg(x_0 + \tfrac{1}{2}h,\ y_0 + \tfrac{1}{2}k_2,\ z_0 + \tfrac{1}{2}k'_2),$$
$$k_4 = hf(x_0 + h,\ y_0 + k_3,\ z_0 + k'_3), \qquad k'_4 = hg(x_0 + h,\ y_0 + k_3,\ z_0 + k'_3),$$
$$y_1 = y_0 + \tfrac{1}{6}(k_1 + 2k_2 + 2k_3 + k_4), \qquad z_1 = z_0 + \tfrac{1}{6}(k'_1 + 2k'_2 + 2k'_3 + k'_4).$$

Note that for the solution of the simultaneous linear differential equations $\frac{dx}{dt} = Ax$ it is better to treat the problem as an algebraic eigenvalue problem (see chapter 9, exercise 2).

20. $y_{r+1}(1 + \frac{1}{2}\alpha h) - (2 - \beta h^2)y_r + y_{r-1}(1 - \frac{1}{2}\alpha h) = 0.$

22. The exact solution is $y = 2e \sinh x/(e^2 - 1)$.

23. The exact solution is $y = \frac{1}{24}x^4 - \frac{1}{16}x^3 + \frac{1}{48}x$.

24.

x	0·2	0·4	0·6	0·8	1·0
y	0·0555	0·1115	0·1693	0·2313	0·3008

x	1·2	1·4	1·6	1·8
y	0·3824	0·4826	0·6101	0·7771

26.

x	0·2	0·4	0·6	0·8	1·0
y	0·2313	0·4455	0·6259	0·7575	0·8269

27. Let $y^{(s)} = \frac{1}{2}a_0^{(s)} + a_1^{(s)}T_1 + a_2^{(s)}T_2 t \ldots$. Show that

$$2ra_r^{(s)} = a_{r-1}^{(s+1)} - a_{r+1}^{(s+1)}.$$

For the given differential equation show that

$$\tfrac{1}{2}(a_{r+1}^{(2)} + a_{r-1}^{(2)}) + a_r^{(1)} + 8(a_{r+1} + a_{r-1}) = 0.$$

Hence deduce that

$$a_{r-2} = a_{r+2} - \tfrac{1}{8}r(a_{r+1}^{(1)} + a_{r-1}^{(1)}).$$

Now take $a_{10} = \tau$, $a_r = 0, r > 10$, and evaluate a_r ($r = 0, \ldots, 9$) in terms of τ. Use the fact that $y(0) = 1$ implies

$$\tfrac{1}{2}a_0 - a_2 + a_4 - \ldots = 1,$$

to obtain τ and $y(1)$.

30. $$y = A\left(1 + 3x^2 + \tfrac{3}{5}x^4 - \frac{1}{15}x^6 + \frac{1}{65}x^8 + \ldots\right)$$
$$+ Bx^{\frac{3}{2}}\left(1 + \frac{3}{8}x^2 - \frac{3}{128}x^4 + \frac{5}{1024}x^6 - \frac{45}{32\,768}x^8 + \ldots\right), \qquad 0 \leqslant x < 2.$$

Make a change of variable, say $x = u + 1$, and find an expansion in terms of u.

32. The solutions are

$$u = x\left(2 - \frac{1}{4}x^2 + \frac{1}{4^2.6}x^4 - \frac{1}{4^2.6^2.8}x^6 + \ldots\right),$$

and

$$u \log x + x^{-1}\left(1 + \frac{1}{2^2}x^2 - \frac{1}{2^2.4}\left(\frac{2}{2} + \frac{1}{4}\right)x^4 + \frac{1}{2^2.4^2.6}\left(\frac{2}{2} + \frac{2}{4} + \frac{1}{6}\right)x^6 - \ldots\right).$$

33. Make the substitution $x = 1/z$ and then solve by method of Frobenius.

35. $$y = e^{-x}\left(1 + \epsilon\frac{x^2}{2} + \epsilon^2\frac{x^4}{2\cdot 4} + \ldots + \epsilon^n\frac{x^{2n}}{2\cdot 4\ldots(2n)} + \ldots\right).$$

36. A and B are symmetric tridiagonal matrices such that $a_{ii} = 2$, $a_{i,\,i+1} = -1$, $b_{ii} = 10$, $b_{i,\,i+1} = 1$ (all i).

Exercises on chapter 12

1. (*a*) Linear; hyperbolic if $x < 0$ and elliptic if $x > 0$.
(*b*) Quasi-linear; elliptic.
(*c*) Quasi-linear; hyperbolic.
(*d*) Linear; hyperbolic if $x^2 + y^2 > 1$ and elliptic if $x^2 + y^2 < 1$. What happens when equality holds?
(*e*) Quasi-linear; if $v = 0$, parabolic; $v \neq 0$, hyperbolic. See answer to exercise 6 for method of determining characteristics.

2. The characteristics are given by $y^2 + 4x = \text{constant}$; the values of u are given by the formula

$$u = -4 + 2(y^2 + 4)^{\frac{1}{2}}.$$

This can be deduced by making a change of variable, putting $\xi = y^2 + 4x$ and $\eta = y$.

4.

x	0·0528	0·1557	0·2558	0·3551	0·4544
y	0·0113	0·0164	0·0267	0·0423	0·0633
u	0·2267	0·3380	0·5567	0·8852	1·329

x	0·5538	0·6533	0·7529	0·8526	0·9523
y	0·0899	0·1222	0·1603	0·2043	0·2544
u	1·893	2·584	3·407	4·368	5·473

5. We find $(a^2 - p^2)\left(\frac{dy}{dx}\right)^2 + 2pq\left(\frac{dy}{dx}\right) + (a^2 - q^2) = 0$. This equation has two real roots if $a^2 < p^2 + q^2$, i.e., the equation is hyperbolic if the motion is supersonic.

6. To obtain the characteristics differentiate the first equation with respect to x and t and the second with respect to x and eliminate second order derivatives in ρ.

9. $B = 2I$, $C = -A$, where A is the symmetric tridiagonal matrix $a_{ii} = \rho + 1$, $a_{i,\,i+1} = -\frac{1}{2}\rho$.

15. Show

$$\mu^2 + 8\rho\mu \sin^2(\tfrac{1}{2}\beta\Delta x) - 1 = 0,$$

which implies the stated result.

16.

y \ x	0·0	0·1	0·2	0·3	0·4	0·5
0·01	0·9757	0·9637	0·9279	0·8694	0·7894	0·6902
0·03	0·9288	0·9174	0·8836	0·8283	0·7537	0·6626
0·05	0·8848	0·8742	0·8429	0·7926	0·7260	0·6480

$y \backslash x$	0·6	0·7	0·8	0·9	1·0
0·01	0·5744	0·4464	0·3143	0·2004	0·1784
0·03	0·5614	0·4594	0·3707	0·3080	0·2874
0·05	0·5651	0·4859	0·4199	0·3759	0·3606

19.

$y \backslash x$	1	2	3	4
1	0·4401	0·7853	1·001	1·074
2	0·2754	0·5000	0·6443	0·6938
3	0·1613	0·2952	0·3825	0·4128
4	0·0746	0·1369	0·1779	0·1921

Tabulated values are those of $u \times 10^{-2}$.

22. $4(u_{-1,0} + u_{1,0} + u_{0,1} + u_{0,-1}) + (u_{-1,-1} + u_{1,-1} + u_{-1,1} + u_{1,1}) - 20u_{0,0} = O(h^6).$

24.

$$V = E\left\{e^{-\rho x/v} + \frac{\sigma x}{v}\int_{x/v}^{t} e^{-\rho\tau}\, \frac{I_1[\sigma(\tau^2 - x^2/v^2)^{\frac{1}{2}}]}{(\tau^2 - x^2/v^2)^{\frac{1}{2}}}\, d\tau\right\} H\left(t - \frac{x}{v}\right),$$

where

$$v = (LC)^{-\frac{1}{2}}, \qquad \rho = (R/2L) + (G/2C), \qquad \sigma = (R/2L) - (G/2C),$$

$I_1(x)$ is the modified Bessel function of first order and $H(t)$ is the Heaviside unit step function ($H(t) = 0$, $t < 0$; $H(t) = 1$, $t > 0$) – see Carslaw and Jaeger [**13**], p. 197 *et seq.*

Exercises on chapter 13

1. If $g(x)$ is a function such that $\left(\frac{d^r g}{dx^r}\right)_{x=a} = 0, r = 1, 2, \ldots, 2n-1$ then $g(x)$ has a minimum value at $x = a$ if $d^{2n}g/dx^{2n} > 0$ and a maximum value at $x = a$ if $d^{2n}g/dx^{2n} < 0$.

3. 1·1462.

4. $x = 1{\cdot}0637$, $y = 0{\cdot}9350$.

5. (0·4472, 0·8944), (0·8944, 0·4472)
(−0·4472, −0·8944), (−0·8944, −0·4472).

7. $R = 0{\cdot}10878$ ohm, $V_2 = 10{\cdot}0013$ volt.

8. There is a maximum value of $\frac{1}{4}$ at (2, 1) and a minimum value of $-\frac{3}{4}$ at $(-\frac{2}{3}, -\frac{1}{3})$.

9. The solutions are $(\frac{5}{3}, -\frac{2}{3}, \frac{4}{3})$ and (1, 0, 2).

10. (48·125, 44·25).

11. The maximum occurs where $x_1 = x_2 = \ldots = x_n$. The minimum occurs where $z_1 = \frac{1}{8}$, $z_2 = \frac{1}{4}$, $z_3 = \frac{1}{2}$.

15. 4·5.

16. 8.

17. $a = (0{\cdot}9998, 0{\cdot}1589)$, $b = (0{\cdot}1589, 0{\cdot}9998)$, $|\det \Lambda| = 0{\cdot}9744$.

18. The minimum value is 0 at (1, 1).

20. $y = 8{\cdot}44 + 3{\cdot}2X'$

$y = 5{\cdot}51 + 3{\cdot}2X' + 0{\cdot}733X'^2$.

23. Minimize $\sum\{(y_i Q(x_i) - P(x_i)\}^2$. The data was generated from $(0{\cdot}95 + x)/(2{\cdot}35 - x)$.

24.
$$\sum X_1 = \hat{a}_1 N + \hat{a}_2 \sum X_2 + \hat{a}_3 \sum X_3,$$
$$\sum X_1 X_2 = \hat{a}_1 \sum X_2 + \hat{a}_2 \sum X_2^2 + \hat{a}_3 \sum X_2 X_3,$$
$$\sum X_1 X_3 = \hat{a} \sum X_3 + \hat{a}_2 \sum X_2 X_3 + \hat{a}_3 \sum X_3^2.$$
$$X_1 = 3{\cdot}6512 + 0{\cdot}8546 X_2 + 1{\cdot}5063 X_3.$$

25. Form the sum of squares

$$\sum_k (Z_k - a_0 - a_1 X_k - a_2 Y_k - a_3 X_k^2 - a_4 X_k Y_k - a_5 Y_k^2)$$

and differentiate with respect to the coefficients $a_0, \ldots, a_5$ to get the required formulae.

26. $x = 0{\cdot}4638$, $f(x) = 1{\cdot}130\,33$.

27. $2360 e^{-0{\cdot}1601x} + e^{-0{\cdot}3046x} (640 \cos 0{\cdot}4935x + 2621 \sin 0{\cdot}4935x)$.

Answers to the miscellaneous exercises

1. For the given quintic equation, with exact arithmetic,

$$|P| = 121.$$

2. The expansion is

$$z = \sum_{n=1}^{\infty} \eta^n \frac{(0{\cdot}1n)^{n-1} e^{-0{\cdot}1n}}{n!}.$$

3. The minimum depth of the seam is 53·8 at $x = 2{\cdot}566$, $y = 2{\cdot}733$. The volume is 3341·7 units. At (2·5, 3·5) the depth is 76·5.

4. 4·9350.

7. See Anderson, D. G., *Math. Comp.*, **19** (1965), 477–81, and Abramowitz and Stegun, p. 920.

8. 14·134 725 14. See Haselgrove, C. B. and Miller, J. C. P., *Tables of the Riemann Zeta Function*, Royal Society Mathematical Tables No. 6, Cambridge University Press (1960).

9. [0·46393, 0·53902].

10. $$\int_{x_0}^{x_0+h} f(x)\,dx = hf(x_0) + \frac{h^2}{2} f'(x_0) + \ldots + \frac{h^k}{k!} f^{(k-1)}(x_0) + \frac{h^{k+1}}{(k+1)!} f^{(k)}(X).$$

11. $$\int_{-1}^{1} f(x)\,dx = \sum_k A_k f(X_k) + R,$$

where A_k is an interval estimate of the weight α_k, X_k is an interval estimate of the abscissa x_k, and R is an interval estimate of the remainder term. The interval estimate R can be obtained using Stirling's formula.

12. Approximate the integral by the trapezium rule to obtain the equation,

$$\sum_{i=1}^{n} \tfrac{1}{2}[k(x, y_{i-1})f(y_{i-1}) + k(x, y_i)f(y_i)] = \lambda f(x) + g(x), \qquad y_0 = a,\ y_n = b.$$

Set $x = y_0, y_1, \ldots, y_n$ and solve the resulting linear equations in $f(y_i)$. Use the technique of deferred correction to improve your results. (The exact solution to the given integral equation is $-6x-4$.)

13. $1, 1, 4/3\pi$ are the exact answers.

14. $$a_{11} = \left(1 + \sum_{i=2}^{n} x_i^2\right)^{-1}, \qquad a_{i1} = -x_i a_{11},\ i \geqslant 2;$$

$$a_{1j} = x_j\left(1 + \sum_{i=2}^{n} x_i^2\right)^{-1}, \qquad a_{jj} = 1 - x_j a_{1j},\ j \geqslant 2;$$

$$a_{ij} = -x_i a_{ij}, \quad i \neq 1, \quad i, j \geqslant 2.$$

15. $$A^{-1} = \begin{bmatrix} \frac{4}{5} & -\frac{3}{5} & \frac{2}{5} & -\frac{1}{5} \\ -\frac{3}{5} & \frac{6}{5} & -\frac{4}{5} & \frac{2}{5} \\ \frac{2}{5} & -\frac{4}{5} & \frac{6}{5} & -\frac{3}{5} \\ -\frac{1}{5} & \frac{2}{5} & -\frac{3}{5} & \frac{4}{5} \end{bmatrix}.$$

18. We need two other conditions to obtain the spline fit, e.g., the values of the derivatives of e^{-x} at the end points.

23. $k_5 = \frac{2}{113}$, $a_2 = \frac{-108}{113}$, $a_3 = \frac{54}{113}$, $a_4 = \frac{-64}{339}$, $a_5 = \frac{16}{339}$.

The maximum error is less in modulus than k_5. For the second part extra conditions are obtained by setting x to be the non-zero roots of $T_5(x) = 0$ and optimizing so that $|k_5| + |k_6|$ is a maximum.

26. $\omega \doteqdot 9{\cdot}9$, $E \doteqdot 1822$.

28. $a_0 = 2$, $a_1 = 3$, $a_2 = -2$, $b_1 = 4$, $b_2 = 0$.

Bibliography

Abramowitz, M. and Stegun, I. A., *Handbook of Mathematical Functions*, Dover Publications Inc., New York (1965).

Birkhoff, G. and Maclane, S., *A Survey of Modern Algebra*, rev. edn., MacMillan Co., New York (1953).

Bodewig, E., *Matrix Calculus*, North-Holland Publ. Co., Amsterdam (1959).

Box, M. J., Davies, D., and Swann, W. H., *Optimization Techniques,* I.C.I. Monograph, No. 5, Oliver and Boyd Ltd., Edinburgh (1969).

Buckingham, R. A., *Numerical Methods*, Pitman, London (1957).

Carslaw, H. S. and Jaeger, J. C., *Operational Methods in Applied Mathematics*, Oxford University Press, London (1941).

Celia, C. W., *Introduction to Numerical Analysis*, McGraw-Hill Publ. Co. Ltd., Maidenhead (1969).

Conte, S. D., *Elementary Numerical Analysis*, McGraw-Hill Book Co. Ltd., New York (1965).

Davies, O. L. (ed.), *The Design and Analysis of Industrial Experiments*, 2nd edn., Oliver and Boyd Ltd., Edinburgh (1956).

Davis, P. J., *Interpolation and Approximation*, Blaisdell, New York (1963).

Davis, P. J. and Rabinowitz, P., *Numerical Integration*, Blaisdell, New York (1967).

Forsythe, G. E. and Wasow, W. R., *Finite Difference Methods for Partial Differential Equations*, J. Wiley and Sons, New York (1960).

Fox, L., The Use and Construction of Mathematical Tables, *N.P.L. Mathematical Tables*, Vol. 1, H.M.S.O., London (1956).

Fox, L., *The Numerical Solution of Two-Point Boundary-Value Problems in Ordinary Differential Equations*, Oxford University Press, London (1957).

Fox, L., *An Introduction to Numerical Linear Algebra*, Oxford University Press, London (1964).

Fox, L. (ed.), *Numerical Solution of Ordinary and Partial Differential Equations*, Pergamon Press, Oxford (1962).

Fox, L. and Mayers, D. F., *Computing Methods for Scientists and Engineers*, Oxford University Press, London (1958).

Fox, L. and Parker, I. B., *Chebyshev Polynomials in Numerical Analysis*, Oxford University Press, London (1968).

Froberg, C- E., *Introduction to Numerical Analysis*, Addison-Wesley, Reading, Mass. (1965).

Gillespie, R. P., *Partial Differentiation*, Oliver and Boyd Ltd., Edinburgh (1954).

Goursat, E., *A Course of Mathematical Analysis* (translation), Dover Publications Inc., New York (1945).

Hammersley, J. M. and Handscomb, D. C., *Monte Carlo Methods*, Methuen, London (1964).

Handscomb, D. C. (ed.), *Methods of Numerical Approximation*, Pergamon Press, Oxford (1966).

Hardy, G. H., *A Course of Pure Mathematics*, 10th edn., Cambridge University Press, London (1952).

Hardy, G. H., *Divergent Series*, Oxford University Press, London (1949).

Hartree, D. R., *Numerical Analysis*, 2nd edn., Oxford University Press, London (1958).

Hildebrand, F. B., *Introduction to Numerical Analysis*, McGraw-Hill Book Co. Ltd., New York (1956).

Hildebrand, F. B., *Advanced Calculus for Application*, Prentice-Hall Inc., New Jersey (1962).

Householder, A. S., *Principles of Numerical Analysis*, McGraw-Hill Book Co. Ltd., New York (1953).

Interpolation and Allied Tables, H.M.S.O., London (1956).
Isaacson, E. and Keller, H. B., *The Analysis of Numerical Methods*, J. Wiley and Sons, New York (1966).
Jeffreys, H. and Jeffreys, B. S., *Methods of Mathematical Physics*, 3rd edn., Cambridge University Press, London (1956).
Kacmarz, S. and Steinhaus, H., *Theorie der Orthogonalreihen*, Chelsea, New York (1951).
Kendall, M. G. and Stuart, A., *The Advanced Theory of Statistics*, Griffin, London (1961).
Kopal, Z., *Numerical Analysis*, J. Wiley and Sons, New York (1955).
Lanczos, C., *Applied Analysis*, Pitman, London (1957).
Mangasarian, O. L., *Non-linear Programming*, McGraw-Hill Publ. Co. Ltd., Maidenhead (1969).
Milne, W. E., *Numerical Calculus*, Princeton University Press, New Jersey (1949).
Modern Computing Methods, N.P.L. Notes on Applied Science No. 16, H.M.S.O., London (1957).
Moore, R. E., *Interval Analysis*, Prentice-Hall Inc., New Jersey (1966).
Moursund, D. G. and Duris, C. S., *Elementary Theory and Application of Numerical Analysis*, McGraw-Hill Book Co. Ltd., New York (1967).
Nef, W., *Linear Algebra*, McGraw-Hill Publ. Co. Ltd., Maidenhead (1967).
Pearson, E. S. and Hartley, H. O., *Biometrika Tables for Statisticians*, 3rd edn., Vol. 1, Cambridge University Press, Cambridge (1966).
Rall, L. B. (ed.), *Error in Digital Computation*, I and II, J. Wiley and Sons, New York (1965).
Ralston, A., *A First Course in Numerical Analysis*, McGraw-Hill Book Co. Ltd., New York (1965).
Ralston, A. and Wilf, H. S., *Mathematical Methods for Digital Computers*, J. Wiley and Sons, New York (1960).
Richtmeyer, R. D., *Difference Methods for Initial Value Problems*, Interscience, J. Wiley and Sons, New York (1957).
Schneider, Yu. A., *The Monte Carlo Method*, Pergamon Press, Oxford (1966).
Smith, G. D., *Numerical Solution of Partial Differential Equations*, Oxford University Press, London (1965).
Szego, G., *Orthogonal Polynomials*, Vol. 23, American Mathematical Society Colloquium Publications, New York (1959).
Titchmarsh, E. C., *The Theory of Functions*, 2nd edn., Oxford University Press, London (1939).
Todd, J. (ed.), *Survey of Numerical Analysis*, McGraw-Hill Book Co. Ltd., New York (1962).
Traub, J. F., *Iterative Methods for the Solution of Equations*, Prentice-Hall Inc., New Jersey (1965).
Varga, R., *Matrix Iterative Analysis*, Prentice-Hall Inc., New Jersey (1962).
Walsh, J. (ed.), *Numerical Analysis. An Introduction*, Academic Press, London (1966).
Whittaker, E. T. and Watson, G. N., *A Course of Modern Analysis*, 4th edn., Cambridge University Press, London (1962).
Wilkinson, J. H., Rounding Errors in Algebraic Processes, *N.P.L. Notes on Applied Science*, No. 32, H.M.S.O., London (1963).
Wilkinson, J. H., *The Algebraic Eigenvalue Problem*, Oxford University Press, London (1965).

The bibliography contains additional material not specifically mentioned in the text. Details of new methods and techniques and their appraisal and comparisons with established methods may be found by consulting journals such as

Journal of the Association of Computing Machinery,
Mathematics of Computation,
Numerische Mathematik,
S.I.A.M. Journal of Numerical Analysis,
S.I.A.M. Reviews,
The Computer Journal,
Journal of the Institute of Mathematics and its Application.

A number of these journals also contain ALGOL procedures for performing many of the computations referred to in the text.

Index of notations

Symbol	*Meaning*
$\subset$	Is contained in
O	Order of a quantity, as in $O(h^4)$
$\sim$	Is asymptotic to
$\doteqdot$	Has approximate numerical value
$\in$	Belongs to
$\propto$	Is proportional to
(a, b), $[a, b]$, I, J, K	Intervals
$a \,.\, 10^b$	Floating point representation (decimal)
$a \,.\, 2^b$	Floating point representation (binary)
$[a]$	Integer part of a
$\{a\}$	Fractional part of a
$\{I_n\}$	Sequence with general term I_n
$\bar{x}$	Computed value of x
fl	Floating point arithmetic operation
fi	Fixed point arithmetic operation
ϵ, η, ξ, E	Errors
$E, E^{-1}, \delta, \Delta, \nabla, \mu, D$	Difference operators
$I_M\,(2^k, 2^{k+1}, \ldots, 2^n)$	Notation for Romberg integration
A, B, C, X, Y	Matrices
L	Lower triangular matrix
U	Upper triangular matrix
I	Unit matrix
O	Zero matrix
$\mathbf{x}, \mathbf{y}, \mathbf{z}$	Vectors
$\mathbf{o}$	Zero vector
$\mathbf{r}_i(A)$	Row i of A
$\mathbf{c}_i(A)$	Column i of A
Λ	Spectral matrix
$\rho(A)$	Spectral radius of A
λ, μ	Eigenvalues
A' or A^T	Transpose of A
$\|A\|$	Norm of A (specified by suffix)
$\lvert A\rvert$ or $\det A$	Determinant of A

Symbol	*Meaning*
A^{-1}	Inverse of A
$\text{tr}\,(A)$	Trace of A
$f(\mathbf{x})$	$f(x_1, x_2, \ldots, x_n)$
$T_n(x)$	Chebyshev polynomial of first kind
$T_n{}^*(x)$	Shifted Chebyshev polynomial of first kind
$U_n(x)$	Chebyshev polynomial of second kind
$P_n(x)$	Legendre polynomial
$H_n(x)$	Hermite polynomial
$L_n(x)$	Laguerre polynomial
$\Gamma(x)$	Gamma function
$x!$	Factorial function $(= \Gamma(x+1))$
$\binom{n}{r}$	$n!/r!(n-r)!$
$\psi(x)$	Psi or digamma function
B_{2n}	Bernoulli numbers
$J_p(x)$	Bessel function of first kind
$I_n(x)$	Modified Bessel function of first kind
$C(t), S(t)$	Fresnel functions
γ	Euler's constant ($\doteqdot 0{\cdot}577\,215\,664\,9$)
π	$3{\cdot}141\,592\,653\,6 \ldots$
e	$2{\cdot}718\,281\,828\,5 \ldots$
$\text{ei}\,(x)$	Exponential integral function
$\bar{I}(s)$	Laplace transform of $I(t)$
$\text{Si}\,(x)$	Sine integral function
Fox [1]	Reference to author (given at end of current chapter)
(8.11)	Numbered equation or inequality (the 11th numbered reference in chapter 8)
Exercise 18	Refers to exercise at end of current chapter.

General index